H O T
J A Z Z
A N D
J A Z Z
D A N C E

HOT JAZZ AND JAZZ DANCE

Selected and Edited by Pryor Dodge

ROGER PRYOR DODGE

COLLECTED WRITINGS

1929 - 1964

NEW YORK OXFORD
OXFORD UNIVERSITY PRESS 1995

OXFORD UNIVERSITY PRESS

Oxford New York Toronto
Delhi Bombay Calcutta Madras Karachi
Kuala Lumpur Singapore Hong Kong Tokyo
Nairobi Dar es Salaam Cape Town
Melbourne Auckland

and associated companies in
Berlin Ibadan

Published by Oxford University Press, Inc.,
198 Madison Avenue, New York, New York 10016

Oxford is a registered trademark of Oxford University Press

Library of Congress Cataloging-in-Publication Data
Dodge, Roger Pryor.
Hot jazz and jazz dance / Roger Pryor Dodge.
p. cm. Includes index.
ISBN 0–19–507185–9
1. Jazz—History and criticism.
2. Jazz dance—History.
I. Title.
ML3507.D63 1995
781.65′09—dc20 94–3477

Frontispiece: Portrait of Roger Pryor Dodge, Paris, 1929.
Dodge presented "Lilies of the Field" at the Paramount Theatre in Paris
under the title "Roger Dodge et Ses Cinq Vagabonds"
to the music *Marche Slav* by Tchaikovsky.
Photograph by Studio-Toroy.

9 8 7 6 5 4 3 2 1

Printed in the United States of America
on acid-free paper

Introduction

Back when the publication of a book on jazz was a rare event, I snapped up a slim volume entitled *Frontiers of Jazz* as soon as I became aware of it. Edited by Ralph De Toledano, it was an anthology of important articles—Ansermet on Bechet, Hobson on Ellington, Morroe Berger on Bunk Johnson, William Russell on Boogie Woogie—but to me, just about eighteen and enamored of early Ellington, the centerpiece was Roger Pryor Dodge's "Harpsichords and Jazz Trumpets," with its incisive comments on the work of Bubber Miley.

This was in 1947. Dodge's piece had been written in 1934 but seemed not at all dated, and its musical literacy set it apart from most of what I'd read about jazz. (I had not yet discovered Winthrop Sargeant's *Jazz, Hot and Hybrid.*) My next encounter with Dodge was in the pages of *Jazzmen,* that seminal 1939 work which concluded with his "Consider the Critics," a vigorous overview of much of what had been written about jazz up to then.

Both these books are still in print, and it is likely that most serious students of jazz are acquainted with them, and thus with the name of Roger Pryor Dodge. But few, if any (including this writer), had a real notion of the scope and breadth of his writings until they were lovingly collected by his son. After the publication of this book it will be impossible to omit Dodge from any serious discussion of jazz criticism, a field to which he brought a unique perspective—that of a professional dancer and choreographer who had worked with jazz music and jazz musicians, who had a profound knowledge of and real love for baroque and pre-baroque Western Classical music, and who did not follow fashion in any of his artistic or intellectual pursuits.

Central to Dodge's ideas about jazz was his conviction that "all great music, even church, leans upon and is developed by the dance," and that, potentially, jazz could evolve into "the greatest music in the best classic mode since Bach and his predecessors." From Dodge's own experiences as a dancer, and from his work with improvising jazz players (notably Bubber Miley), Dodge knew that "when warmed up to improvising, an artist can do things impossible to re-create, even if adequately notated." Yet he was convinced that the best in jazz could only be preserved—and allowed to develop—if a way could be found toward a compositional process that would employ material from transcribed improvised solos (taken from recordings) and put them together in a manner that would add up to something more than the individual parts. This idea preoccupied Dodge to the very end, and he never lost hope that such a method could be found.

It will seem to some readers of this book that Dodge's rejection of most of what we understand by the term "jazz composition" is unreasonable. He tells us that he has no use for arranged jazz—in the main because the devices used by jazz arrangers are derived from nineteenth-century classical music (i.e., the romantic era), for which he has no sympathy. Even Ellington, whom he rates higher than any other jazz arranger-composer-orchestrator, falls short of Dodge's standards, as we can see from his review of the recording of *Black, Brown and Beige.* Hand in hand with this goes his dislike of popular songs as material for jazz improvisation. Jazz, he felt, needed to develop its own thematic material, and some of that could be found in the New Orleans repertory.

It would be a mistake, however, to view Dodge as a traditionalist. For instance, in the polemic triggered by his negative review of the first recordings of the Lu Watters band (the beginning of the traditional revival), he attacks "the purist attitude of disregarding all the best jazz since [New Orleans], and restricting ourselves to that style alone, including what was good as well as bad." And his excellent analysis of a Charlie Christian solo certainly reveals an open mind. But we must see Dodge in the context of his times.

One of the great paradoxes of jazz history is that the music's rich past was being rediscovered at the very time when its present began to undergo its most radical change. While Dizzy Gillespie and Charlie Parker were jamming in Harlem, Bunk Johnson was being recorded in New Orleans. Within a few years, both modern and traditional jazz had vociferous adherents, and they were soon at each other's throats in the pages of *Down Beat, Metronome,* and the smaller jazz magazines. Critical strafing soon escalated into a full-scale war between self-styled progressives (stigmatized as "dirty boppers") and traditionalists (stigmatized, more imaginatively, as "moldy figs"). In the course of this schism much mischief was wrought, and it lingered on for decades. Dodge never stooped to name-calling, nor did he spout any party line, but there is no doubt that the contentious climate had some effect even on him. (Certainly his enthusiasm for the French revivalist efforts of Claude Luter seems triggered by what was taking place in the modern realm.)

That Dodge is no dogmatist is clearly revealed by his record and concert reviews, where he can even admit that *Talk of the Town* is a "likeable old tune." Most importantly, these reviews, even when one might disagree with certain opinions, are free from cant and idolatry. He does not fall into the trap of reflexive reaction; even his great favorites, such as Johnny Dodds, are not sacrosanct. And throughout this book, he is always at pains to clarify terms, such as "commercial," here as applied to Louis Armstrong: "Armstrong sells very good. Maybe he went a little commercial in order to do so. But what do we mean by 'going commercial'? It is very possible that the change in Armstrong's playing made him sell well, or will it be said that in order to sell well he changed? If to sell well, one changed, then it is 'going commercial,' but if one changes with none but a musical motive and because of it sells well, I cannot see that we are justified in calling that person commercial."

Though his love for New Orleans jazz made him identify with the traditionalists, he transcended such categories. Thus, in 1955, he could state that "it is the attitude that we *must* benefit by proliferating modes held by both the modern academic and progressive jazz schools, which is fallacious. There may be artistic determinants requiring certain turns in art, but the artist should always feel free to curtail any practice no matter how arrestingly resourceful, newly arrived, or venerable it may be."

To Dodge, jazz at its best represents an antidote to the decadence and "sweetness" of romantic and post-romantic art. Even in his earliest piece, written in 1924 (but not published until 1929), he speaks of jazz as "virile, non-emasculated," and virile remained one of his favorite adjectives applied to jazz. He feels that too many listeners are motivated by sensuousness (tone) rather than esthetics (melodic line), but that "music's sensuousness is a static sort of pleasure." He says that the melodic line is the "essential element in jazz." It is not always easy to follow Dodge's reasoning in these matters, but it pays off to try. At the very least, it takes us, like a strong melodic line, into areas far removed from the usual jazz writing, with its emphasis on history or sociology or "begats."

Roger Pryor Dodge came to jazz with sensibilities very different from those of our time, but what he has to tell us is very much worth hearing. He was convinced of the value of this strange and different new music, and he was concerned about its survival and future. He did not think the music could grow artistically without a compositional methodology of a new kind, but this was something he could not bring about—he was, after all, not a composer. What he could do, however, was to contribute to an esthetic of jazz. "If hot jazz is as good as most of us believe it is (unless we are simply confining ourselves to expressions of personal pleasure), it is worth establishing why it is so valuable," he wrote in 1942, and this was what he wanted to do for "the most significant music of recent centuries."

To follow him in his search is always worthwhile. He often forces us to reconsider comfortable assumptions, and best of all, he helps us not just to listen, but to think. In this age of repertory jazz, Dodge's ideas about jazz notation and the nature of the transcribed solo as a medium for re-creation are of more than historical interest. Much more.

Dan Morgenstern

Portrait of Roger Pryor Dodge, New York, 1966. *Photograph by Pryor Dodge.*

Preface

My father, the late Roger Pryor Dodge, a dancer, choreographer and writer, was fervent about the performing arts and fortunate to discover jazz at a time when its folk element was still hot. Although the importance of his work was not entirely apparent during his lifetime, he now holds a prominent place in jazz history as one of the first serious writers to recognize jazz as a great music. This collection of his writings forms a seminal contribution to the understanding of the relationship of jazz to classical music and to dance, three art forms of passionate interest to my father and the subject of his lifelong creative and critical work.

His writing career perhaps took even him by surprise. A classically trained and practiced dancer, he was nonetheless capable of analytic and impassioned criticism. It all began in 1924 when he attended the second New York performance of "Rhapsody in Blue," which had opened to rave reviews. Upon hearing this reputedly new jazz, he bristled from the experience and felt compelled to enlighten the public that the music was not in fact real jazz. He was unable to publish his first article, "Jazz Contra Whiteman," written in 1925, until 1929, when re-titled "Negro Jazz," it appeared in *The Dancing Times,* an English review devoted to ballet. The article argued that "the word 'jazz' is being used too loosely and too indiscriminately by persons who have little perception of the true nature of the embryonic form now developing amongst us." This article was the prelude to my father's lifetime exploration of elements common to the development of both classical music and jazz.

The son of the academic muralist William de Leftwich Dodge, Roger Pryor Dodge was born in Paris in 1898 while his father was on one of his extended visits to paint and exhibit his work at the Salon. When the Dodges returned to New York, my father was raised in the artistic milieu of Greenwich Village and in the vicinity of Stony Brook, Long Island, where his father designed a neo-Greek villa with caryatids.

After quitting high school, he became interested in social dancing, a pursuit which brought him to the Ritz Carlton Ballroom in New York, where a dancing partner introduced him to Les Ballets Russes starring the great Vaslav Nijinsky in 1916. This proved to be a momentous experience for my father, and in 1920, at the age of 22, he began studying classical ballet and soon left for Paris to continue his training with Nijinsky's teacher, Nicholas Legat, and the maître de ballet at the Paris Opera.

Nijinsky had stopped dancing in 1917. My father realized that photography was the only available means to experience his greatness, and he had the foresight to preserve a photographic record of one of the greatest dancers of our time. He proceeded to order prints from the photographers who had taken studio portraits of Nijinsky in the various roles he created. His efforts resulted in the most comprehensive collection of photographic images of Vaslav Nijinsky, donated years later to the Dance Collection of the New York Public Library and eventually published in a widely acclaimed book, *Nijinsky Dancing.*

Upon returning to New York in 1921, my father entered the Metropolitan Opera corps de ballet, an engagement he held for six seasons while continuing his ballet training with Michel Fokine. Desiring a greater freedom of expression, he began exploring more modern techniques, including the Dalcroze School, Duncan technique, and classes with Michio Itow, who introduced him to vaudeville in his Pin Wheel Revue. His excursions in vaudeville continued, including a tour with the Marx Brothers and performances at the Shubert Theater among others, all the while dancing at the Met, and eventually with the Adolph Bolm Company, with which he went on tour to Argentina.

Although he was featured in John Alden Carpenter's "Skyscrapers," the first "jazz ballet," in 1926, my father's taste in jazz had been formed two years earlier when he first heard several jazz recordings including one by Bessie Smith. His continued fascination with jazz in the twenties led him eventually to choreograph dances to "East St. Louis Toodle-Oo," "Black and Tan Fantasy," "Yellow Dog Blues," and "St. Louis Blues," which were performed in various theaters including the Roxy. These dances were accompanied by Bubber Miley in Billy Rose's revue *Sweet and Low* on an extended engagement. He also performed to "The Mooche," "Call of the Freaks," and "King of the Zulus," with a number of these dances eventually recorded on film. He took one of his pieces to Paris for performances at the Paramount. But owing to back problems, his dance career ended in 1942.

My father's urge to revive his favorite blues solos from the twenties even extended to my own personal involvement at a young age. In 1960, when I was ten, he had me transcribe the trumpet, clarinet, and trombone solos he needed to have notated for the musicians with whom he was working. This project lasted a few years and was an integral part of the musical education he had begun with me four years earlier at the age of six, when he sat me down daily for a recorder lesson.

It was with these lessons that I was first introduced to Bach chorales and movements from dance suites. Later, after I had progressed to the flute, he brought out his hand-written, hand-bound selection of "Hot Solos" and opened it to Armstrong's solo in "Potato Head Blues." What for my father was one of the great solos of all time became my introduction to jazz. I played it almost every day until I had it memorized in different keys, an exercise my father had me do with most pieces.

The transcriptions were also "ear training" exercises, and at the rate of a nickel a measure, there was an added incentive for me to participate. Determined to be as accurate as possible, I often had to slow passages down to 45 or even 33 rpm in order to capture the intricate rhythms. Some of these rhythms were syncopated with dotted sixteenth and thirty-second notes or runs with sixty-fourths. Pitch could be tricky, as musicians often bent notes and recording speeds were not always in synch with playback.

With my musical transcriptions in hand, my father would meet and rehearse with his musicians. Later he would venture forth with his Wollensak to record them, usually returning dissatisfied, as the musicians improvised in a modern rather than period style. This important distinction was indicative of certain fundamental principles that guided my father in his pursuits. I remember the story he repeatedly told regarding the harpsichord revivalist Wanda Landowska when she claimed to play Bach "his way." According to my father, this style, based on dance, was in contrast to the accepted romantic approach to baroque music. He was so moved by Landowska's playing that he had his first wife, a classical pianist, attend her classes.

I never knew my father the dancer, for his professional career ended about seven years before I was born. However, at the age of 75, when I was 23 and living and studying flute in Paris, my father choreographed five baroque dances for himself that he intended to film. He designed an 18th-century costume for himself based on a print of Faune in *Le Triomphe de Bacchus,* to be worn with a papier-mâché mask of his own making. Three of the pieces were dances selected from Gluck's *Orpheus,* Bach's Wedding Cantata and a Handel Organ Sonata. The other two were a sarabande by Rameau and a dance by C. P. E. Bach. Although he didn't live to see the project completed, my mother did witness a very moving private performance.

My father also performed these last dances for Mura Dehn, who had been his principal dancing partner in the 1930s. In addition, he performed a new version of an early jazz dance composition. Later, Mura Dehn wrote that these pieces were a summation of his knowledge of and reflections on dance. She stated that my father wanted to "emphasize a style as known and felt by the average people of each epoch, performed by a dancer of middle age with artistry and taste—not a virtuoso. Watching him I thought George Washington could have danced that way. But to his jazz piece he gave more brilliance, more wildness and abandon than ever."

Fortunately, through the films my father made of his jazz dance compositions in the thirties, I too can remember him as a dancer. Besides enabling me to appreciate his creativity and his unique angular style, seeing my father perform in these films has helped me to understand why dance was fundamental to his concept of musical history and development. He experienced his own most vital expression in dance and in the music dance brought to life, an experience that informed the ideas he shared in his writings.

P. D.

New York
June 1995

Contents

HOT JAZZ AND JAZZ DANCE

Negro Jazz

Ernest Newman, the eminent musical critic of the London *Times,* some short time since allowed his keen judgment to be overthrown in the violence of a controversy for and against jazz. A reading of the remarks of Mr. Newman and the answers of his opponents convinces me that jazz needs protection not so much against its enemies as against its friends.

The word "jazz" is being used too loosely and too indiscriminately by persons who have little perception of the true nature of the embryonic form now developing amongst us. It is no wonder that critics are unable to agree when no two of them are discussing the same thing. The word "jazz" as it is currently used seems to cover both true jazz and popular music in general. It covers Paul Whiteman, George Gershwin, and Irving Berlin, none of whom I consider as belonging to the ranks of jazz at all. But if these men are not exponents of jazz, who is? And what is jazz?

It has been said that there is no such thing as jazz music and that what is commonly called jazz is only a manner of playing music. This is partly true. Any composition can be transformed into a sort of imitation jazz, either by an instrumental technique that applies to the melody, stunts that cannot be notated, or by subjecting it to a jazz-like rhythmical treatment that is susceptible of notation. Critics lay stress on the appearance in jazz of instrumental stunts that cannot be notated as evidence of its ephemeral nature; but this, it seems to me, is a perfectly natural condition. Because these effects cannot be notated now is no proof that they will always be incapable of notation.

The Dancing Times (London), October 1929.

Those that prove to have value will so develop that in time they will come to be written down, either within our own notation or by creating one of their own. In fact, it seems almost inevitable that a type of music arising from the folk and not from the academies should give vent to sounds that have not yet been foreseen by our present system of notation.

There are two styles of jazz as we have it now which combine to form the most perfect type of jazz—the rhythmic treatment which can give a jazz flavor to any composition and the "blues." It is this rhythmic treatment which is generally referred to as jazz, and it is a technique which, once acquired, can be applied to any type of music. The most familiar variety is that which adds the thinnest possible veneer of jazz to the original composition, which, as a rule, is a popular song or a mediocre classic. The melody of the selected composition either undergoes a slight rhythmical distortion to obtain the jazz effect, or is provided with a background of jazz accompaniment; but in both cases enough, unfortunately, of the original is left so that the result of this treatment, as applied to the lesser classics, is the loss of whatever small value the original possessed.

The blues have developed from the spirituals, which had their genesis in the simple but powerful four-part harmonies of our hymns. They are the result of straining a formal and highly cultivated music through the barbaric and musical mind of the Negro. The spirituals are the Negroes' digestion of our hymn tunes. The blues, in turn, are another step in the development, as Handy's book does not neglect to explain, and it is this material that constitutes the basis of true jazz. Into the Negroes' singing or playing of the spirituals crept the savage rhythms that had shaped or been shaped by the ancestral dances of the tribe and these formed in time a definite playing style; and in recognizing this style we recognize jazz.

As I have implied above, the point at which pseudo-jazz stops and jazz begins is largely a matter of degree only. The Negro bands often take music foreign to their own culture and base their jazz on the very popular songs or classics which form the foundation of certain of the Whiteman performances. The result is most interesting. But it is very different from the civilized and elegant versions of the symphonic jazz band. For the Negro has taken the least possible contribution from the notes of the melody. He distorts it beyond recognition, makes of it an entirely new synthesis and his product is a composition—whereas that of the symphonic band is no more than a clever arrangement. This is indeed interesting, but vastly more so are the two forms entirely Negroid, entirely born of the jazz impulse, the blues themselves and that faster form, the stomp.

A common criticism of jazz is that it is nothing but rhythm; that if the drums were removed there would be nothing left. This is an argument that will hardly hold water. The rhythmical peculiarity of jazz is not superimposed upon it by the drums; instead the whole orchestra is engaged in creating the rhythm; and though this rhythm may have been introduced at first by the drums, it is not their sole property. Take away the tom-toms and the steady tom-tom beat can still be heard in the rhythmical pulsations of accompaniment and melody. The development from the hymn and spiritual to the blues and stomp is paralleled in the history of classical music by the development of the fundamental chant as influenced by the rhythms of the dance tunes; a movement which culminated in the rhythmic, but certainly not drum-like, works of Bach.

Up to the time of Bach, music was an art of the people which found expression through individual geniuses, of whom Bach was one and the last. Jazz is certainly, as was the contrapuntal music, a music of the people. And like the music of the contrapuntal period, it is distinguished by the same bare melodies, stripped to fundamentals, driving with a continuous flow of musical thought to a natural and inevitable conclusion. There is no pause, no turning to one side, no attempt to lift the music to uncertain heights, as in romantic music of the pre-decadent period. Jazz melodies, like contrapuntal melodies, inspire both melodic and contrapuntal development and do not depend on full-throated orchestration to cover a lack of fundamental virility. Just as a fugue carries in its melody the rhythmic beat that never ceases to pulsate until the end of the composition is reached, so a jazz melody contains a rhythm that carries it through to its conclusion without pausing for false emotional effects.

This old contrapuntal music is marked by a steady pulsating rhythm that holds it to its purpose of musical development. The compositions are masterpieces of structural beauty; they have an inevitability about them the Romantic school has never attained, due to its constant vacillation in both tempo and content—its melodies that refuse to stand alone without the aid of expressive "interpretation." Since the time of Bach, music has ceased to be a high development of folk art, and become merely a vehicle for self-expression. It is a decadent art and as is the general rule in arts that are in a decadent state it has created for itself forms that are at once the result and the proof of its decadence.

Let us take the old chant as a beginning, and carry the academic line to Bach, its highest contrapuntal development. Down from this strong and vital line fell constant little lines to the people. It is these adulterated offshoots which the people seized upon, revitalized and finally sent back to the academy. Thus the new school was slowly seeded, rising contemporaneously with the old school which was nearing its highest point of development. This impregnation of the people by the contrapuntal school resulted in the birth of the Romantic school; the second beginning. It was around these folk tunes Haydn built his little variations—these tunes which have a long musical line, and are not to be confused or associated with the primitive chants. From Haydn through Strauss the Romantic school has thrived upon and now practically exhausted Occidental folk tunes. From Haydn through Strauss men have been expressing themselves by means of variations on a folk tune, be they impromptu or symphonic—variations never intrinsically musical, depending as they do on expressive "interpretation." This school has tapered out to its logical and decadent conclusion. It was ripe under Beethoven. It is now and has been for a long time rotten. Consequently jazz which was born out of the first beginning and has been kept by the Negro in the same stern school, is of an altogether different caliber than the themes that are treated by the Romantic school. To subject jazz to Romantic treatment is an outrage comparable only to romanticizing the pre-Bach music.

Now what have our jazz reformers done? The writers in America have not tried to develop jazz, but have applied it to other ends than its own. Gershwin selected that stalest of decadent forms, the rhapsody; an episodic form that allows the furthest possible departure from the logical carrying to completion of a single musical idea, and hence, probably the form more remote from the genuine jazz ideal than any other. He produced an episodic and disjointed composition of which the separate parts resembled jazz, but which had not, as a whole, the continuity and directness of true jazz music. It was no more than we should have expected, however, from a composer who is not, whatever his publishers may think, a jazz writer, but simply a first-rate popular song writer. The jazz contribution of Whiteman's famous orchestra consists in the application of little jazz rhythms or syncopation to popular songs and popular classics. This method is in direct contradiction to that of a certain composer who used two blues as themes for a symphonic work. Both took a part of jazz to use for other ends; but neither of them wrote jazz. Some moderns who have been influenced by jazz, as almost all of them have been, have shown better taste in avoiding classical forms when they approached jazz. Stravinsky, it is true, wrote a concerto very much in the medium of jazz, but as a rule he makes no attempt, when he composes a jazz piece, to drape it on the structural form of a symphony or an opera.

Between those who seize upon such superficialities as syncopation to mean jazz to them, and those who give the word a broader definition we discern a decided difference in viewpoint. We may suspect that the term "lively art" as applied to jazz, means for the first, "frivolous art," and for the second, "vital art." Whiteman, Gershwin, Berlin, and others are exponents of frivolous art; if there is anything bogus about jazz it is certain to be found in their type of jazz. Taking popular songs that are hand-me-downs from the romantic classics and "heart songs," Whiteman's orchestra paints them with a jazz rhythm, and is forthwith acclaimed a pioneer of jazz. To these jazz artists who confine themselves to syncopating the classics, I have only to say that their conduct is a mild outrage. I do not care for the jazz players who do it, but it is difficult to get excited if the "Anvil Chorus" from *Il Trovatore,* the "March" from *Aida,* or even the "Moonlight Sonata" are jazzed. Those classics which would be in any sense desecrated by being jazzed have never been touched.

It is this syncopation of jazz which has led so many of its imitators astray. There is a great deal of syncopation in jazz, but it is all subordinate to the more important steady two-four or four-four beat. Composers imitating jazz—Stravinsky is again an example—produce syncopation a great deal more intricate and less dependent on the main beat than that found in true jazz. The treatment is the result of the natural impression of the novice in a new type of music, who has seized upon one of its features and exaggerated it. As we become better acquainted with jazz rhythms, our new jazz compositions will undoubtedly incorporate a more natural type of syncopation and abandon tricks of rhythm that may be momentarily amusing but are without permanence. It seems that the development of jazz lies not so much along the lines of further syncopation as it does in the possibility of harmonic and contrapuntal advance—counterpoint, a device which is already used abundantly in jazz. Much as jazz is supposed to dominate our modern music, it is really rare in its pure state. Since its appeal is still to the few, except in adulterated form, the big cities in America are not rich in fine jazz orchestras. Instead, the chief jazz orchestras are scattered all over the country and the only feasible way to hear good jazz in quantity is through phonograph records.

What, then, is the condition and status of jazz? To my mind the creative playing found in *low-down* jazz is establishing a stronger form than any that has arisen for centuries. It is a musical form produced by the primitive innate musical instinct of the Negro and of those lower members of the white race who have not yet lost their feeling for the primitive. It is appreciated not only by these primitives but by those who can participate in the enjoyment of a stern school and can appreciate its vitality while it is still in the process of development. It is disliked by those who know it only in its diluted form and who, often under the impression that they are defending it, desire to bring about its fusion or confusion with the windbag symphony or the trick programme closing rhapsody. There is no important similarity between the orchestras of Paul Whiteman, Jack Hylton, etc., and such organizations as Ted Lewis and His Band, Fletcher Henderson, and His Orchestra, Mound City Blue Blowers, King Oliver's Jazz Band, Thomas Morris and His Seven Hot Babies, Red Nichols and His Five Pennies, Duke Ellington and His Orchestra, Louis Armstrong and His Hot Five, and Jimmy O'Bryant's Famous Original Washboard Band.

Also, among the many soloists who have contributed to the development of virile non-emasculated jazz may be mentioned the singers, Bessie and Clara

Smith; the pianists, Seger Ellis and Jimmy Johnson; the clarinetists, Boyd Senter and Wilton Crawley; the cornetists, Miley, Swayzee, Johnny Dunn, and Louis Armstrong; the trombonist, Joe Williams; the guitarist, Ed Lang and the team Bobby Leecan and Robert Cooksey playing the harmonica and guitar. These are a few of the pioneers, scattered far and wide over the country, only to be brought together through the medium of the phonograph. But their music will grow; the confusion between true jazz and its bastard children of the polite orchestras will become less; musical publications will begin to recognize the existence of the true jazz; and it will no longer be necessary to defend jazz from its defenders. Not that it really requires defense; for it will exist so long as the Negro lives *in* our civilisation but not *of* it.

Serge Lifar: A Study

With the passing of Diaghileff, I think it proper to say something about his last great dancer, Serge Lifar. He is not only the last great dancer who was with the company, but also stands beside Nijinsky, Bolm and Massine as one of the four great dancers whom Diaghileff brought forth. He brought something new to the company, a something I would call fine *emotional expression through movement*—not to be confused with pantomime. In this he recalls somewhat the later style of Isadora Duncan, the difference lying chiefly in the fact that she breathed the music, while he moves with the music as a background. There would not be this unfortunate difference, aside from the disparity in their degrees of greatness, if it were not for the antagonism between the two schools. The school of ballet still considers that she had no technique whatsoever, while she always felt that ballet was stilted and unexpressive. The fact is, she got along without the old technique while the Diaghileff company, though its great artists were undoubtedly influenced by her, was much in need of her technique in many ballets. If there had only been more of an understanding between them, I am sure, in the case of Lifar, he would have been a much greater dancer than he is.

In his return home in "The Prodigal Son," uninteresting as a ballet, but particularly happy as a vehicle for the emotional expression, he held the stage for an extremely long time in a way no other dancer of Diaghileff's company could have done. Here there was enough choreographic movement to make the scene quite different from the actual home re-

The Dancing Times (London), May 1930.

turning of a son, and also enough choreographic movement to put this scene beyond the ability of even a very good actor. In "Apollo," wearing a gold wig and a little red tunic, his figure not unlike that of a Greek athlete, he managed by means of his personality and a stylistic modern choreography to recapture more of the Greek ideal than either Fokine or Nijinsky were able to in "Daphne and Chloe" and "Narcissus." Always in "Daphne and Chloe," and especially "Narcissus," I felt a great breach of aesthetics. The over-developed legs of the dancers, so much a part of tights and ballet seemed to lose their aestheticism in these productions, whereas Lifar, disregarding the bounding style of ballet, walked the stage in simplicity, giving one the feeling of a youth of that period.

In "The Cat," Lifar's facial intensity, notably when he came down stage and did a step in place, giving the impression that he was running, was a very good example of the just amount of self-expression to add to such abstract movement. But even his ability could not make up for the lack of modernism which this ballet was so obviously trying to express. When anything modern slips into the style of the department store cubist, lacking a true feeling for modernism, it seems to lose whatever value it might have had otherwise. Also the music, such an important part of a ballet, was especially mediocre. In fact, considering most decor is done in that experimental spirit, which, if necessary, on the face of it excuses itself, I should attribute the failure of "The Cat" principally to the lack of even an experimental modernism on the part of the music composer. It was in "The Fox," the most modern ballet of Diaghileff's last season, simple but true modernism in both music and decor, that Lifar revealed himself as an accomplished choreographer. He choreographed the ballet for four male dancers, three to be duplicated by acrobats—the acrobats alternating with the dancers. For my part I found the tumbling a little tiresome, and would have preferred seeing Lifar take one of the leading parts himself. Nevertheless, the inventive movements he created to carry out this amusing fable, the well-balanced even line of *tempo* maintained through the entire string of episodes and a fine sense of design, made "The Fox" by far the most interesting choreographic work of the season.

Strictly speaking, Lifar is not a ballet dancer. He *uses* ballet, but does not *do* it; and he uses it because the Diaghileff company was built on it, and this was the only technique they studied. In fact, the corps de ballet, when called to rehearsal, say for a Spanish ballet, would get the first inkling of this new technique at rehearsal only! There was a complete lack of regard for the previous study or training such a technique requires. Some of the dancers, it is true, studied character techniques privately, but few made anything of them except in so far as to give local color to a production. It seemed to be taken for granted that a dancer equipped with an entrechat six and a turn in the air, could, when called upon, sustain long passages where ballet footwork was being ignored entirely. It is surprising that such an arm technique as is possessed by any Oriental school, or even such a technique as Michio Ito has evolved out of his native dance, never seems to have been considered worth the experiment. Of course, when Lifar does ballet dancing it is, as far as it goes, exact enough—but it is stiff and unfeeling. He does not work within the character of his technique. Now this is a great handicap, not working within the character of any technique; of merely using a school for the strong physical foundation it can give. In great techniques such as ballet, natural movement, some national dancing and oriental dancing, the aesthetic appeal of the character of the technique alone, can reach out into space—even though the exponent may not be great. There is a vast difference between he who has only a superficial knowledge of the character of a technique, however great an artist he may be, and one who, besides having the technique, also feels intently its special character. Lifar works within no vital dance form. He lacks the special vitality that can only come out of a school. If we consider that dancing covers anything from purely artificial dance form to the emotional expression of a Duse, we will find that Lifar, at his best, is far removed from the dance form and well along the line towards Duse. Nevertheless, even though he may not have this solid character of any school, he brings to a stage presentation such a personal spirit of the dance, that his movements are of a real dancing quality—a quality not easy to acquire outside of a technique.

In thinking over what Lifar has done, it seems to me that in individuality he stands close to Nijinsky and Bolm. Massine is certainly the best all round dancer, ranking particularly high in his Spanish work, but I feel his attitude towards the dance has always been more that of the choreographer than the dancer. Considering the power Lifar has shown in the choreography of "The Fox," and in his dancing in general, we can guess that his strength is hardly tapped; though it remains to be seen what he can do free-lancing. Other Diaghileff dancers did not do much after they left the company, and if anything of value was done, it was never a step forward. Gala concerts now and then do not speak very loudly for the dancer who wishes to be considered a creative artist. Lifar's position is more hopeful in that he was not given the chance to run dry in the company. Though one personal handicap I have observed is that so far he has seemed unable to resolve a whole concept into himself and send it out again with a new and vital understanding. One might compare him to a distinguished soloist of an orchestra ensemble, as opposed to a recital artist. He holds the stage but he never takes it. This would appear to make it necessary for him to work in a ballet ensemble. However, if he has anything in him he must show it now, as he is no longer a part of the Diaghileff tradition; circumstances have forced him into competition with the recital artist. It is not an easy job for a beginner.

Harpsichords and Jazz Trumpets

Hound & Horn, July–September 1934.

In the history of jazz we find that the immediate result of the bringing together of the four-part hymn and the Negro was the spiritual. The spiritual, though concededly the most original any one thing the Negro has contributed *outright,* seems to me chiefly significant as containing the first seed of jazz. Unfortunately a great deal more critical interest is expended on the spiritual out of the *church,* than on the jazz out of the *dance hall.* In fact, quite aside from my personal conviction that jazz is by far the most important music of the two, the spiritual is so well taken care of that new collections are constantly appearing, whereas jazz, taken for granted as contemporary dance music is scarcely acknowledged, let alone notated. For we can hardly consider a popular song publishing company's issue of the simple ground bass, or harmonic vamp accompaniment with occasional uninspired instrumental suggestions, as the written counterpart of that extraordinary and highly developed music. This present lack of adequate notation can be compared very simply to the similar musical situation in Europe during the 16th, the 17th, and even the 18th century. In this connection a few lines from a letter written by a certain André Maugars in 1639 upon the occasion of a visit to Rome, give an exciting picture of the times:

> I will describe to you the most celebrated and most excellent concert which I have heard. . . . As to the instrumental music, it was composed of an organ, a large harpsichord, two or three archlutes, an Archiviole-da-Lyra and two or three violins. . . . Now a violin played alone to the or-

> gan, then another answered; another time all three played together different parts, then all the instruments went together. Now an archlute made a thousand divisions on ten or twelve notes each of five or six bars length, then the others did the same in a different way. I remember that a violin played in the true chromatic mode and although it seemed harsh to my ear at first, I nevertheless got used to this novelty and took extreme pleasure in it. But above all the great Frescobaldi exhibited thousands of inventions on his harpsichord, the organ always playing the ground. It is not without cause that the famous organist of St. Peter has acquired such a reputation in Europe, for although his published compositions are witnesses to his genius, yet to judge of his profound learning, you must hear him improvise.*

He also adds "In the Antienne they had . . . some archlutes playing certain dance tunes and answering one another. . . ." (Which, by the way, helps bear out the theory that all great music, even church, leans upon and is developed by the dance.) Now, as then, there is such a musical bustle and excitement in the air that no jazz musician needs more than a harmonic base or a catchy melody, to play extempore in solo or in "consort." Improvised jazz is comparable to such music as Maugars heard at St. Peter's, and though the distorted (to our ears) dynamics and instrumental tone quality of the negro brass and woodwind sound harsh and fantastic at first, like Maugars, one gets "used to this novelty" and finally takes "extreme pleasure in it." Moreover its start and development occurring during our lifetime, we should feel its power tremendously and have a definite emotional reaction; a purely contemporary enthusiasm, which can never be experienced for a bygone music, no matter how great it was.

When we consider that not only Frescobaldi but Handel, J. S. Bach, Haydn, Mozart and even Beethoven were all great improvisers, we realize it was intellectual superiority which made them write down what they could improvise more easily—not the limitations of a modern academic composer. We realize that such individuals who could improvise the most difficult and inventive counterpoint and fugue on a keyboard, needed only to push their minds a step further to dispose parts to an orchestra. On the other hand when we consider that the Negro instrumentalist is apparently uninterested and incapable of writing down his own real improvisations, that his inspiration is absolutely dependent upon harmonic progressions provided by other instruments than his own and that though he takes great pleasure in it, his counterpoint is the happy accident of a confrere "getting off" at the same time, then, we can understand perhaps, why this structure of jazz, this musical development by the instruments themselves (and the different musical styles that implies) is at a standstill as far as native, written composition for solo or symphony goes.

*Arnold Dolmetsch's *The Interpretations of the Music of the XVIIth and XVIII Centuries.*

To appreciate the significance of the act of improvisation, we must not overlook the fact that improvisation is absolutely imperative to the development of an art form such as music and dancing. On the other hand we must not overvalue the ability itself, as at the time, this resource must be so commonplace that every performer can avail himself of it with perfect ease. It is when the spirit of a folk school of music so excites the folk artist that it is the most natural thing in the world for him to make variations on every melody he hears, or to invent new melody on a familiar harmony or to extemporize in general, that we find a real freedom of invention. It isn't essential that the individual accomplishments be masterpieces, but it is essential that the whole group experience the same improvisatory spirit. This is the only way, in my opinion, to insure telling change and growth. When this atmosphere does not prevail, the creations of the solitary individual, no matter how revolutionary, always will lack the force of those of a much less important man who has the basic, group impetus back of him. The creations of the former, after the first shock, become more old-fashioned than the most common material of folk improvisation. Richard Wagner is a good example. Moreover it must be understood that by group feeling I do not mean the will to organize a group. A Dalcroze, a Mary Wigman, Les Six, or a Picasso may impose rules on a train of satellites, but instead of receiving back new force and inspiration from contemporary fellow artists, on the contrary, they are run into the ground by a coterie of pupils and imitators. Also it must be understood that what passes for improvisatory art in our exclusive little studios of both dancing and music, where the girls and boys find new freedom in expressing the machine-age or the dynamic release of the soul, or in musical combinations in the manner of the written works of Liszt, Scriabin, Milhaud, or Gershwin, is not the art of improvisation that I discuss.

Whereas the academic child prodigies of today content their masters and their public with nothing more than a mature reading of a score, even a child in early days was expected to improvise, and the quality of his extemporaneous playing was the criterion by which he was judged. When Mozart held an audition for the child Beethoven, he fell half asleep listening to what he presumed were prepared pieces. When Beethoven, greatly vexed, for he had been improvising, insisted the Master give him a theme and then made countless variations on it, Mozart is reported to have jumped up crying, "Pay attention to him: he will make a noise in the world some day." And when Mozart as a child had played for Papa Haydn, he had shown the same prodigal invention. And remember Haydn, Mozart, and Beethoven were at the close of a great musical period and were improvising as that tradition had demanded, even though they themselves were the founders of the Romantic virtuoso movement. It is for that involuntary, impersonal connection with the past that

I mention them, not for the new avenue they opened up for the 19th century.

If we turn to the musical literature of the 17th and 18th centuries we find that no two artists were supposed to play identical variations and ornaments on the same piece; on the contrary, the artist was expected extemporaneously to fill in rests, ornament whole notes, and rhythmically break up chords. The basic melody, as in jazz, was considered common property. If the player exactly imitated somebody else or faithfully followed the written compositions of another composer, he was a student, not a professional. However, to the student we owe the inspiring textbooks written by such masters as Couperin, Ph. E. Bach, Geminiani, Mace, Quantz, etc. This is a literature which from all sides presses the fact that if the pupil has no natural inspiration and fantasy in melody, no feeling as to how long to trill, or where to grace notes, or in what rhythm to break up a figured bass, he had better give up all hope of pleasing his contemporaries. It was not the contemporary virtuoso, professional or semi-professional musician who benefited by the few notations in circulation. It was the student. The fact is, that in a healthy school of music it is a drawback to have to read music. It is unnecessary to write it for your own convenience and too much trouble to take the infinite pains necessary to notate a fellow artist's daily compositions. In such a school it is the well-balanced composer not depending on written notes himself, but with an eye on posterity or with pupils to interest, who takes the pains to notate more than the simple harmony or melody.

At that time one listened first, as one does now in jazz, for the melody, then recognized the variations as such and drew intense enjoyment from the musical talent familiarly inspired. Instead of waiting months for a show piece to be composed and then interpreted (our modern academic procedure), then, in one evening, you could hear a thousand beautiful pieces, as you can now in jazz. Instead of going to a dance hall to hear Armstrong, in earlier times you might have gone to church and heard Frescobaldi; or danced all night to Haydn's orchestra; or attended a salon and listened to Handel accompany a violinist—with his extemporaneous variations so matter of course; or sneaked in on one of Bach's little evenings at home, when to prove his theory of the well-tempered clavier he would improvise in every key, not a stunt improvisation in the manner of someone else, but preludes and fugues probably vastly superior to his famous notated ones. Academicians of today can improvise in the styles of various old schools but the result is commonplace, not only because of the fact of improvising in a school that is out of date, but because such an urge is precious and weak in itself, limiting the improvisor to forms he has already seen in print. Even in contemporary modern music, the working out is so intellectual that the extempore act does not give the modernist time to concoct anything he himself would consider significant.

Contrary to the modern academician and similar to the early composers, jazz musicians give forth a folk utterance, impossible to notate adequately. For even if every little rest, 64th note, slide, trill, mute, blast, and rhythmic accent is approximately notated and handed back to them, it is impossible to get them to read it. To read with any facility these extremely difficult improvisations takes a highly developed academic training, a training which is not general, usually, till the best part of an improvising period is past. The great Negro musicians are not pianists or harpsichordists consciously contrapuntal. The very range of the keyboard which stimulated the 17th- and 18th-century European mind to great solo feats of combined polyphonic and harmonic invention only suggests to them, for the most part, a simple harmonic, rhythmic accompaniment. The Negro is par excellence an instrumentalist: a trumpeter, a trombonist, a clarinetist. He is still musically unconscious of what he has done or what he may do. But, do not conclude from this that jazz music is still at the simple folk-tune stage. Far from it. For though the birth of the spiritual was the birth of a new folk song containing the seed of jazz, jazz itself is something more than just another folk tune. Jazz has reached the highest development of any folk music since the early Christian hymns and dances grew into the most developed contrapuntal music known to history.

To understand even better the source of jazz, remember the spiritual is a song, a highly developed hymn if you will, compared to which the blues, the seed manifestation, is really a step backward in the direction of the chant: a step altogether natural and necessary for a new art form to take, as witness the retrogression of early Christian music surrounded by Greco-Roman culture. Although the simple sing-song monotonous way of both blues and spirituals reveals a lack of depth, comparatively foreign to the old chant, this is to be explained, I think, by an appreciation of the vast difference in direct antecedents. The Negro received his little bit of the greatness of a choral, from the Protestant hymn mixed with Moody and Sankey. Whereas the Gregorian chant came out of the austere Greek modes mixed with the passionate Semetic plain song. Also, the tunes which skimmed along in the drawing-rooms and music halls during the whole of the 19th century were principally polkas, Irish reels, jigs, or the schottische. These the Negro was quite naturally exposed to. So as he was breaking down the four-part hymn into the spiritual, so was he, through his own heritage of *rhythm,* twisting this music of the marches and jigs into first, a cake-walk, later—ragtime. To play this early American dance music he had to accustom himself to the white man's musical instruments; and it was this familiarization which laid the foundation for his extraordinary instrumental development into jazz.

Ragtime, we now perceive, was the rhythmical twist the Negro gave to the early American dance tune. Here, the different instruments were finding their places in the musical pattern and already daring to add their own peculiar instrumental qualities. But—suddenly, the whole breadth of melodic and harmonic difference between the *folk-tune* stuff ragtime was made out of, and the *chant* stuff the racial blues were made out of, touched something very deep in the Negro. He found himself going way beyond anything he had done so far. For he had now incorporated his own melodic blues within his own syncopated dance rhythms and miraculously created a new music—a new music which moved him so emotionally that jazz bands sprang up like mushrooms all around him. The blues, retrogressed hymn, secular spiritual, had fathered itself by way of the clarinet, trumpet, trombone, banjo, drums, and piano into a rebirth, and christened itself JAZZ!

Now, the many things that go into making a *playing style* suit one instrument rather than another, are usually taken for granted. As a matter of fact they are the result of an experimental development which takes time and is very interesting to trace. The playing style of the harpsichord was not evolved in a day and neither was the playing style of the jazz valve trumpet. Both started by emulating the human voice (the harpsichord by way of the organ); that is, they took their melody from the singers; and both twisted this song into a stronger instrumental form. Taking these two examples as broadly representative of their respective cultures, we can see of what little importance, after the first vital impulse, the human voice was in the development of these two musics. I doubt if the voice could ever carry the development of a music very far without the advent of a composer, as there always seems to be such satisfaction for a folk singer in repeating verse after verse and letting the words alone be inventive. Though the harpsichord seems a very complicated instrument to compare alongside the single-noted valve trumpet, or a slide trombone, nevertheless I feel more of a true basis of comparison here than with, say, the trumpet and violin. We moderns only know the present virtuoso violin, an instrument without any real inventive playing style of its own—an instrument merely swinging back and forth between the imitative sweet singing of a tune and the highly developed musical figures lifted out of keyboard music. But a careful, lively carrying out of all the turns and graces of old music on the harpsichord can give us a fairly clear outlook on the playing style two or three hundred years ago; and only the extreme artificiality of the harpsichord has made this possible. Any instrument with the dynamic range of the modern piano and violin possesses, possibly, a clean crystallized style at the outset, but the traditional playing style can be absolutely swept away by one little wave of romanticism. The mechanical construction of the harpsichord itself has stood in the way of any such collapse into smooth and suave decadence.

If in the harpsichord music of the 17th and 18th centuries we find crystallized the various styles of the other instruments, it is due to the fact that while that instrument lasted, a fairly traditional way of playing persisted. Since the birth of the augmented symphonic orchestra, one hundred years or so, we have been listening to lukewarm instruments, some forgetful of a playing style originally belonging to them and others unaware of a playing style possible to them. They have no bite in performance. They are completely swamped under the arbitrary dictum of a conductor reading the arbitrary dictum of a composer. Such a clever, dramatic juxtaposition of instruments, as indicated by Stravinsky seems to me no more than clever, and absolutely no more than one man can do without help from the instruments themselves. And I think, whatever his followers may be still doing, he himself is aware of this in some sort of a way and feels that he is tired of exploiting the folk tune, horizontally, vertically, atonally, seriously, or comically. I do not call intellectual messing around with the tone colors of atrophied, academic conservatory instrumentalists composition in significant instrumental playing style. And I know, merely intellectually, one could never invent such a style. A *playing style* does not spring out of subjective interpretation or subjective composition. It springs out of primitive group feeling spreading itself deliriously, growing and feeding upon itself. Out of this feeling may or may not appear conscious, composing artists. Their appearance is, however, the beginning of the end, for from then on the group tends towards *listening,* not *participating,* and it is not long before the composer becomes one of a small class, forced to fall back on himself and his kind for nourishment. In this connection consider the most natural of instruments, the human body. In 17th- and 18th-century ballet, we know there existed a highly significant and artificial movement and posture. Now we see its complete romantic disintegration—the brief spurt of modern ballet being more of a healthy modernist criticism than actual healthy, artificial dancing. Of the old technique of ballet all we have left is disintegration; and a revolt against disintegration as intellectually manufactured as Schönberg's revolt against harmonic accord.

The shaping of melody by the instruments involved is something I feel accounts not only for the character of old European music but for the character of jazz; and the development of the latter I have had the opportunity of observing. First, the trumpet, piercing and high pitched, dominating the whole orchestra as it could, took over the principal presentation and variations of the melody, something hitherto left to the violins. So the importance of the first violinist vanished, the first trumpeter taking his place. In a limited way the trumpeter already held this position in the military brass band, but there he was either traditional *cor de chasse, cor de bataille,* or simply playing violin or voice music relegated to the trumpet. In a jazz orchestra he is inventing his own music and doing things previously considered impossible on such an instrument. In order to satisfy a wild desire to play higher and higher he blew harder and harder and in the process made unavoidable squawks and fouled notes.

These not only surprised but delighted him, and now, though he has a trumpet technique inferior to none, he still blasts and blows foul notes in beautiful and subtle succession, and with his extraordinary manipulation of mutes gets a hushed dramatic intensity that entails harder blowing than ever before. This difficulty of performance has kept the trumpet, so far, melodically inventive in the hands of the inventors, plus a few white imitators. But lately even the great artist Louis Armstrong has fallen into a florid cadenza style, induced, I fear, by excess technical ease. Armstrong has always favored the "open" manner of trumpet playing and a melody of the wide, broken chord variety, seemingly impossible in range. Even when he sings he is really playing trumpet solos with his voice.

The natural playing style of the slide trombone quite obviously would be dictated by the rhythmic movement of the arm sliding back and forth, something no trombonist has been allowed to feel heretofore. The Negro trombone player has become a sort of dancer in the rhythmic play of his right arm. He makes this instrument live, by improvising solos as natural to a trombone as the simplest of folk tunes are to the voice. This cannot be said of the trombone in any other music save jazz. However, as important as this instrument is, I find because of its low register and a certain cumbersomeness in size that the trumpet has gone beyond it in inventiveness, even carrying further the trombone's own newly created rhythmic and melodic twists. This copying of one another's style we meet constantly, and in order not to detract from the original creative importance I wish to distribute amongst the various instruments, it is well to understand how that which one instrument creates another may incorporate to more complete advantage. In other words, many trumpets can play in trombone style but I've heard only one trombonist create trumpet melodies, and he is Joseph Nanton (Tricky Sam) of Duke Ellington's orchestra. Theirs is a lazy style, which the trumpeter has seized upon with the rhythmic instinct of his mind and transposed to the trumpet for variety's sake, but which they, having conceived, are confined to. At this stage of the game if we had to choose between trumpet and trombone we would find the trumpet more of an all around instrument.

The faculty of imitation is possessed to a high degree by the pianist and almost to the exclusion of originality. This of course has not tended to make him an important contributing factor to jazz. The piano has been pretty generally relegated to the position of harmonic and rhythmic background and its occasional excursions into the foreground are not noticeably happy in inspiration. Owing to the conspicuous commonplaceness of our virtuosi, the Negro pianist only too easily slips into the fluid superficialities of a Liszt cadenza. This tendency of the Negro to imitate the florid piano music of the 19th century which he hears all around him has kept the piano backward in finding its own jazz medium. It takes a very developed musical sense to improvise significantly on the piano; a talent for thinking in more than one voice. The counterpoint that jazz instruments achieve ensemble is possible to a certain extent on the piano alone, but this takes a degree of development jazz has not yet reached. The best piano solos so far, in my opinion, are the melodic "breaks" imitating trumpet and trombone. Lately the pianist has found some biting chords, and felt a new desire to break up melody, not only rhythmically as inspired by the drum, but rhythmically as a percussion instrument fundamentally inspired by its own peculiar harmonic percussion. This perhaps, will lead him to contribute something no other musician has. Claude Hopkins is a notable example of such a pianist, but on the other hand Bix Beiderbecke (white) is doing this, and seems to have lost all sense of melody in the process. His intellect has gone ahead of his emotion, instead of keeping pace with it. That is the main trouble I find with our "hot" white orchestras. All they have they got from the Negro, and they are a little too inclined to fling back the word "corny." Until their own output proves more truly melodic or the Negro has completely succumbed to the surrounding decadence, they are still melodically "catching up" to the Negro.

The clarinet has been the instrument of very inventive players but somehow its facile technique has inspired florid rippling solos only too often. The saxophones have found their place, playing the background harmony in threes and taking the "sweet" choruses. Though I dislike the way the saxophone is generally played, it can be as "hot" as the clarinet when it is "jazzed up"; however, it is mostly used for soft, sentimental passages. The drums on occasion have qualified the playing style of the entire orchestra, as in the old washboard bands. Nobody who has heard a clarinetist used to playing in conjunction with a washboard, can have missed noticing the persistently syncopated and galloping style induced by the incessant rubadubadubadub of the washboard. Amazing things have been done with kettle drums, and any good drummer can take a "break" and off-hand crowd into it more exciting rhythms than a modernist can concoct after the lengthiest meditation. The violin has not been favored by the hot jazz bands although when "jazzed up" it falls in very readily with the spirit. Here again we can see the same musical result of the arm movement as in the case of the trombone, although not quite as pronounced. The style of playing, I should judge, is similar to that in vogue before the advent of the tight bow. They even get an effect of the old bow by loosening the hair and wedging the violin in between the hair and the bow and then playing all the strings of the violin at once. The remaining instruments, bass fiddle, guitar, banjo, and tuba are in very versatile hands but for the most part contribute solos in styles similar to the ones I have already discussed. They have contributed, however, a few individual elements such as the "slap" of the bass or the contrapuntal, inventive accompaniment of the guitar.

A jazz composition is made up of the improvisations of these players, and although the arranger is coming more and more into prominence, his work is still very secondary. The arranger takes a fragmentary rhythm from an improvisation on a given melody, and applies this rhythm over and over again throughout the natural course of the melody. Maybe this rhythm is given to the three trumpets, and a counter theme given to the other musicians; but in any case it is usually musically very uninteresting and only saved by the piercing interpretation of the men in the better bands. Though most written music of moderato tempo or faster, from Haydn on, has had such rhythmic patterns applied throughout, the composers have been conscious of their best melodic phrases and have woven them into the piece with intellectual skill. But this being one of the last stages in folk development, the jazz arrangers are not equal to the task and their output is very tiresome. Even if any highly inventive improvisation should be orchestrated into the score it would lose its original appeal through the self-consciousness of the reader, for the actual melodic richness is unappreciated both by the jazz player as well as the listeners, only the robust playing style of the instrumentalist exciting their admiration.

It is a fact that the many jazz orchestras playing under the name of a star performer or leader mislead the public into thinking that the leader or performer creates the music. This brings up the question as to how we should rate benefactors, managers, impresarios, and the like of all large organizations promoting art, whether opera, ballet, symphony, or jazz orchestra. The Diaghileff ballets are a good example, for these productions included not only the dancers and choreographers, painters, and musicians, but Diaghileff, upon whom depended everything from the securing of backing to the choice of ballets. He managed the financing, he controlled and selected the great artists, smoothing out their differences, and he proved himself a rare man; probably more rare than any single one of the artists. But important as he was to the life of the organization, students will always pick out the important separate creators and give them their due credit, for these were the people who were the backbone of the organization and upon whom all lasting significance depends. To attribute their art significance to Diaghileff is as foolish as it would be to attribute a Beethoven symphony to any one of his benefactors. I think the jazz orchestra is placed in about the same situation. There are players, arrangers, and a personality in every orchestra, and the leader can be anything from the best player to no practicing musician at all. He can be everything from a great personality to a shrewd businessman, and whether he is a musician or not he is always able to take the role of a master of ceremonies. He usually sets the policy of the orchestra, and is responsible for the quality of players and arrangements. Also we find the leader, more often than not, employs orchestrators. As a rule he *can* orchestrate, but because of the tiresome routine of public performance and the running of the organization, he employs, let us say, the saxophonist to do this tedious job, though since all except a few pieces are simply arrangements of passing popular tunes, any outsider does just as well. When we see the names of two composers on a piece of music, one well known and the other unknown, we can guess who did the composing. Always look for this other name before attributing the work to the well-known name. And of course on a record of this piece where the melody is entirely changed into a new, vastly more important synthesis by an unknown individual in the orchestra, we see only the name of the original composer, as no musical importance is attached to these variations. In jazz as in ballet, the public must discover for itself who is responsible for the various works.

The four solos that follow are notated from records made by Bubber Miley. He said they were variations on a spiritual his mother used to sing, called *Hosanna,* but the spiritual turns out to be a part of Stephen Adams's *Holy City* commencing at the seventeenth bar. There the tune is in four-four time and eight bars long, but Miley's version has about two bars taken out next to the last bar and the remaining six bars drawn out to twelve by dividing each bar into two. In the composition *Black and Tan Fantasy* this theme is announced in the minor, but his hot solos (variations) are on the original major. As he improvises these solos the orchestra simply plays a "vamp" rhythmic accompaniment. I have written them out under each other as in an orchestral score, with the

theme on top, in order to facilitate intercomparison; but it must be understood that these are pure improvisations out of a folk school, with no idea of adequate notation. All of these solos are by Bubber Miley (see last lines of p. 253) except the first twelve bars of No. 2, which is by Joseph Nanton, a trombonist in Duke Ellington's orchestra. As I have said, Negro improvisations are either on the melody or on the harmony, and it would appear that Miley paid no attention to the melody, so far removed are his variations; but by playing certain parts of the theme, then the corresponding part in any one of the hot solos, you will find that many times he did have the theme in mind.

In No. 1, all through the first twelve bars, there is a vague resemblance to the theme. The thirteenth and fifteenth bars are exactly the same as the theme but his treatment of these bars takes the startling form of blasts. In No. 3, if we play the fifth and sixth bars of the theme, and then his corresponding variations, we again see a melodic resemblance, but it is curious how this is his first melodic attack after the four-bar hold; that is, instead of continuing with the melody, he is starting one.

In the thirteenth bar of No. 3 he wonderfully distorts the B flat in the theme, to an E natural. Here is a take-off of the most extreme kind, and accordingly he followed the harmony until he could catch up with the melody. This he did at the twenty-first bar, finishing with a jazzed-up version of the theme.

A typical jazz distortion of the given melody is to lengthen the time value of one note by stealing from another, thereby sometimes reducing the melody to an organ point. Though this was practiced prior to Miley, it is interesting to see how the whole note in the first bar of the theme is held longer and longer until in No. 4 he is holding it seven bars. Notice how the improvisations do not have any break between the twelfth and thirteenth bars, that is, where the theme begins its repeat.

It seems to me the little phrases in bars eight, nine, and ten of No. 1, where he plays with his melody at either end of the octave, can only be found elsewhere in such music as Bach's *Goldberg Variations.* For example:

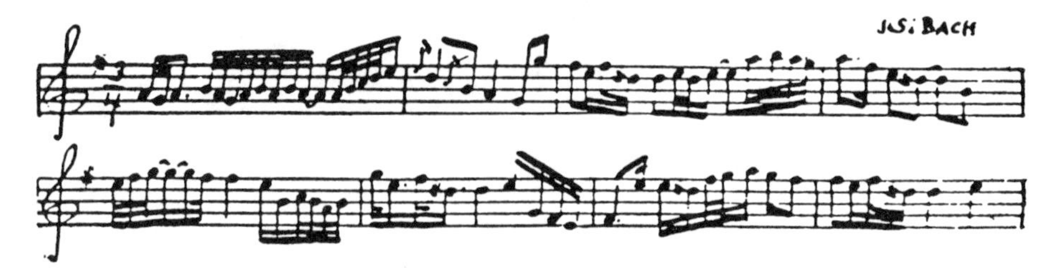

Joseph Nanton's twelve-bar variation in No. 2 seems to be on the harmony. It is followed by Miley's beautiful entrance, a slow trill on the original B flat. In

No. 4, which he made for me, the little coda to the long note is the purest music I have ever heard in jazz. I speak of purity in its resemblance to the opening of the Credo for soprano voices in Palestrina's *Missa Papae Marcelli.*

You will observe that the thirteenth bars of both No. 3 and No. 4 are the same. The freedom leading up to the C in the fifteenth bar is amazing, and the A flat in the nineteenth bar, after all the agitation, is no less surprising.

There were two elements essential for this freedom of thought: harmony and rhythm. Miley told me he needed the strictest beat and at least a three-part harmony. Though the piano could give him this, he was always better, however, with the orchestra and its background of drums, etc. Whereas the academy now might be able to compose parts like this, write them down and with a little shaping make something very inventive, no folk artist could do so, as his improvising in such a manner that rhythm and melody are torn apart really demands these two elements. We can now understand how a person like Duke Ellington was indispensable to Miley—"When I get off the Duke is always there." The Duke's co-operation, in fact, inspired Miley to the best work he ever did and neither of them sustained very well their unfortunate parting of the ways. The Duke has never since touched the heights that he and Bubber Miley reached in such records as *East St. Louis Toodle-Oo, Flaming Youth, Got Everything But You, Yellow Dog Blues,* etc., etc.—and of course the many *Black and Tan Fantasys*. The sudden and tragic death of Bubber Miley put a stop to his career before he was thirty—though without the guidance of the Duke, who is a real Diaghileff in a small way, perhaps he would have slipped backwards too.

In an article I wrote seven years ago entitled "Negro Jazz" I held out high hopes for the art, though at the time those critics who deigned to notice jazz were in no undecided terms announcing its complete extinction. Since that time jazz has not only persisted but advanced way beyond my expectations. It is only now that I have my doubts; it is the present tendencies that seem to be spelling doom in the near future. For the Negro is tired of the blues and likes to write the popular tunes which are a sort of compromise between his former music and Tin Pan Alley, and fairly eats up any like compromise of a white person. There are few real blues singers left like Bessie Smith. Her inventive way of singing does not seem to have been contagious. When I hear an early record of Bessie Smith and then listen to a Cab Calloway and see how much more the Negro now enjoys the latter, I realize that the blues have been superseded and white decadence has once more ironed out and sweetened a vital art. At the moment, through the arrangers and the more conscious players, jazz is in process of being crystallized into a written music, but the gulf is too great between this and what I consider good jazz for such a crystallization to have any significance in the future.

One would suppose that academic composers would jump at this medium, but the little that has been done in the field is of less value than those arrangements I have spoken of. With the awareness of an academic education, the composers have combined the simple side of jazz with the complex side of modern music, and the public, with a similar viewpoint towards the treatment of such music, finds it interesting. But even for them this music does not seem to wear well and probably is no more than their standard of novelty. Our composers may have the craftsmanship of Bach still sticking in their craws, but they lack even a taste of his melodic significance—it is an already embellished melody of the 17th and 18th centuries which Bach has so vitally mixed with his craftsmanship. These moderns give us musical mathematics and acrobatics applied to any and every folk tune, but their work lacks as true a line of melody as we might find in the most obscure trombone solo. As I am in complete accord with other moderns who theoretically object to using jazz, it must be understood that I am not urging composers to *use* it in the same sense that Dvořák *used* folk tunes. What I propose for consideration is, that as this whole period is permeated with jazz, it cannot be such a precious or out-of-the-way attitude to become *part of it.* Though a fine modern jazz music may still be written, frankly, the best I think we can hope for is that this eating decadence of jazz be a slow process, and that in the meantime the Negro will crystallize more of his work on records, such work as Duke Ellington and Bubber Miley turned out in the old days.

Of the many American writings on jazz, both pro and con, few are knowledgeably critical, none of any instructive value. There are magazines and articles in Europe with an attitude towards jazz as serious music, that we haven't approached. And there is Prunières, the one important critic, to my knowledge, who has an appreciation of the improvised solo in jazz. The American criticisms on the subject seem confusedly to hover around on the one hand, the spirit of America, the brave tempo of modern life, absence of sentimentalism, the importance of syncopation and the good old Virginia cornfields; and on the other hand, the monotonous beat, the unmusical noises, the jaded Harlem Negro, alcoholism, and sexual debauch. These solos in the *Black and Tan Fantasy* may not have the significance I attribute to them but they could at least be a premise for criticism. As notated music it certainly is not just noise, squawks, and monotonous rhythm: nor do vague favorable praises seem appropriate. Such solos as I have printed demand musical investigation.

Negro Jazz as Folk Material for Our Modern Dance

National Dance Congress, 1936.

If jazz music has permeated the civilized world, jazz dancing is not far behind. Although recognized by everybody as a "sort of" folk art of this period, and by some as a very vital folk art, I attribute its careful avoidance by modern dancers to a mistaken idea of wherein its art form lies. In fact, a good number of our aesthetic confusions may be traced to vagueness in all critical discussions involving race idiosyncracy and folk vitality and the part they play in advanced forms of art.

Probably every critic, grudgingly or otherwise, has acknowledged folk art as a source of the more extended forms now known as *fine* art and *classical* art, but it is not always clear from their writings exactly how contributory they believe the folk to be—or how necessary. Nor is it easy to decide whether they distinguish between pure folk material and that nebulous stuff conveniently called "spirit of the times." In any event, there seems to be a loose agreement that art comes from somewhere in life and that a valid art should be an art expressing its time. I would qualify this by saying: when the folk form is used, the resultant art depends more upon the significance of this folk form than on any outside influence of the times; that is, "spirit of the times," no matter how revolutionary, cannot build high on a non-significant folk basis, or, worse still, take the place of non-existent folk material.

There is little to be said for creating forms for ourselves in the "spirit of the times" and then trying to build an art upon them. Any such conscious, intellectualized approach to the creation of the very forms we are to work within infuses them with a highly detrimental, esoteric

Roger Pryor Dodge with Mura Dehn, 1937.

quality from the start. I find it quite noteworthy that only in the dance do we find so complete a satisfaction expressed with forms having no roots in the past. This may be due to the fact that the human body involves many activities releasing emotion, any one of which can lead the neophyte, bent upon wish-fulfillment, to a complete conviction that he is engaged in art activity, when any thoughtful analysis of art values might assume quite the contrary. No other art save the dance provides opportunity for such an unblushing display of gap between the folk forms, or origins of art, and that material which the conscious artist invents.

To clarify my distinction of valid folk material as opposed to expression of the "spirit of the times," let me draw an analogy in the field of music. Folk melody, when consciously used by a composer, is easily recognized as coming from the people, likewise, though perhaps not so easily, is the complicated superstructure built around the folk tune, called form. Just as the folk tunes are extended (or cut up) by the composer into melody quite different from their folk parentage, so are all musical forms, through the progressive freedom of variation, extensions of the early, simple structures. If we thus analyze dancing, the very opposite may be said of the movement inaugurated by Isadora Duncan. Regardless of her personal genius and whether or not her art expression, as is said, represented the "spirit of the times"—by no stretch of the imagination could she be said to have extended folk dance forms as I have defined the process. The freedom that she stood for, and everything springing from her, is not only *not* part of an aesthetic evolutionary process but is possible to any new social order revolting against the old. Such self-expression might have organized itself during the Age of Reason following the French Revolution; or the Age of Pericles; the Cromwellian era; or as it actually did, the Victorian era. Isadora Duncan is situated in any history of fact but in the history of the evolution of dance art her personal divergence and all subsequent ramification have no significant place, save as a *qualifying element,* extra to the development of lineal descent in art form.

Negro jazz, the folk form I propose for American consideration, is an actual new folk form susceptible of extension. It is a valid basis for American art. Neither jazz music nor jazz dancing is pure African art, but Negro development of our art—and by our art I mean the forms a combined Western culture has been evolving for hundreds of years. The Negro has used our instruments, our musical scale, our four-part harmony, our shoes to predicate his dancing feet, our clothes to qualify his movements, and our couple dancing, and with these has opened up a new era. The result is no actual melding of his art with ours, nor superimposed African melody and African tribal dance (in a manner comparable to Bali folk tune in European arrangement or Spanish dance in ballet) but a fine, true, and electrifying art development, suggesting in scope the birth of modern music out of Italian polyphony.

Negro jazz has so infused all our popular music and dancing that, in some degree, this entire generation exercises itself in it. Each does it in his own way. The confusing element is that the Negro forms, when presented to us by Negroes, naturally carry with them racial idiosyncracy. Although most art is the distillation of many personal and racial idiosyncracies, nevertheless in the time arts, where re-creation constitutes the performance there are certain idiosyncracies which cannot be transferred from one race to another—or, for that matter, from one person to another. These are surface idiosyncracies which do not develop into any very significant form but merely go along with the original performing of the more vital forms. Unfortunately, a time art such as dancing can be, and very often is, made up of just such surface idiosyncracies. But jazz music and dancing do not depend upon superficial racial tricks. If it were so, jazz would never have spread beyond the Negro, or have moved white Europe and America to such general adoption. Where the blackface comedian was an imitation of the Negro idiosyncracy, white jazz music and dancing, on the other hand, is our reaction to the Negro reaction to our culture. We do not see everyone playing blackface, but we do see everyone dancing and singing jazz in one form or another. We actually live jazz in this doing; in blackface we play the mimic. In other words, blackface is imitation of racial idiosyncracy, whereas modern jazz is doing something vital in our own way.

There is a great difference between expressing a natural impulse within our own medium (or somebody's else) and making an effort to imitate another in either his personal or racial idiosyncracy. Participating in someone else's art can become a part of us—we can even create something within this medium, making it as natural and personal a basis as it was for the one who first created it. But to imitate is to copy, without the least divergence, something someone else has created—something emotionally foreign, but which can be thus rendered fairly satisfactorily. Certainly, most of the music suites played in concert during the 17th and 18th centuries had a 15th- and 16th-century dance origin quite foreign to the 18th century—not to mention our 20th-century revival. But even though original racial and personal idiosyncratic interpretations do not necessarily carry over into the succeeding generations, the actual body of the art is not less vital. That is to say, the music that survives is in itself composed of such solid material that long after the special idiosyncracy is lost the music can be played. If too precious a concern were taken regarding the undisputed passing of the original and authentic idiosyncracy, and for that reason modern performance of old music were shunned, then I think we would have lost the music in our effort to avoid inevitable anachronisms of performance. Fortunately we recognize great musical forms and present them for what they are worth to us. We can only guess how the 17th-century violins were bowed, or what were the accents of harpsichord playing, but the notated music is none the less great.

Likewise, we must not confuse dance style having significant and true dance movements, in which the human race unconsciously habituates itself, with weak movements, which, deprived of their racial or personal idiosyncracies, are void of dance meaning. Currently, jazz is thought to be too light a material when the average chorus girl practices it and too Negroid when the Negro does it, but I believe underneath both manifestations there is movement upon which anyone can build to suit his own nature—but he must forget the inimitable idiosyncracies. He must forget for a moment the showmanship of the Negro performer wherein he brings everything characteristic of his race to the fore, and instead watch him go through his paces in a relaxed mood. Then and only then can be seen what natural drive the raw material has in itself. This drive is constant, and does not *depend* upon the misleading, heightening effects that we are led to expect from timed, dramatic performances.

In improvisation, which is the folk medium, there is a vitality, a liveness, an exciting spontaneity which is never present in a set work of art. When warmed up to improvising an artist can do things impossible to recreate, even if adequately notated and offered anew for interpretation. It is these very things which give improvisation its vitality, etc. This could be carried over into a direct analogy by imagining ourselves in the 18th century, called upon to judge between any *written* variation of J. S. Bach and the white heat performance of an improviser of the same period. The impromptu performance would have a vital performing style—electrify us upon first hearing, yet very possibly lack the inventiveness with which a Bach would have infused the same theme. Vital performance can be felt immediately—whereas only constant repetition makes the inventive work grow in our eyes. In a growing folk school, either music or dancing, it must be remembered that the performers are constantly, joyously, doing their art at white heat. On the other hand, a conscious artist spends most of his time on perfecting smoothness, and spends very little time on movement at concert pitch. For one thing he hasn't the time, as his pursuits are so diversified. Moreover, he doesn't want to do over and over the same work of art. He is constantly *creating* new works of art.

The folk are constantly *practicing* for personal entertainment the same thing in slightly different sequences. Folk performers, who earn their living in the theatre, however, have to have a set work. This they repeat day after day, over and over, until it becomes unsalable, audience excitement making them do it at pitch. Three or four routines stolidly sweated out will do them year in and year out, until death or the times overtake them. Moreover, as the nervousness a recital carries with it is completely absent from their work it is often possible to observe an unbeatable performance if you can catch their show on a day when they feel at their best. At such a performance, technical brilliancy plus natural idiosyncrasy can so cover up whatever art form may or may not be there, that the casual onlooker will neither look for nor perhaps miss what makes up the solid, root material.

To the neophyte, jazz is just an extraordinary amount of noise and rhythm; jazz dancing, a vulgar form of exhibitionism. Moreover to a person simply looking for two qualities, i.e. group or personal idiosyncrasy and technical brilliance, all else seems calculated and drab. Concededly, folk art in its short day of creation holds this upper hand of idiosyncrasy and technical brilliance over the slowly crystallizing academy. Probably folk art loses immeasurable qualities in its transformation into an academy, but the measurable qualities added are there to stay—and because of their measurability. What folk art loses in its metamorphosis into academy is a tragedy, but the stuff is too ephemeral to transplant. What it gains is of measurable value and can be everlasting.

In jazz dancing, as probably in all folk dancing, there are two forms of movement. One comprises a series of juxtaposed movements which, except for the addition of dynamics, depend entirely upon repetition for adequate presentation. The other form of movement is as natural as walking. When a person walks we do not think of him as merely repeating himself. After he has taken two steps we do not say "Oh, all he is doing is repeating the placing of first his left foot and then his right." No. The function of walking is thought of as a thing in itself which might take too long, but is never too repetitive. Jazz has in its various forms of the strut and particularly in its latest development, trucking, accomplished a natural function of movement that is as repetitive as walking, but like walking, still remains a flowing thing in itself. All such fundamental movements are highly charged with the style out of which they came, and especially is this so of trucking. For such movement to burst into actuality, a whole school of dancing must be, and is, eminent. The many converging forces present in trucking which give it its pure, kinetic vitality make it inconceivable that trucking could come from a dictating minority group. American jazz provokes the thought that we do *not* have to invent art forms for the American people; that on the contrary, the people of our contemporary Negro dance halls—the *folk*—are offering us a strong vital and flavored medium in both music and dancing.

On Nijinsky Photographs

Photography is of paramount importance to the dance, whether motion picture or "still." It is indispensable for any record of the dancer and the key to his final evaluation. Lately "action" photography has presented vivid photographs of dancers who in general take very insipid photographs in the studio. As dancing becomes more and more *kinetic self-expression* and less and less *movement in art form,* the man who can express his innermost dance feeling in controlled pose is not only a rarity but a consummate artist. Nijinsky, even in the old days of bad studio photography, achieved as much life and vitality in his poses as modern action photography catches during performance.

Although action photography is important in taking pictures of great dancers who are not able to pose in the studio, it is definitely misleading as an indication of dance values when it catches a dynamic movement of a bad dancer—movement which only took on value at the moment of extreme muscular exertion and which in terms of dance art is no more significant than the football player caught in the air straining for the ball. The direct opposite of this approach is in one photograph of Nijinsky, here his attitude is so alive, so as though he were leaping high in the air, that although his foot is firmly on the ground the photograph has been retouched in order to give a photographic effect of mid-air leaping!

If we analyze the difference between a dead pose and a lively one full of vitality even in repose, we find it is a case of the amount of *lift* given to the arms and the upper torso. Within dance style, Nijinsky had this *lift* just up to point where he

The American Dancer, March 1938.

Vaslav Nijinsky in "Le Spectre de la Rose," Paris, 1911.

wished the hand or arm to hang, and there he controlled it, permitting the rest of the limb to fall naturally in place. The composition he gave his whole body was the balance of parts, in which head, torso, and limbs were complementary to each other in creating a greater whole. Moreover, no matter how ethereal, twisted, brutal, or reposed the character, Nijinsky always achieved genuineness of facial expression. Falsity of expression predominates in practically all other modern dancers.

The dance has had a long history of contribution to the space arts. Throughout Eastern art in general, more particularly in Bali and Cambodia, we find in painting, print, or sculpture the dancer's pose influencing and formalizing the art style. An equivalent in Western culture is the great school of ballet and its contribution to the French art of the 18th century. This French period draws most of its style and significance from 18th-century ballet posture—not the craftsmanship of the artist. It is only the hangover of the 19th-century art criticism which keeps us from properly distinguishing between a painter's craftsmanship and his borrowed use of dance pose. Nijinsky's photographs are great dance pictures, far more significant as pictures than any paintings made of him at the time. Reviewing a large collection of Nijinsky photographs not only leaves little doubt as to his genius but fixes him for all time in poses of which even his former audiences retain but fleeting memories.

Dance in the Cinema

Cinema is of two-fold importance to the dance: as an auxiliary to script notation and as a means of preservation of dance art. In the first instance we must remember that all systems of dance notation are limited; whether the method whereby the single symbol stands for the completed conventional movement, or the method which tries to notate the history or (start to finish) of a movement. If the Feuillet and De Leon methods were simply useful to their own period as an indication of the contemporary forms then familiar to ballet dancers, the modern Von Laban system, through its very effort to re-create the all-time *how* of a movement, is overwhelmingly cumbersome for the professional dancer as an everyday recording device. On the other hand neither is it easy to reconstruct a rhythmic dance from film alone. It is a task analogous to transcribing to paper an intricate piece of recorded music—a piece a good sight reader will take the actual four minutes to play from paper but perhaps four or five hours to transcribe from the disc. The difficulty of cinema transcription lies in the fact that all dancers have a way of so qualifying or barely indicating a step that it is next to impossible to reconstruct it from the screen without first knowing the original source of the step. The solution lies in combining a simple choreographic notation with the dance film. Such notation tells us what the original step was. The dance style, mood, precision in space, which up to the moving picture era were so inaccessible to successful reconstruction, are accurately presented on the film. In conjunction with dance notation, the dance film is invaluable.

Dance Herald (Monthly publication of the American Dance Association, New York, N.Y.), (April 1938).

When we consider the dance film in its second character, that is, as the only *lasting* specific indication of the art of the individual dancer, the film becomes a work of art itself—not an art of cinema, but a *living* facsimile of the art of dance. In the so-called art of cinema, dances have been torn to shreds—sacrificed to the contemporary applause handed a doubtful camera trick. Definitely, from my point of view, the moving pictures have yet to prove their intrinsic art worth. When the time comes that a photoplay, say twenty years old, is not merely amusing but as absorbing as a contemporary photoplay, then only can we be sure there is art significance in cinema itself. The present fad for old "Westerns," Theda Bara tragedies, and Mack Sennet comedies does not indicate we are refreshing the memory of a former cinema art emotion. It indicates we are amused in the light of our present-day sophistication. On the other hand, dance films can perpetually give us (however non-contemporaneous either dance or cinema technique may be) that constant emotion we feel at all times in the presence of all significant art. But in dance films, dance must be treated as an art that has proven itself, and must not have imposed upon it the few camera tricks so recently invented and so likely to be demoded next year.

It is incredible that we dancers of today should be still backward in preserving our art. The two greatest dancers of our time, Vaslav Nijinsky and Isadora Duncan, although living well into a period of advanced cinematography, failed to take advantage of this technical device which would have preserved their art for all time. More unfortunate still, these two dancers were not only the greatest artists of their own time but apparently the two greatest in the past hundred years and, who knows, perhaps for many years to come. This apathy towards preserving their art now leaves them in the same ephemeral position in art history as Vestris and Taglioni.

To those who believe the dance cannot be successfully filmed owing to the three-dimensional quality present in actual dance performance and the admitted two-dimensional character of moving pictures, I submit the following thought: we must distinguish between expression and that substance which is dance form proper. That is, in spite of the fact the mechanics of our present-day theatre may excite us by their dramatic and heightening effects, the actual dance art is precisely the same whether the dance is produced with or without them. A machine which misses a few overtones does not fail to record the salient features of the art. The first wax impression was successful in recording music, notwithstanding it has taken forty years to actually achieve perfection of recording; the musical value of these early, mechanically imperfect recordings of singers and instrumentalists long since passed away is now beyond dispute. In fact a dance that is dependent upon three-dimensional presentation might be said to lack the qualities that go to make up dance art. In any event, color films are a great advance towards the accomplishment of a presentation suggesting depth and reality—although I am suspicious when excessive praise of the color film is expressed by dancers. The invention of pigment photography seems to be the satisfying element—not the fact that cinema, color or no color, is the only way to dance-art preservation.

Dancers! Let us not pass up the only known means of dance preservation because it is not yet mechanically perfect. It is an act of short-sightedness to work hard, push ahead, build up personal prestige in the name of dance and then not have faith enough in the dance art to preserve it. Future generations cannot take into account the value of an art which they have heard about but not seen. From now on, even if the individual dancer does risk a lower level in history than the attitude of his times would lead him to expect, nevertheless, by filming himself, he may feel confident that he has assisted dance to a standing in art history heretofore denied it. In the name of common sense let us concentrate on what Hollywood has neglected—the filming of dance art as it is today.

From Spirituals to Swing: An Evening of American Negro Music

Conceived and produced by John Hammond
Directed by Charles Friedman

H.R.S. Society Rag, January 1939.

Preamble

The general precarious position of hot jazz music makes any attempt to segregate the good wheat from the commercial chaff something to be thankful for. John Hammond and the *New Masses* brought us a variety of talent not in the usual run of whipped-up jazz concerts. If Hammond leaned a trifle heavily upon the "at least it is all sincere" angle, the concern was none the less a good deal better than the professionally managed swing concert. "From Spirituals to Swing" was a cross between an art lecture and a concert—the art lecture consisting of progressive demonstrations of what fine stuff makes up the basis of jazz, the concert presenting some rarely heard piano compositions in jazz, played by the outstanding virtuosi of our time. Personally I feel we can do more for jazz by stressing the concert side. Not only is the concert angle much more interesting for those of us who know jazz, but virtuosi playing *outstanding works* turn a jazz concert into a more exciting experience for the average concertgoer as well. In any event it is probably out of place to expatiate on how badly run the concert was, especially when we know that some of the artists were not free to perform at the times good program-making might demand.

Africa

The concert started with an African recording. I question the degree of importance Hammond attaches to it. I have heard many African records in this country and for the most part far more exciting ones.

Ruby Smith

The inclusion of Ruby Smith can readily be understood on the grounds of sentimental relationship, but if it is true that this was her first solo experience before an audience, she should not have opened the program. Moreover the selections she was allowed to sing were not of the best blues type. Blues written back in 1924 were half popular song. It is true Bessie Smith sang these blues as well as the best type, but it is the avoidance of this very circumstance of poor selection on the part of the folk artists themselves that will encourage critical attendance at jazz concerts.

James P. Johnson

James P. Johnson is a great pianist, but I am sorry to say one would not always know it. His piano accompaniment of Bessie Smith on the recording of *Back Water Blues* alone is enough to prove his superiority, but his accompaniment of Ruby Smith might have been equalled by any professional pianist. What he produced in his solo *Carolina Shout* is an example of the kind of piano playing that has *always* disappointed me. Juxtaposition to boogie-woogie puts it to a test it can't stand.

Big Bill

The singing of the blues, a much commoner activity than instrumentalizing them, falls into either one of two categories: one, the intoning of couplet after couplet to an unaltered monotoned melody; two, the same lengthy round of lyrics, *but* in conjunction with a subtle and varied intonation of the melody. Big Bill falls well within the first category, whereas Bessie Smith was supreme master of the second category. Big Bill was featured as a sharecropper and the son of a sharecropper. This was probably to fall in line with a *New Masses* audience.

Meade "Lux" Lewis and Albert Ammons

There is hardly enough room in this review to write fully about the importance of Meade "Lux" Lewis to piano jazz. He and Albert Ammons are two extraordinary pianists. We might say that Albert Ammons has more virtuosity—consisting in a stronger left hand and (in this concert) has more invention in his melodic improvisations. However, it is not easy to evaluate invention at one hearing and Meade "Lux" Lewis has more than proved his creative ability on his outstanding discs. If piano jazz in the Teddy Wilson manner is always a mild disappointment, I doubt very much whether in our wildest dreams we could imagine a style so apt as the boogie-woogie, and a piece so significant as *Honky Tonk Train Blues.* Although Albert Ammons has also made a recording of his own *Boogie Woogie,* with orchestra, he most decidedly does not play long enough on it. His *Boogie Woogie* piano solo should be recorded in full.

Joe Turner

Joe Turner's singing was another case of the turning over of blues for the sake of their lyrics. His melody is on the style of Lonnie Johnson's. Peter Johnson, his accompanist, played a very good *Boogie Woogie.* In fact the whole concert could have been billed *A night of boogie-woogie—also some singers and orchestras.*

Sanford "Sonny" Terry

Sanford "Sonny" Terry with a limited jazz vocabulary hit a strange note in his combination of harmonica and singing. His contribution demonstrated once again how significant is the basis of jazz—so significant that even on such a bypath as chosen by Terry, the music arrives.

Mitchell's Christian Singers

Mitchell's Christian Singers kept us well within the confines of the archaeological jazz lecture. They were not, thank God, the Negro-spiritual-Toscanini-choir type, but, in a jazz concert very little spiritual goes a long way! The Christian Singers were as "sincere" as the day is long, but perhaps if they had sung the *Lord's Prayer* as the Mills Brothers sing *Miss Otis Regrets,* the *Lord's Prayer* would have approached boogie-woogie pianoforte playing in interest.

Sidney Bechet and Kansas City Six

The Kansas City Six played some very mediocre swing. We have very few functioning links between the pre-1920s in jazz and our own day. Sidney Bechet, the soprano saxophonist, whose full, rich tone is heard on many records since that time, played some very fast jazz in the New Orleans style.

This kind of a band is at a disadvantage on a large stage, as they are only at their best when jamming in a small club. The publication *Hot Jazz* informs us that Ansermet, the eminent French conductor, praised Sidney Bechet as far back as 1919. Of his own art, Bechet could only say that he follows his "own way." But Ansermet's critical discernment could exclaim: "And when one thinks that this 'own way' is perhaps the road the whole world will be treading tomorrow."

Sister Tharpe

Sister Tharpe is an entertaining mixture of Broadway and back-country singing. The musical result, however, overlooking her charming personality, is the *patter* type of hymn-spiritual.

Boogie-Woogie Trio

The unique event of the evening was the three boogie-woogie pianists, Meade "Lux" Lewis, Albert Ammons, and Peter Johnson playing in trio. Bad staging squeezed Meade Lux in back of the other two but nevertheless their ensemble performance was dynamic. There was an outburst of justly timed boos. We may possibly hear any one of these pianists again or secure solo records, but I doubt whether we will ever have another opportunity to hear all three in concerto. Such improvising as they were easing themselves into should have been nurtured along—not thwarted.

Count Basie

Count Basie's orchestral selections were certainly not contributory to the general excellence of the concert. In spite of a capable enough band he seems to have nothing to say. His ensemble arrangements are pretty poor stuff, on the whole, when we consider what the modern ensemble can do. It is about time that orchestras on concert platforms gave us more than the latest tune—necessary as such vaudeville may be at the Paramount or 125th Street Apollo. Basie's *One O'Clock Jump* is excellent and might be called a classic, but nothing much can be said for his other numbers. There are classics in jazz, which if played adequately would arouse a concert audience more than any snappy, up-to-the-minute popular tune arrangement. Outstanding pieces should be in the concert repertoire of every leading orchestra; such pieces as *Black and Tan Fantasy, East St. Louis Toodle-Oo, The King of the Zulus, Call of the Freaks, Mr. Bach Goes to Town,* etc. Page played a good trumpet. He is a fine soloist—something which is lacking in Basie's band.

Summary

Both in variety and distinction of soloists it was the best concert I have ever heard. Unlike the one-band concerts, it resorted to no padding, each effort having its own sustained value in varying degrees. But performance demands the guiding hand of a program-maker. Even a good band cannot be left to its own resources. Obviously there is an unexplored field for a jazz promoter.

Consider the Critics

As soon as jazz became disturbingly identifiable as something more than "our popular music," countless uninformed commentators sprang up with something to say about it. In what the era might have called "the spirit of the thing," they made a jocose offering of a great part of the early recognition of jazz.

In general, symphonic jazz was considered a progressive advance upon primitive improvisation, and critics were anxious to see an art form blossom divorced from the dance and comparable to nineteenth-century concert music. Even throughout the most sympathetic critical writing we find jazz tackled as a problem-child whose significant development is dependent upon immediate separation from the untutored musician.

Unfortunately, such premature white-collar meddling with jazz not only cut off the music public from following the slow but determined development of jazz by jazz musicians themselves, but induced academic-minded composers to leap headlong into vast, pretentious jazz works. Pretentious folk-art extension, in the word's best sense, can never be seriously entertained unless the vital elements of the folk-art have been first seriously considered by the ambitious composer. There is no doubt that changes are in order when a folk-art is taken into an extended form; that is, there are elements which the critic might like to cling to but which the composer is defiantly aware cannot be carried along into his extended form. But, such an admission hardly covers the mayhem perpetrated by the first so-called jazz composers in their course of romantic, rhapsodic, European

Jazzmen, 1939.

folk extension. All sensitive development of the form inherent in folk-jazz itself was studiously avoided. Perhaps the difference between folk material and the first advanced work of art, should be no more than that in the one the art is scattered and in the other there is an apt concentration into one composition. For such an advance, significant improvised hot solos were scattered about any number of recordings before 1924. However, a conscientious America, anxious to promote American music and secure in the knowledge that any written art work superseded improvised phonograph recordings, gratefully settled for the *Rhapsody in Blue!*

In this article I have drawn out for extended comment three categories of critics: one, important men in any field; two, men who seem to have said the right thing; and three, men who have taken time and care to write at serious length upon the subject. As regards those contemporary, bright but forgotten music commentators, devoted to consideration of the "jazz age," "jazz morals," and "jazz haircuts," time and space have not been so generous to me for purposes of refutation as they were to the original observers.

One of the first outstanding critics I have been able to discover is the writer, Carl Van Vechten. And this is in spite of the fact that the underlying generalizations he made in 1917 pointed towards something quite different from that which he later particularized upon with evident satisfaction. It is strange how many sincere and sympathetic pioneer discoverers of jazz shied away from their own first premises the moment the "refining" element was introduced and the "major" work appeared! In 1917 we find him saying:

> Popular songs, indeed, form as good a basis for the serious composer to work upon as the folk-song. . . . If the American composers with (what they consider) more serious aims, instead of writing symphonies and other worn-out and exhausted forms which belong to another age of composition, would strive to put into their music the rhythms and tunes that dominate the hearts of the people a new form would evolve which might prove to be the child of the Great American Composer we have all been waiting for so anxiously. I do not mean to suggest that Edgar Stillman Kelley should write variations on the theme *Oh, You Beautiful Doll!* Or that Arthur Farnell should compose a symphony utilizing *The Gaby Glide* for the first subject of the allegro and *Everybody's Doing It* for the second with the adagio movement based on *Pretty Baby* in the minor key. It is not my intention to start someone writing a tone-poem called New York. . . . But, if any composer, bearing these tendencies (jazz) in mind, will allow his inspiration to run riot, it will not be necessary to quote or to pour his thought into the mould of the symphony, the string quartet, or any other defunct form, to stir a modern audience.[1]

1. Carl Van Vechten, "The Great American Composer," *Vanity Fair,* April, 1917.

In 1925 we find him still generalizing in fine terms:

> real American music (*Alexander's Ragtime Band*)—music of such vitality that it made the Grieg-Schumann-Wagner dilutions of MacDowell sound a little thin, and the saccharine bars of *Narcissus* and *Ophelia* so much pseudo Chaminade concocted in an American back-parlor, while it completely routed the so-called art music of the professors.[2]

But suddenly we are brought up short by the remark:

> February 12th, 1924, a date which many of us will remember henceforth as commemorative of another event of importance besides the birth of our most famous president, George Gershwin's *Rhapsody in Blue* was performed for the first time by Paul Whiteman's orchestra with the composer at the piano.[3]

The *Rhapsody in Blue* probably violated all of Van Vechten's modern art stipulations. His long-awaited "running riot in new form" was no more than a rhapsodic bastardization of what since the day of Haydn has been called the symphonic sonata form: the sonata form with a subject, counter-subject, development, tempi changes, etc. Curiously enough, five months later we find him saying that the blues[4] deserve, from every point of view, the same serious attention that has been tardily awarded the spirituals. If he found melodic beauty in the blues and looked forward to a new form of music out of the jazz idiom, his enthusiastic recognition of Gershwin was a sad jumbling of theory.

Carl Van Vechten allied himself to the exponent of one of the most decadent and at the same time most trying of forms; a form whose no one part is strong enough to bridle bad taste. Moreover, the hot jazz recordings made prior to February 12, 1924, are concrete witness to the actual new style of variation already in use by the jazz folk themselves. If, today, jazz has shown no more interest than twenty years ago in progressing beyond the basis of a tune played over and over to encourage variation, from one point of view it is presumptuous to consider anything further as *progressive.* Let us consider how securely the mighty *Goldberg Variations* of Bach, or the variations in the last movement of Brahms's *Fourth Symphony* are wedged in the contemporary repertoire! It is the romantics, including Gershwin, who are never satisfied with such a timeless form as the variation.

As early as 1918 we find Olin Downes making this statement; "'Ragtime' in its best estate is for me one of our most precious musical assets."[5] In thus taking

2. Carl Van Vechten, "George Gershwin," *Vanity Fair,* March, 1925.

3. *Ibid.*

4. Carl Van Vechten, "The Black Blues," *Vanity Fair,* August, 1925.

5. Olin Downes, "An American Composer," *Musical Quarterly,* January, 1918.

cognizance of the raw material of jazz of that date and then not following through with earnest investigation, Downes has distinctly avoided a consistent course as music critic. However, he tried to pierce through the tone of the 1924 Whiteman concert:

> Thus the *Livery Stable Blues* was introduced apologetically as an example of the depraved past from which modern jazz has arisen. The apology is herewith indignantly rejected, for this is a gorgeous piece of impudence, much better in its unbuttoned jocosity and Rabelaisian laughter than other and more polite compositions that came later.[6]

One might suspect that perhaps he was holding himself just a little above all jazz music, but in 1937 such a suspicion is confounded:

> He [Gershwin] looked into the promised land, and pointed a way—one way—that a greater musician might follow.[7]

The keenest insight into early jazz coupled with the most sympathetic understanding of the necessarily slow development of folk instrumental music, comes from the eminent European concert conductor, Ernest Ansermet. He wrote as early as 1919:

> The first thing that strikes one about the Southern Syncopated Orchestra is the astonishing perfection, the superb taste, and the fervor of its playing. . . . It is only in the field of harmony that the Negro hasn't yet created his own distinct means of expression. . . . But, in general, harmony is perhaps a musical element which appears in a scheme of musical evolution only at a stage which the Negro art has not yet attained.[8]

It seems Ansermet knew how little to expect, manifested his delight when he got more than he expected and in every way came to his subject with an erudition equal to the solution of, or temporary toleration of, the apparently insuperable obstacles which all new art carries in its wake. When faced with Sidney Bechet's clarinet solos his comment was not only musically sure but esthetically sensitive:

> they gave the idea of a style, and their form was gripping, abrupt, harsh, with a brusque and pitiless ending like that of Bach's second *Brandenburg Concerto* . . . what a moving thing it is to meet this very black, fat boy with white teeth and that narrow forehead, who is very glad one likes what he does, but who can say nothing of his art, save that he follows his "own way," and when

6. Olin Downes, New York *Times,* February 13, 1924.

7. Olin Downes, "George Gershwin." Editor, Merle Armitage, Longmans, Green, 1938.

8. Ernest Ansermet, "On a Negro Orchestra," *Revue Romande,* October 15, 1919. Translated by Walter E. Schaap for *Jazz Hot,* November–December, 1938.

> one thinks that this "own way" is perhaps the highway the whole world will swing along tomorrow.[9]

When we consider that the style of playing in 1919 merely pointed the way to significant jazz solos of a later day, our admiration of Ansermet for visualizing the mature style which actually materialized, grows by leaps and bounds.

In this same year, 1919, from George Jean Nathan, editor of the *American Mercury,* we get such an incredible dismissal as:

> The Negro, with his unusual sense of rhythm, is no more accurately to be called musical than a metronome is to be called a Swiss music-box.[10]

Carl Engel was one of the first American music critics of note to be intelligently receptive to our folk-art of jazz. In 1922 we find him trying to dispel the notion that jazz must be ostracized because it is vulgar:

> To a great many minds the word "jazz" implies frivolous or obscene deportment. Let me ask what the word "sarabande" suggests to you? I have no doubt that to most of you it will mean everything that is diametrically opposed to "jazzing." When you hear mention of a "sarabande," you think of Bach's, of Handel's slow and stately airs. . . . Yet the sarabande, when it was first danced in Spain, about 1588, was probably far more shocking to behold than is the most shocking jazz today.[11]

With great sense, Engel warns the twentieth century not to indulge in the sort of eighteenth-century criticism which was proved so contrary to fact by the subsequent good standing of eighteenth-century dance music. He decries, as uninformed, the point of view which denies the use of dance material for what is popularly called "serious art." He pertinently quotes the following comment of Karl Spazier in the *Musikalischer Wochenblatt* in 1791, to show how ridiculous such a stand has always been:

> I furthermore hold that minuets are contrary to good effect, because, if they are composed straightforward in that form, they remind us inevitably and painfully of the dance hall and abuses of music, while, if they are caricatured—as is often done by Haydn and Pleyel—they incite laughter.[12]

Unfortunately, Engel stretched his attitude to include any and all popular music, and it must be agreed that he let a laudable generalization lead him into unconsidered particularization when we find him denying that

9. Ernest Ansermet, "On a Negro Orchestra," *Revue Romande,* October 15, 1919.

10. George Jean Nathan, *Comedians All,* Knopf, 1919, p. 133.

11. Carl Engel, "Jazz," *Atlantic,* August, 1922.

12. Carl Engel, "Views and Reviews," *Musical Quarterly,* April, 1926.

> to borrow material for a piano concerto from Mr. Flo Ziegfeld and the American Beauty Chorus should be thought more incongruous than to ask it of Gregory the Great and the Roman Antiphonary. . . .[13]

It just happens that Flo Ziegfeld and the American Beauty Chorus are pretty poor stuff, not because of their moral level of functioning, but for legitimate art reasons. Much of the strength of his comment is weakened by his unnecessary confusion between what is jazz and what is popular song. Actually, Engel never completely identified jazz for himself. On the other hand, like Ansermet, he recognized the strength of improvisation and said as early as 1922: "For jazz finds its last and supreme glory in the skill for improvization exhibited by the performers."[14] His admiration of European composers for their prompt recognition of jazz did not cloud him from observing that in their practical use of jazz "they do not throw me into ecstasies." He rightly felt it was more the "spirit" of jazz that affected Europe than an appreciative desire to write within the music itself.

In 1926 he admired Henry Osgood's book *So This Is Jazz,* devoted to uniform eulogy of Gershwin and Whiteman. But six years later we find him saying:

> let us further state that we are by no means an admirer of everything Mr. Gershwin has written. His *Rhapsody in Blue* and his *American in Paris,* except for a few isolated measures leave us cold . . . With M. Goffin's estimate of Paul Whiteman and Jack Hylton we are in full accord.[15]

It can be seen that as soon as Goffin's book appeared (1932) Engel's point of view altered. He should have been able to dispose of Henry Osgood by himself. Perhaps, if Engel, constantly concerned with the published opinion of the early twenties, had had a more extended acquaintanceship with hot jazz itself, he would have quickly and emotionally responded to it. As it is, his contribution to jazz criticism has been more that of the erudite music critic, highly sympathetic to the idea of jazz.

George Antheil, the *enfant terrible* of Paris, in the limelight for some time both by order of his own compositions in jazz and by what he has seen fit to say about it, started off in 1922 with the musically meaningless statement:

> Jazz is not a craze—it has existed in America for the last hundred years, and continues to exist each year more potently than the last. And as for its artistic significance, the organization of its line and color, its new dimensions, its new

13. *Ibid.*

14. Carl Engel, "Jazz," *Atlantic,* August, 1922.

15. Carl Engel, "Views and Reviews," *Musical Quarterly,* October, 1932.

> dynamics and mechanics,—its significance is that it is one of the greatest landmarks of modern art.[16]

Such confused statements from a musician discussing music would hardly suggest that when we hear a certain kind of music and call it jazz, the word "jazz" stands for that kind of music; that when we hear another kind of music and call it ragtime, the word "ragtime" stands for that kind of music! Antheil's effusion on "the greatest landmark in modern art" would seem to be a sincere enough acknowledgment for the time, but in the light of his subsequent entanglement I am led to believe that this early statement was more in line with the "Paris group" smartness than a sincere presentiment of what was to come. Six years later he seemed to think that the validity of the point of view that jazz had great art value, could be easily tested:—"The development of a great composer out of jazz is the only really clinching argument."[17] This, of course, means anything or nothing, depending upon the terms established for great composers. However, further along in the article the following remark seems clear and sensible:

> Jazz is her own way out to the future. But until jazz finds its way a little more clearly, let us [the composers] not take it into the concert hall.[18]

But in further discussion of the concert hall, which he admits has often been the scene of bitter and bloody conflict, he suggests that jazz must quickly attain "a dignity that the mere serving out of a parade of popular and clever melodies in trick orchestra garb can never attain." Has jazz moved Antheil or has he confused the issue? Four months later the secret is out with his published comment:

> The works of Vincent Youmans are pure clear, and extremely beautiful examples of jazz that is a pure music.[19][!]

Two great European composers, Darius Milhaud and Igor Stravinsky, composed music under the influence of jazz. Stravinsky, speaking of his piece called *Ragtime,* composed in the early twenties, comments:

> Its dimensions are modest, but it is indicative of the passion I felt at that time for jazz, which burst into life so suddenly when the war ended. At my request, a whole pile of this music was sent to me, enchanting me by its truly popular

16. George Antheil, "Jazz," *Der Querschnitt,* Germany, Summer, 1922.

17. George Antheil, "Jazz is Music," *The Forum,* July, 1928.

18. *Ibid.*

19. George Antheil, "American Folk Music," a letter to the editor of *The Forum,* December, 1928.

> appeal, its freshness, and the novel rhythm which so distinctly revealed its negro origin. These impressions suggested the idea of creating a composite portrait of this new dance music, giving the creation the importance of a concert piece, as, in the past, the composers of their periods had done for the minuet, the waltz, the mazurka, etc.[20]

It is ever a pity that Stravinsky, unlike the minuet composers of the past, merely toyed with this new dance form upon slight acquaintanceship; that his material was limited to popular sheet music.

Darius Milhaud was publishing small and large so-called jazz compositions at an early date. In an interview in 1923, he is quoted as saying:

> One thing I want to emphasize very particularly and that is the beneficial influence upon all music of jazz. It has been enormous and in my opinion, an influence of good. It is a new idea and has brought in new rhythms and almost, one might say, new forms. Stravinsky owes much to it. It is a pity that it is limited at present, practically to dance music, but that will be remedied.[21]

There was something in the best of jazz—blues singing, hot playing—that stirred the emotions, but such is the fallaciousness of human judgment that the intellectual musicians did no more than meddle with the phenomenon, and the popular song writers no more than wallow in the shallowest imitation!

Although the Whiteman-Gershwin concert was not the first of its kind, it came at the time when fever of musical expectancy was at its height. American critics had been indefatigable in building up a case for American music. They insisted that this folk-art, which they had been critically nurturing, respectably arrive via a piano concerto and a symphony orchestra. And for them the Whiteman affair was wholly satisfying. February 12 now stood for two births, that of Abraham Lincoln and the *Rhapsody in Blue.* Critics, whose business was sharp musical observation, succumbed to the reasoning that *something* vital must have occurred since a concert crowd roared. Even those sympathetic to Gershwin's bathos, but trained enough to recognize and comment upon the inept handling of the completely familiar concerto form, followed the line of least resistance to the concert public's will. So as we reach the year 1924, we find more writers voicing their opinion of jazz than at any other time until the advent of the particularizing "swing" critics ten years later.

A popular book of great importance was published in 1924—*Seven Lively Arts* by Gilbert Seldes. In this book, Seldes bravely justified the importance of the so-called *minor* and *lively* arts. He was outspoken in his attack upon those who are ill at ease before great art until it has been approved by great authority.

20. Stravinsky, *An Autobiography,* Simon & Schuster, 1936.

21. John Alan Haughton, "Darius Milhaud," *Musical America,* January 13, 1923.

But in the following quotation (the italics are mine) Seldes not only shows he makes no distinction between jazz and popular song but confuses what otherwise would be a fine art attitude:

> there is no difference between the great and the lively arts . . . both are opposed in the spirit to the middle or bogus arts. . . . The characteristic of the great arts is high seriousness—it occurs in Mozart and Aristophanes and Rabelais and Molière as surely as in Aeschylus and Racine. And the essence of the minor arts is high levity which existed in the *commedia dell' arte* and exists in Chaplin, which you find in the music of Berlin and Kern (not "funny" in any case). . . . We require, for nourishment, something fresh and *transient.* It is this which *makes jazz* much the characteristic of our time.[22]

If we admit that the "high seriousness" of major art, and the "high levity" of minor art, is an excellent distinction to make between that which we acknowledge as major art and the work of Berlin and Kern, the inclusion of the *commedia dell' arte* and jazz, as art forms to be equally and categorically linked with "fresh and transient" Tin Pan Alley, is highly objectionable. Seldes fails to tell us that running along with minor and lively arts, and even nourishing them, we often find the new vital folk-art—in its later metamorphosis to be known as the great art of "high seriousness!" In 1924, jazz had already shown qualities which distinguished it from the "transient" art of popular song. But suddenly his artistic boldness turns fainthearted.

> I say the negro is not our salvation because with all my feeling for what he instinctively offers, for his desirable indifference to our set of conventions of emotional decency, I am on the side of civilization. To any one who inherits several thousand centuries of civilization, none of the things the negro offers can matter unless they are apprehended by the mind as well as by the body and the spirit.[23]

The Negro may have been indifferent to our emotional decency, but he was certainly far from indifferent to our harmony and musical form. Into Negro jazz has gone the vital part of our centuries of musical experience. Seldes continues in the same vein:

> Nowhere is the failure of the negro to exploit his gifts more obvious than in the use he has made of the jazz orchestra; for although nearly every negro jazz band is better than nearly every white band, no negro band has yet come up to the level of the best white ones, and the leader of the best of all, by a little joke, is called Whiteman.[24]

22. Gilbert Seldes, *Seven Lively Arts,* Harper, 1924, pp. 348–49.

23. *Ibid.,* p. 97.

24. *Ibid.,* p. 99.

The concluding remark, admittedly slight in witty intent, will ever strike back at its author as a more humorous error in art judgment! As late as 1934, in an article on George Gershwin, Seldes still has no comment to make on hot jazz solos.

In 1922, Clive Bell wrote an essay entitled *Plus de Jazz.* This famous English art critic found it convenient to express the opinion that art

> is a matter of profound emotion and of intense and passionate thought; and that these things are rarely found in dancing-palaces and hotel lounges.[25]

But in 1924 he found it convenient to deplore the similar snobbism which maintains that

> the *Last Judgment* by Michelangelo, is something essentially nobler and more important than a picture painted by Watteau.[26][!]

Again, apparently finding it hard to explain why *I'm Just Wild About Harry* is not as great as the *B Minor Mass,* he hazards the opinion that

> maybe the only difference between a comic song by Mr. Irving Berlin and a comic song by Mozart is that one stylishly expresses Mr. Berlin, and the other stylishly expresses Mozart.[27]

But almost immediately he feels constrained to explain some difference in magnitude and reverts to his own brand of snobbism in declaring that those who like Mozart

> might have understood and been intimate with Mozart himself, whereas the latter [jazz musicians] could have been for him [Mozart], only objects of curiosity, surprise, amusement, or distaste.[28]

In 1924, the magazine *Etude* published a revealing collection of jazz comment headed by an editorial expatiating exclusively upon the art of Berlin, Confrey, and Gershwin. It concluded with the exclamation:

> But who knows, the needs of Jazz may be Burbanked into orchestral symphonies![29]

This same thought has been advanced time and time again. Art is to be "Burbanked" from Berlin to symphonies! A symphony, apparently, is the puri-

25. Clive Bell, *Since Cézanne,* Harcourt, Brace, 1922, p. 215.

26. Clive Bell, "There is an Art in Drinking a Cup of Tea," *Vanity Fair,* July, 1924.

27. *Ibid.*

28. *Ibid.*

29. "Where the Etude Stands," *Etude,* August, 1924.

fication of what would otherwise be dross—never the culminating synthesis of simple but pure elements!

In this same collection of comment, John Alden Carpenter, composer of the jazz ballet *Skyscraper,* goes out of the way to make the invidious distinction:

> I am convinced that our contemporary popular music (please note that I avoid labelling it "jazz") is by far the most spontaneous, the most personal, the most characteristic, and by virtue of these qualities, the most important musical expression that America has achieved.[30]

Here also, Will Earheart, director of music, Pittsburgh, tersely and amusingly enough disposes of jazz:

> Bach fugues, Beethoven's symphonies, . . . are heard in certain places and received by a certain clientele gathered there. They seem appropriate to the places in which they are heard, and to the people who are gathered to hear them. So does "jazz."[31]

Yet Sir C. Hubert Parry, professor of music at the University of Oxford, and director of the Royal College of Music, in a most important contribution to *Grove's Dictionary of Music,* deplored the aloof attitude of composers towards popular dance hall music and maintained that such an attitude had no basis in reality; that on the contrary, all concert music owes an incalculable debt to dance music. He wrote, in part:

> Dance rhythm and dance gestures have exerted the most powerful influence on music from prehistoric times till the present day. . . . The connection between popular songs and dancing led to a state of definiteness in the rhythm and periods of secular music . . . and in course of time the tunes so produced were not only actually used by the serious composers of choral music, as the inner thread of their works, but they also exerted a modifying influence upon their style, and led them by degrees to change the unrhythmic vagueness of the early state of things to a regularly definite rhythmic system. . . . In fact, dance rhythm may be securely asserted to have been the *immediate origin of all instrumental music.* [Italics mine.][32]

However, it is a truism, that when professors are brought upstanding with a new and *living* art fact, only too often their own contemporary moral compulsions force a retreat from their previous art premise so easily arrived at on the basis of historically dead issues!

30. *Ibid.*

31. *Ibid.*

32. Sir H. C. Parry, "Dance Rhythm," in *Grove's Dictionary of Music and Musicians.*

Let us clear up a little point that still induces much argument and was brought up by George Vail in this same issue of *Etude:*

> Mozart, Haydn, and Chopin, were they alive today, would write foxtrots as naturally and inevitably as they once composed gavottes, minuets and mazurkas.[33]

This is not necessarily true. Although the attitude of *their times* towards art was different from ours, the artists themselves could not be expected to hold a different professional point of view—given the same conditions—from that of our own academic composers. It was only the *times* that made them write gavottes and mazurkas—not their better judgment.

In 1924, Lawrence Gilman, a thoughtful yet conservative music critic, put his finger, with rare honesty, on the weakness of the whole business surrounding the Whiteman-Gershwin faddism:

> We have before expressed our conviction that the trouble with Jazz—the best Jazz, according to the showing of the Palais Royalists themselves [Whiteman's band] is its conformity, its conventionality, its lack of daring . . . it seems to us that this music is only half alive. Its gorgeous vitality of rhythm and of instrumental color is impaired by melodic and harmonic anemia of a most pernicious kind. Listen to Mr. Archer's *I Love You* or to Mr. Kern's *Raggedy-Ann,* or to Mr. Gershwin's *Rhapsody in Blue.*[34]

Mr. Gilman's point of view was one of the few examples of contemporary considered American opinion which questioned whether the *Rhapsody in Blue* should be properly termed progressive modern art. He speaks of Gershwin as "lacking daring." He was obviously right. For Mr. Gilman could think in terms of the daring of Stravinsky's *Le Sacre du Printemps.* However, if Gershwin was unbelievably old-fashioned, jazz was *always* daring—but only in its hot solos. For the time, that was enough. Gilman, not seeing as clearly as Ansermet, was oblivious to the great possibilities latent in this new folk music. Moreover, although a large number of hot solos were and are daring, there are also any number of hot solos that are significant, but not startlingly novel. They are no more daring than a melody of Mozart is daring. The whole confused attitude toward Modern Art at that time hung on one hook, and still hangs on it to a degree, that *shock* must prevail: that it is only from him who shocks that we may expect Modern Art!

33. George Vail, "Would Mozart Write Foxtrots If He Lived Today?", *Etude,* September, 1924.

34. Lawrence Gilman, "Music," New York *Tribune,* February 13, 1924.

In 1924, the modern composer Virgil Thomson decided that it was impossible to use jazz material for "serious" composition:

> [jazz] rhythm shakes but it won't flow. There is no climax. It never gets anywhere emotionally. In the symphony it would either lose its character or wreck the structure. It is exactly analogous to the hoochee-coochee.[35]

Thomson obviously believes that modern music should continue along latter day sonata-symphonic lines rather than make a new beginning on a new dance impulse. In a later article he concludes:

> For after all, America is just a collection of individuals. . . . The idea that they can be expressed by a standardized national art is of a piece with the idea that they should be cross-bred into a standardized national character, 100 per cent North American blood.[36]

The year before, Charles Buchanan made much the same comment:

> Possibly, the most peculiar and arresting phenomenon that the heterogeneous art activities of this country have brought forth is the wide-spread idea, amounting almost to an obsession, that American painting and music must create and express themselves through the medium of an unmistakeable national idiom . . . to prescribe that the American composer renounce the heritage of four hundred years of musical development . . . is disaster breeding nonsense.[37]

These two similar statements incite equal parts of agreement and disagreement. That is, jazz is *not* the medium for an American composer to express himself within, if he feels that there is still some vitality to the European art-type handed down to us. But the American, or any other national composer, who believes that the old art feeling has been worn threadbare, who feels that some new folk infusion, such as jazz, has vitality, should be urged to study jazz. Buchanan, in further expressing the opinion that "art is the expression of an individual, not of a nation" avoids the fact that music direct from the people makes up 75 per cent of the work of our great seventeenth- and eighteenth-century composers.

B. H. Haggin, an outstanding young music critic, has been occupied with jazz for some years. In 1925 he sympathetically acknowledged it. In 1935 he regularly devoted a portion of his music column to discussing the new manifestation—"swing." Nevertheless, in 1925 he seriously stated that jazz was

35. Virgil Thomson, "Jazz," *American Mercury,* August, 1924.

36. Virgil Thomson, "The Cult of Jazz," *Vanity Fair,* June, 1925.

37. Charles Buchanan, "The National Music Fallacy," *Arts and Decoration,* February, 1924.

> lacking the most important source of rhythmic variety in serious music, namely, variation in the length and shape of phrases, with artistic use of figuration.[38]

This was true of the fox-trot sheet music but most emphatically not so of the improvised hot solos of 1925. In another article, after stating that if the "pluck-pluck" were dispensed with there would be little or nothing left of jazz, Haggin makes plain his distress over the monotonous foundation beat of modern dance music. He states flatly that this beat is not emphasized in the old *bourrées;* that the bar line is only a "convenience" of notation for musicians.[39] It may be truthfully said that the bar line can be forgotten in, say, a Wagnerian opera. But, if the simple beating out of a *bourrée* inevitably makes one out of every four beats very strong, then I maintain that not convenience of notation, but a very good musical reason marks a bar line before this strong beat. Of course, this *foundation beat* of the bar is not to be confused with an accented counter *melodic* cadence which may cross it at any time. Haggin goes on to state that because of the monotony of rhythmic folk music, such material is never used save for the lighter movements of symphony. But any discussion of former folk music and its extension, in relation to possibilities for extension latent in jazz, obviously demands a skip of the nineteenth-century attitude and a return to at least that of the eighteenth—if not previous centuries. Moreover, his remark that jazz, in the light of modern academic composition, cannot be characterized as syncopated, seems unconsidered.

There are many ways of writing *technically* syncopated music, but it is not always a profusion of these different ways that leads the listener to identify the result as syncopation. In *Sacre du Printemps,* Stravinsky has been able to make the listener definitely conscious of syncopation; whereas Antheil, in his *Aeroplane Sonata,* although running the gamut of tempi changes from 17–8 time to 1–8 in order to transfer the rhythmic accents, merely leaves the listener with a vague feeling of slightly syncopated music. Obviously, on paper, we may show what is technically known as syncopation, but the music, as heard, may lack that peculiar displacement of expected rhythm we identify as syncopation. The exaggerated syncopation found in many a symphonic score may make transcribed jazz recordings appear comparatively uncomplicated, but a glance at the *melody line* of hot jazz solos gives evidence of intrinsically syncopated melodies. If we grant that Stravinsky can far outstrip the jazz orchestrator (or arranger) in the clever disposition of mass, rhythmic, chordal shocks of accompaniment, nevertheless such calculated effect does not deny the actuality of the syncopated melodic line of hot jazz solos.

38. B. H. Haggin, "The Pedant Looks at Jazz," *The Nation,* December 9, 1925.

39. B. H. Haggin, "Music, Two Parodies," *The Nation,* January 13, 1926.

So This Is Jazz, by Henry C. Osgood, already mentioned as occupying the attention of Carl Engel, was a widely read book in 1926. It went into the worst aspects of popular music with obvious admiration. At one point Osgood approvingly quotes a reported interview with a Negro boatman to illustrate how a completely new tune is introduced:

> Some good sperichils are started jess out o' curiosity. I been a-raise a sing myself once. We boys went for tote some rice and de nigger-driver, he keep a-callin on us, and I say, "O, de ole nigger-driver!" Den anudder said, "Fust ting my mammy tole me was, 'Nothin' so bad as nigger-drivers.'" Den I made a sing, jess puttin' a word and den anudder word.[40]

I quote this as an example of a false but common assumption. The story may be quite correct insofar as the *lyrics* are concerned but it is doubtful whether any folk remember the circumstance of spontaneous creation of *melody.* When a Negro sings the blues, he may for the occasion make up the words in entirety. But the melody he sings is never solely his creation; his individual musical contribution is no more than a slight variation on the already familiar form. Only through a slow metamorphic process does one folk tune, in time, become a definite *new* tune.

Although Osgood in championing Zez Confrey's *Kitten on the Keys* definitely felt that "You can't compare Confrey with Beethoven . . . And you don't need to argue with me that the 'Fifth Symphony' is better music and more important to you and me than anything Confrey ever wrote or may write,"[41] he nevertheless identified *Kitten on the Keys* as a masterpiece in jazz! How often have we found this confused championing of the wrong thing under the right name—with the actual circumstance of hot jazz making such patronage laughable!

In 1926 there appeared the very important book *Blues,* by W. C. Handy with an introduction by Abbe Niles. In his foreword, Niles is by far most competent when analyzing the actual blues. His incorporation of the music of Gershwin, accompanied by such comment as, ". . . the *Rhapsody in Blue* without doubt conveys . . . a rowdy, troubled humor as marked as that of the best of the old blues,"[42] demonstrates the usual confusion of a critic when faced with the first extension of a new folk form. Niles continues with the absurd statement (absurd in the light of the rhythmical background of much so-called "serious" music) that there is

40. Henry C. Osgood, *So This Is Jazz,* Little, Brown, 1926, p. 56.

41. *Ibid.,* p. 80.

42. W. C. Handy, *Blues,* A. and C. Boni, 1926, p. 23.

> One valid objection to the idea that jazz is timber for serious writing . . . that of rhythm, especially, of an unvarying tom-tom beat which after a time, must become intolerably monotonous.[43]

Niles is familiar with his subject, but the aesthetic conclusions he draws are the weakness of what is otherwise an excellent treatise.

In 1927, Aaron Copland, the well-known composer, expressed the opinion that jazz should become basic material for modern composers and that it was not as outmoded as Darius Milhaud, having finished his *Création du Monde,* now seemed to think. Copland appeared to believe that if only jazz were adequately defined, such definition would be of aid in composition. I doubt whether musicians in the past depended upon adequate verbal definition of their medium in order to compose within it! He expressed great annoyance with jazz musicians who merely

> can tell jazz, from what isn't jazz and let it go at that. Such vagueness will do nothing toward a real understanding of it.[44]

He attributes to Don Knowlton, the author of an innocuous enough little article,[45] the ridiculous statement that the written rhythm:

is a deceptive notation for a far more complicated polyrhythmic performance, i.e.,

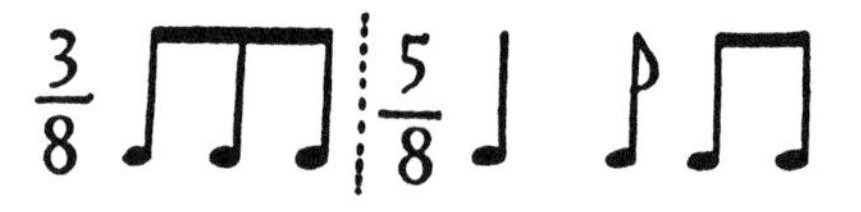

Actually, Knowlton said nothing of the kind, merely confining himself to the general remark that publishers have left out polyrhythmic effects in their printed sheet music. But Copland presses the point into such a false sequitur as:

> He [Knowlton] was the first to show that this jazz rhythm is in reality much subtler than in its printed form. . . . Therefore it [jazz] contains no syncopation.[46]

43. *Ibid.,* p. 47.

44. Aaron Copland, "Jazz Structure and Influence," *Modern Music,* January–February, 1927.

45. Don Knowlton, "The Anatomy of Jazz," *Harper's,* April, 1926.

46. Copland, *loc. cit.*

Once more we meet with a theory limiting the amount of syncopation in jazz! But this time it is buttressed by inexcusable misquotation. Copland's flimsily built-up theory that the foregoing rhythm is not syncopated is absurd. There is no basis for the assumption that the jazz player thinks in such polyrhythmic terms; that he consciously alters a normal syncopated 4–4 time into 3–8 plus 5–8 time. Moreover, Knowlton's article was too slight in content—too patently deficient in critical acumen, to hazard erecting a theory upon it.

Jazz, written in 1926 by Paul Whiteman and Margaret McBride, is amazingly naïve; it lets us in on the ground floor of Whiteman's aesthetics:

> I still believe that *Livery Stable Blues* and *A Rhapsody in Blue,* played at the concert by its talented composer, George Gershwin, are so many million miles apart, that to speak of them both as jazz needlessly confuses the person who is trying to understand modern American music. . . . When they laughed and seemed pleased with *Livery Stable Blues,* the crude jazz of the past, I had for a moment, the panicky feeling that they hadn't realized the attempt at burlesque—that they were ignorantly applauding the thing on its merits.[47]

That panicky feeling has been ultimately justified, for the *Livery Stable Blues,* burlesqued as it was in spots, contained the style which encourages musicians to improvise in jazz; it has outlived the symphonic jazz orchestra. What is now known as a "swing band" is a continuation of what Whiteman failed to explain away!

In 1927, we find a book written by an Englishman, Robert Mendl, titled *The Appeal of Jazz.* Mendl, explaining how dance movements were incorporated into suites during the eighteenth century, similarly tried to justify the continued use of jazz. But he slipped only too often into such curious comment as:

> You cannot play jazz music as a pianoforte solo: if you perform syncopated dance music on the pianoforte it is ragtime, not jazz. It only becomes jazz when it is played on a jazz orchestra.[48]

Although admiring Paul Whiteman and Jack Hylton for their "refining" influences, at the same time he decried the current attitude that the jazz musician should be rebuked for using other tunes than his own. He justly observed:

> "The greatest genius is the most indebted man," and it is the use which he makes of his legacy that counts.[49]

47. Paul Whiteman and Margaret McBride, *Jazz,* J. H. Sears & Co., Inc., 1926.

48. Robert Mendl, *The Appeal of Jazz,* P. Allen & Co., Ltd., London, 1927.

49. *Ibid.,* p. 134.

In 1927, André Levinson, the well-known European dance critic, made an observation on rhythm that should be taken to heart by the jitterbugs:

> We should not, however, jump to the conclusion that because of his extraordinary rhythmic gift alone the Negro dancer and musician should be taken seriously as an artist. Rhythm is not, after all, an art in itself.[50]

This I feel is a timely rebuke to those whom we find exclusively in the thrall of "swing" rhythm for its own sake—regardless of the value of the melodic line.

Ernest Newman, England's eminent music critic, writes in 1927 as though he were commenting on jazz in 1922. He occupies himself unnecessarily with the jazzing of the classics. However objectionable such jazzing may be, it certainly is a minor phenomenon. It becomes very clear that Newman's well-known antagonism towards jazz is based on no real knowledge of jazz. Newman probably has in mind such ineffectual pieces as Paul Whiteman arranges for a concert, when he says:

> The jazzsmiths, however, speaking generally, are not clever enough to make their manipulations of the classics tolerable.[51]

This is very true but very unimportant. Another time, he states flatly that:

> There is not, and never can be, a specifically jazz technique of music, apart from orchestration.[52]

This is followed by the astounding comment:

> Jazz is not a "form" like, let us say, the waltz or the fugue, that leaves the composer's imagination free within the form; it is a bundle of tricks—of syncopation and so on.[53]

A waltz is hardly a "form." It is primarily a rhythm with a certain style of melody superimposed. A fugue is a form, likewise a sonata—but not a waltz. Jazz is a style in which forms are developing. However, Newman immediately detected in the music of Gershwin what should have been more generally observed:

> when he launches out into "straight" piano concerto music we begin to ask ourselves what all this has to do with jazz. The work was, in fact, though Mr.

50. André Levinson, "The Negro Dance Under European Eyes," *Theatre Arts Monthly*, January–June, 1927.

51. Ernest Newman, "Summing Up Music's Case Against Jazz," London; printed in New York *Times Magazine*, March 6, 1927.

52. *Ibid.*

53. *Ibid.*

> Gershwin may not have known it at the time, a commendable effort to shake himself jazz-free.[54]

It is unfortunate that so keen and knowledgeable a critic as Newman is ever bound to the discussion of the tricked-out banalities of "popular music." Possibly he has never heard any jazz of value. But this should not excuse him from searching it out. In a still later article we find the bitter conclusion that

> (America is) a purveyor of the most dreary, the most brainless, the most offensive form of music that the earth has ever known.[55]

In an article written in 1928, Sigmund Spaeth praises Gershwin and Whiteman, but definitely concludes that jazz is not music.[56] Later, he brings up the matter in detail:

> Jazz is not a musical form; it is a method of treatment. It is possible to take any conventional piece of music, and "jazz it." The actual process is one of distorting, of rebellion against normalcy.[57]

If he means that the treatment is part of a process that results in an actual music called jazz, he would be correct, but it is only too apparent he means that jazz is a tricky attitude we apply to something of value—resulting in nothing. Moreover, it should be obvious that any style, applied to another style, is a distortion. For instance, a current theme is fitted into fugue by way of distortion—the result, however, is a new normalcy. The same is true of the treatment of tunes by dance movement. We can make a minuet, a waltz, a tango, a sarabande, a rhumba, out of any tune and most composers have done so since the sixteenth century! The musician who employs such comment to put jazz in a derogatory light is extending a weak hand to the layman anxious for something technical with which to disqualify jazz.

Further along he remarks that: "Jazz melodies have been mostly simple and obvious, easily remembered after one or two hearings."[58] If there are blues and popular tunes—and for that matter classic tunes—which are easy to remember, in point of fact a hot jazz solo is as comparatively difficult to rerender as a Bach invention. But when he says that "Stravinsky's *Ragtime* and the jazz movement of his piano concerto cannot compare (as music) with the work of

54. *Ibid.*

55. Ernest Newman, "Music and International Amity," *Vanity Fair,* April, 1930.

56. Sigmund Spaeth, "Jazz Is Not Music," *The Forum,* August, 1928.

57. Sigmund Spaeth, "They Still Sing of Love," H. Liveright, 1929, p. 140.

58. *Ibid.*

Gershwin, Souvanie, or Grofé,"[59] Spaeth has left the platform where there is possible room for argument. On no academic platform of standing would such a statement be offered in evidence. No one of these musicians wrote significant jazz; but Stravinsky, unlike our enterprising popular song purveyors, made a most significant contribution to the Modern Academy.

Spaeth's conception of melody is best stated in his own words:

> Melody is the sticky sweetness of music, the cloying jazz which needs a background of nourishing bread before it becomes really palatable.[60]

Quite the contrary is true. A poor melody is unpalatable no matter how lavish the background. More good melodies have been spoiled by sugar arrangements than sugar melodies noticeably fortified by superior background.

As late as 1929 we find a typical literary patronage of jazz expressed by the English author Aldous Huxley. Speaking of a jazz band featured in one of the first moving pictures with sound (*The Jazz Singer*), Huxley says:

> The jazz players were forced upon me; I regarded them with a fascinated horror. It was the first time, I suddenly realized, that I had ever clearly *seen* a jazz band. The spectacle was positively terrifying.[61]

There is never such "hay-wire" comment found to describe a new phenomenon of music as that utilized by those in the strictly literary field! However, perhaps it is no more amusing than that of the group of so-called music Modernists in America, who are no less oblivious to the qualities of jazz than are their own special *bêtes noires,* the old fogies of the Academy! Paul Rosenfeld is a fair example of this sort of critic.

> But the greatest fullness of power and prophecy yet come to music in America, lodges in the orchestral compositions of Edgar Varèse. . . . His high tension and elevated pitch, excessive velocity, telegraph-style compression, shrill and subtle coloration, new sonorities and metallic and eerie effects are merely the result of his development of the search-and-discovery principle in the twentieth century world.[62]

Rosenfeld is continually bathed in a mysticism that must definitely hinder him from listening to any music in terms of music. His identification of great music resolves into the following: a music spelling the space of the planets,

59. *Ibid.*

60. Sigmund Spaeth, "They Still Sing of Love," H. Liveright, 1929, p. 140.

61. Aldous Huxley, *Do What You Will,* Doubleday, Doran, 1929, p. 58.

62. Paul Rosenfeld, *An Hour with American Music,* J. B. Lippincott, 1929, pp. 160–66.

digging deep into humanity, sounding the struggle of masses, rushing on with the immensity of it all. A so-informed public in Mozart's age would have expected a music quite different from that which it got in the *G Minor Symphony;* or in Bach's age, in the *D Major Suite* for orchestra! In fact, upon being given such a definition, only to be faced with a pure example of musical art, every age would have retreated in puzzled disappointment. Truly, if we were convinced that this was the avowed purpose of the composer and an indication of what to look for in the major works issuing from him, how naturally we would hold in contempt any such pure melodic manifestation as jazz! Rosenfeld, like many another critic, thinking in terms of the complex rhythms of Stravinsky and Varèse, is dogged in his search for the bizarre. Let us conclude with his firm statement:

> American music is not jazz. Jazz is not music.[63]

Among the last of the non-jazz critics we find the late Isaac Goldberg.

> The Anglo-Saxon American has no more talent for writing or playing jazz than Europeans. Both of them are bungling at it.[64]

This is a curious statement to find in a book written as late as 1930—an American book which purports to go deeply into the history of popular music; such a statement merely adds to the confusion between popular song and jazz and is of no help at all in the understanding of folk-art. If white folk musicians do not push jazz conspicuously ahead, they most certainly have availed themselves of the medium with great distinction. On one page we read that "at first the jazz band was a crude nuisance, with the emphasis on noise," and on the next, we find him assailing the academic jazz musician as a ". . . decent white lady in her parlor trying to sing a hot jazz number that literally cries for a wild black mamma."[65] However, one thing always stands out in such books as this one: the heated discussion never commences until the jazz chapter is reached!

The year 1930 closes the chapter on early critics who, for one reason or another, felt something should be said about jazz. And jazz, it would seem, was as much a problem child for the music critics as it was for less specialized attention. We rarely hear names other than those of Whiteman, Gershwin, Lopez, Lewis, or Confrey mentioned. Ansermet is really outstanding for his expressed interest in the hot solo playing of Sidney Bechet.

We now approach the era of the Jazz Critic—he who is to jazz, what the concert critic is to classical, academic music. It is no longer the exception to

63. *Ibid.,* p. 11.

64. Isaac Goldberg, *Tin Pan Alley,* John Day, New York, 1930, p. 15.

65. *Ibid.*

find a jazz critic on familiar ground when discussing his subject. He knows all of the heretofore anonymous players by name and handles the subject turned over to him in the same sublimated *shop talk* manner as his established, classical *confrères* handle the Academy! Just as the concert critic, well acquainted with the seventeenth- and eighteenth-century music and full of well-bred enthusiasm, will, however, find no terms enthusiastic enough for Wagner or Debussy, so will the jazz critic talk seriously on Boogie Woogie piano and then proceed with obvious satisfaction to consider a Teddy Wilson!

To my knowledge, Charles Edward Smith published the first article in which separate, improvising players were mentioned by name and detached from the group as a whole, and the symphonic orchestra relegated to the background. In 1930 he wrote:

> It may be said, almost without qualification, that jazz is universally misunderstood, that the men of jazz, those of the authentic minority, have remained obscure to the last. . . . Paul Whiteman developed a symphonic jazz band, for *concert* jazz, Gershwin composed a clever *tour de force* known as *Rhapsody in Blue,* and the grand misconception was off to a glorious start, generously footnoted by writers none too sure of their material, such as the author of *The Seven Lively Arts.*[66]

He made three distinctions which more or less hold true today:

> To the connoisseur there are three classes of dance music. First of all there is jazz which is called *hot*—though all that is *hot* is not jazz. Secondly, there is popular, which the musicians call *sweet;* when ordered to play a *sweet* piece as written they play upon the words in, "Come on, boys, play this one *as rotten!*" Last of all there is that which imitates or plagiarizes real jazz and this brand of stuff is termed *corny,* the last word in disparagement.[67]

In a footnote Smith says—

> My contention is that when *hot* bands turn corny they are much, much worse than sweet bands could be. There is nothing so disgusting to a musical palate as a simulated ad lib. chorus.

I feel, however, that there is something healthier, in spite of itself, in the attitude of the *corny* players, than in the saccharine politeness of the "sweet" bands. Eight years later we find Smith saying even more pointedly:

> Technical embellishment as an end in itself is so closely identified with improvisation against background harmonies (as distinguished from variation on

66. Charles Edward Smith, "Jazz, Some Little Known Aspects," *The Symposium,* October, 1930.
67. *Ibid.*

> theme) that the two tendencies are sometimes confused. . . . It is unadulterated *corn* whether the guilty party is an accomplished musician playing games with his unsuspecting public or a tyro trying to sneak out of the long underwear gang . . . If it seems to have musicianship on its side it is a spurious musicianship, for this word should connote integration of form and substance, not merely a mastery of one's instrument from a technical point of view.[68]

Generally speaking, I find myself in agreement; but I think Smith overlooks how hard it is to create notable melody. Irving Berlin is one of the few who can consistently do so, if only on the popular song level. Mozart could do it. Clementi could not. Bach presented a much longer line of hits than Scarlatti, and Beethoven could turn out more than Brahms. Outstandingly musical jazz solos might be said to be equally rare phenomena also. To call all other solos corn is a misplaced use of the word.

In view of the fact Panassié feels jazz cannot be notated, it is interesting to observe that Smith feels some musicians have "a technical equipment that *gives to almost any kind of chorus* an air of authority, sometimes convincing the musician himself that what he is doing is the real thing."[69] (Italics mine) Such a remark as the "real thing" must mean: music that stands up notated.

One of the most popular books on jazz, and deservedly so, was *Le Jazz Hot,* by Hugues Panassié. Published in Paris in 1934, it not only went into the hot jazz question soberly and authoritatively, but also managed to catch the attention of the "swing" public by way of the numerous Swing Societies. Panassié takes up point by point every feature of jazz; the player, the orchestra, the orchestrator, the different instruments, etc., etc. The breadth of the book is what counts in its favor, and it is only through serious reading in its entirety that its full scope is brought to the fore. However, without meaning to lessen its general importance, there is no doubt it brings out a number of false attitudes towards jazz and art. Panassié tries to establish the difference between jazz and other music by saying:

> in most music the composer creates the musical idea, and the performers recreate these ideas as nearly as possible. . . . They [jazz players] are *creators,* as well.[70]

This is an all-embracing distinction prevalent amongst even the best of jazz critics. It certainly would be a poor way to differentiate jazz from any but nineteenth century music. Although Panassié spends much time stressing the

68. Charles Edward Smith, "Two Ways of Improvising on a Tune," *Down Beat,* May, 1938.
69. *Ibid.*
70. Hugues Panassié, *Hot Jazz,* Witmark, 1936, p. 1. Translated by Lyle and Eleanor Dowling.

thought that jazz *must* have "swing," this whole section is weak because he cannot give a musical definition of swing.

Panassié insists on simply maintaining a disc preservation of jazz. He seems to believe that, although "in classical music, a few sheets of paper are enough to note down a work to preserve it with all its values. In jazz, on the contrary, even when there is no improvisation, the actual performance is itself most important."[71] Even if he insists upon believing that for the first time in history a *playing style* is significant, it is strange he does not see that jazz is also significant on paper, and in this way far more susceptible to considered investigation.

Panassié also seems to believe that merely one or two musicians freed jazz from its early inhibitions. King Oliver and Louis Armstrong were not indispensable to the growth of jazz. However, this should not be taken to mean that they did not exercise great influence. It has always been difficult justly to place the great man in the great school. It is proper that the works of an outstanding individual be segregated for special comment as the best examples of the period, not only for their originality but for the high consistency maintained. Our mistake is in attributing to the individual the entire art phenomenon. This is not his due. A school of art is based on a rank and file, and out of this rank and file rises the outstanding individual. The relation of their respective contributions is one of degree; the waters of genius rise higher, but they are part of the same fountain produced by the rank and file. From this point of view, the degree of his contribution, if he is working within a strong school, is never as great as it seems in isolation. Thus, outstanding as Oliver and Armstrong were in the early days, the school of jazz playing had in it, even then, all those characteristics which these two artists were able to point up so well. If hot solos were shorter, and did not reach the recording studio, nevertheless, they were all about, and full of the extraordinary variation of jazz.

What is his attitude to the actual music of either jazz or Bach? If he has said that jazz cannot be notated, that there are things in the playing of it that are impossible to divorce from the player, why does he quote with evident satisfaction:

> One day I was playing the first measures of his [Teschemacher's] solo in *I Found a New Baby* with one finger at the piano. "Is that Bach?" a composer who was listening, asked me.[72]

Yet here he shows how inventive Teschemacher was, on the basis of the melodic line itself!

It is very evident that Panassié thinks in terms of performance. Although he

71. *Ibid.,* p. 20.

72. Hugues Panassié, *Hot Jazz,* Witmark, 1936, p. 91.

speaks well of Meade Lux Lewis's *Honky Tonk Train Blues* and of Joe Sullivan's *Gin Mill Blues,* he no more than lists these composers along with all the other pianists; apparently judging them by the amount of *swing* they have to their playing style. But he picks out Bix Beiderbecke's published piano solo, *In a Mist,* as the great piano solo contribution to jazz music. Panassié clearly retroverts to the French school of modern music when he picks this superficially mannered music as an example of the best piano jazz. Further along, Panassié seems to contradict himself and acknowledge that jazz may be advancing towards an important notated stage when he says:

> In *Panama,* Lewis Russell has written several ensemble choruses with such hot elaborations that a soloist has only to play the written melodic line to sound as if he were improvising.[73]

It seems to me that such a statement from an author denying the possibility of significant, written jazz, gives cause for reflection.

Panassié dismisses the music of Gershwin as though it were music of foreign origin, having no connection with jazz. As much as we may dislike Gershwin's attempt, no critic should avoid discussion of such development, with its precedent in all art—in all classic music. In fact the suspicion cannot be avoided, that all the classic music we now have notated is rooted in just such a spontaneous music as jazz, and that sooner or later jazz, in spite of all the vital qualities of performance we now link to it, will also take its place in notated form, alongside our great classical music. But, whatever the disagreement with Panassié on these and other points, his book is a landmark in jazz criticism.

One very well-known critic, whose name has become synonymous with hot jazz, is John Henry Hammond, Jr. Otis Ferguson sums him up as follows:

> John Henry Hammond (Junior) is known to practically every one who ever mounted a band stand, or plugged a song, or got on the free list for records and wrote articles using such phrases as gut-bucket and out of this world and dig that stomp-box. He is known as The Critic, the Little Father, the Guardian Angel and the Big Bringdown, of dance music. But the point is he is known.[74]

John Hammond occasionally gives evidence of being a critic definitely detached from such musicians' shop talk, but for the most part he seems to be deeply embroiled in the commercial ambitions of a band—its desire for smoothness, or for more personality in its singers, etc., etc. All of this is purely the business of a publicity agent—not of a hot jazz critic. But Hammond has said:

73. *Ibid.,* p. 211.

74. Otis Ferguson, "John Hammond," Hot Record Society's *Society Rag,* September, 1938.

> To my way of thinking Bessie Smith was the greatest artist American jazz ever produced; in fact, I am not sure that her art did not reach far beyond the limits of the term "jazz."[75]

This is not the talk of the usual up-to-the-minute hot jazz gossip purveyor. And again we find him saying:

> nobody would be more incredulous upon hearing that this [jazz] is art than the throngs of jitterbugs and the hot musicians themselves. Perhaps this is the spirit of the early Renaissance art movement in Italy, of the stone carvers on Romanesque cathedrals, because it is the thing taken for granted and warmly participated in by the people.[76]

Another time he states:

> The hysterical roars of the crowd which once had been sweet music to his [Goodman's] ears, first perplexed, then irked him. He wanted his music to be appreciated for its essential worth and not because of its fortissimo volume and crazy antics. . . . Benny likes good, simple, relaxed music, whether it is by Fletcher Henderson or Mozart.[77]

The remark "simple, relaxed music, . . . by Fletcher Henderson or Mozart" is very interesting. The sympathetic way Hammond handles this interpretation of Goodman's thoughts leads me to believe that Hammond is quite in agreement. Moreover, the way Benny Goodman actually plays his clarinet, his excursions into Mozart, all point to a musician who feels a need for a more refined form of swing; what he might call a higher type of swing; something nobler to be sought after than "fortissimo volume and crazy antics." But does that side of jazz, considered analogous to Mozart by both Hammond and Goodman, really consist of the noble, relaxed simplicity we encounter in Mozart? Some of us who accept hot jazz as we do the sterner kind of classic music, cannot accept Benny Goodman's refined jazz on the same basis as Mozart. In classic music our good taste is varied, and within certain limits we like to hear more than one kind of music, but it does not follow that the large "swing" public which enthusiastically approves genteel "swing" along with hot jazz, is exemplary of a similar good taste.

Let us think back on the long history of Occidental music and observe the invading sweetness that comes sooner or later over each new form of music. It is a sort of decadence which creeps over all art. The strongest academic com-

75. "John Hammond Says," *Tempo,* November, 1937.

76. John Hammond, "The Music Nobody Knows," Program Notes for the concert *From Spirituals to Swing,* December 23, 1938, Carnegie Hall.

77. John Hammond, "Hysterical Public Split Goodman and Krupa," *Down Beat,* April, 1938.

posers are always fighting it; but in spite of their fight, if their music holds any elements of melodic lyricism, the school becomes sweeter and sweeter with each succeeding generation. The result of such a decadence is our present-day American popular music, sweet to the point of puerility. On the other hand, the American Negro jazz phenomenon is a new beginning; and being new it has the earthy quality we find in all new art. America, subjected to both schools, accepts both with equal intensity. We make no real distinction in caliber, and when the two are mixed, even the swing addicts are apparently fooled.

For good jazz, good musicians still have to play at white heat; hence the term "hot jazz." Hot players who cool off in their playing, generally become either sweet or disturbingly banal. This is the way of a folk. Only when the full significance of this jazz material has become apparent to them, will they be able to play their material with a relaxed, simple sweetness comparable to the classic sweetness of a Mozart. The later Goodman clarinet, the Wilson piano, much of the Dorsey and Teagarden trombone work, not to mention the recent Duke Ellington arrangements, are not representative of the best of jazz played with refined distinction but are, rather, an unfortunate demonstration of a superimposed, cheap, classical experience. Neither this experience, nor its derivative, the popular-song-infected mind, can aid us in arriving at "refined, relaxed and simple."

In *Jazz: Hot and Hybrid,* by Winthrop Sargeant, we find an academic book that successfully breaks down the structure of jazz. In rhythmic, scalar and harmonic analysis of jazz, and consideration of both Negro and European background, this book is by far the best to date. But, Sargeant's final and uncomplimentary estimate of jazz is a familiar example of the old truism, that full knowledge of the inner workings of an art form does not necessarily lead to a sensitive awareness of its position in art history.

It is curious to find him dismayed, as the following quotation indicates, because Negro spirituals have been taken into the concert hall:

> This, to the superficial observer, has given them [the spirituals] the status of "compositions," that is, of fixed musical creations designed to be "interpreted" according to the conventions of our concert art. This notion, as has often been pointed out by those familiar at first hand with Negro musical expression, is an essential misunderstanding of their nature and function.[78]

When we reverse the thought and consider how often bawdy, secular tunes have been used for religious purposes, prohibition on such a basis as this seems childish. The original intent of folk music is always lost when it is used in concert—whether religious or dance.

 78. Winthrop Sargeant, *Jazz: Hot and Hybrid,* Arrow Editions, 1939, p. 18.

But if we agree with his statement that the "jazz musician has a remarkable sense of sub-divided and subordinated accents in what he is playing, even if it be the slowest sort of jazz," it is difficult to understand why he believes that this is "*quite foreign* to European music." (Italics mine.) If it is true enough of the playing style of the average, contemporary concert artist, nevertheless the recapture of past dance styles by Wanda Landowska (on the harpsichord) shows us that seventeenth- and eighteenth-century music, at least, was intended to be played with as much awareness of small metrical units as is contemporary jazz. This is the big mistake Sargeant makes. He gives the impression that European music is intrinsically devoid of this rhythmic quality, whereas it is only a lost school of playing that has given us the unrhythmic, lifeless, classical attitude we are now used to.

Moreover, much as we may be interested in what Sargeant can academically analyze for us, in his concluding chapter titled "Jazz in Its Proper Place," he not only brings up a point of view aesthetically repudiating all that which he has previously intimated, but he leaves us in this disconcerting position without one word of explanation. Let us consider the following statement:

> It is not surprising that a society which has evolved the skyscraper, the baseball game and the "happy ending" movie, should find its most characteristic musical expression in an art like jazz. Contrast a skyscraper with a Greek temple or a mediaeval cathedral. . . . Like the jazz "composition" it is an impermanent link in a continuous process. And, like jazz, skyscraper architecture lacks the restraint of the older forms. The skyscraper has a beginning, and perhaps a middle. But its end is an indefinite upward thrust. A jazz performance ends, not because of the demands of musical logic, but because the performers or listeners are tired, or wish to turn to something else for a change. As far as form is concerned it might end equally well at the finish of almost any of its eight-bar phrases. A skyscraper ends its upward thrust in precisely the same way. It might be stopped at almost any point in its towering series of floors. It must, of course, stop somewhere. But the stop is not made primarily for reasons of proportion. Nor does it carry that sense of inevitableness that attaches to the height of the Greek temple.[79]

When Sargeant contrasts the skyscraper with the Greek temple *and* medieval cathedral, I might ask how the medieval cathedral can be categorically linked with the Greek temple! The medieval extravagance of design, its riot of graphic description for what was once a simple device of architecture, might have outraged the ancient Greek if as academic-minded as Sargeant! As far as simplicity is concerned, the skyscraper is the only architectural counterpart of the Greek temple. Moreover, if it is the *undetermined* upward thrust of the skyscraper which limits its art significance, let us not forget that the large Greek

79. *Ibid.*, p. 214.

temple adhered to no particular length of construction. When he ties up this supposedly pertinent architectural analogy to its musical counterpart of *undetermined* termination in jazz performance, we realize that he is simply observing that jazz is based on the variation form—and that sometimes we know when the end is coming and sometimes we do not. But—do Haydn's 64 variations on a theme give any more notice? The same adjective flung at jazz for its uncultured and so-called "formless" state, can be flung at all pre-Bach music.

One thing keeps intruding throughout the classic references Sargeant enjoys making, and that is, he never refers to past art in terms of great *art form.* For instance he will say that the arts of America

> all lack the element of "form" that is so essential, for example, to tragedy, to the symphony, to the great novel or the great opera, to monumental architecture, and to a large amount of Europe's less pretentious expression.[80]

Here, he refers to art in its period of almost sententious importance. Jazz is not at the grandiose stage. The art stuff out of which a Bach passion is tied together *ideologically,* in its original form, had none of the identifying qualities of the great *Passion.* Sargeant has already admitted this himself. Why then, such a conclusive, lofty dismissal of a new art stuff for *that* reason? The following reasons seem hardly satisfactory:

> The attendant weakness of jazz is that it is an art without positive moral values, an art that evades those attitudes of restraint and intellectual poise upon which complex civilizations are built. At best it offers civilized man a temporary escape into drunken self-hypnotism. . . . It is a far cry from the jazz state of mind to that psychology of human perfectability, of aspiration, that lies, for example, behind the symphonies of a Beethoven or the music dramas of a Wagner.[81]

Unfortunately for such a theory—and scratch every academic critic hard enough and you will find it—jazz is no different in spirit from any other folk music in the heyday of creation. In fact, as Engel pointed out, the liveliness and the ribald atmosphere surrounding most of the early classic music, was far more at odds with the more refined culture of the day than is jazz contemporaneously.

American Jazz Music, by Wilder Hobson, is a book seeking to make us more intimately acquainted with the best in jazz. We find on the jacket the following well-expressed comment:

80. Winthrop Sargeant, *Jazz: Hot and Hybrid,* Arrow Editions, 1939, p. 216.

81. *Ibid.,* p. 217.

> . . . [Hobson's purpose] has been, not to prove that jazz is "better than Bach" . . . but simply to tell its story and to make its complex, unprecedented rhythms more understandable and hence a greater source of enjoyment.

This, Hobson does well. Occasionally he feels constrained to point out that jazz is different from *all* other music, but at the same time he admits the circumstances surrounding jazz appear familiar.

> It is often said that jazz cannot be notated. It cannot; and, strictly speaking, of course, *neither can any other music.* Any music is played with a "translation" of the written note values according to tradition for that particular kind of music and the instincts of the performer. [Italics mine.][82]

Hobson truly observes that no music is written without a special school of playing in mind. His comment upon the "steady beat" so many critics find objectionable, is well taken and straight to the point:

> The very regularity of the beat, of course, facilitates the improvising of rhythms, as does the fact that the beat is unaccented. In other words, movements may be planned around a certain base, whereas an uncertain base allows only uncertainty and the constant likelihood of getting lost—or, in this case, of rhythmic confusion.[83]

The importance of this observation can never be overstressed. Altogether this book is highly satisfying: a sensitive approach to the music the author likes.

In 1938 William Russell presented an extremely interesting analysis and review of Boogie Woogie piano records. With great sensitivity he points out that:

> One of the most important developments in hot jazz, the last few years, has been the rediscovery of the Chicago school of Boogie Woogie pianists. . . . In making full use of the resources of the instrument, the Boogie Woogie is the most pianistic of all styles. The piano is treated as a percussion instrument rather than an imitation of the voice or a substitute for an orchestra.[84]

In conclusion, let us say that the early critics occupied themselves with the *extension* of jazz but entertained little consideration for jazz itself. On the other hand the recent critics have concentrated upon the living recording as an end in itself. Without doubt the disc is a priceless innovation towards the preservation of music; it would be a great loss to jazz history and early jazz music had

82. Wilder Hobson, *American Jazz Music,* Norton, 1939, p. 29.

83. *Ibid.,* p. 48.

84. William Russell, "Boogie Woogie," *Hot Jazz* Nos. 25 and 26, June–September, Paris, France, 1938.

recording not existed. Nevertheless, if the musicians themselves are becoming conscious of written music with constant recourse to their own records, such awareness, *built* on the past, will be more lastingly significant than taking off from scratch every day.

My personal attitude might be expressed by the hope that jazz will continue along the path the best of it has taken so far. Inevitably there will be an academy. But the ultimate significance of this academy will be wholly dependent upon the continuance of this virile school of improvisation until such time as the academy materializes. Moreover, the longer we can postpone the arrival of the academy the more profitably will a vigorous school prepare itself for the event. A forced academy, compelled by composers removed from the jazz field, will not do! The jazzmen themselves must crystallize their school of playing. So far, the jazz school is a brilliantly improvisatory one. It is the group of arrangers who set, augment, and redistribute this living art, which has not distinguished itself. The fast "killer-diller" arrangement, the suave hot arrangement, the nostalgic arrangement, although commendable in their effort to break away from the banal, express a tendency to break away from the dance also. It is the dance-inciting, hot, but loosely organized arrangement, that seems safest for the present. As Ezra Pound has so well said:

> The author's conviction on this day of New Year is that music begins to atrophy when it departs too far from the dance. . . . Bach and Mozart are never too far from physical movement.[85]

So let us hope that as jazz crystallizes, one portion will be dance music of the best kind, and the other, music of a caliber worthy of intensified listening in concert. There need be no antagonism between the two nor any distinct separation. A quotation from Mozart helps to make me feel this is possible:

> I saw, however, with the greatest pleasure, all these people flying with such delight to the music of my "Figaro," transformed into quadrilles and waltzes; for here nothing is talked of but "Figaro," nothing played but "Figaro," nothing whistled or sung but "Figaro,"—very flattering to me, certainly.[86]

By this of course I do not mean to suggest the jazzing of another school's concert music, but the happy balance of a unified folk and academy fastidiously giving and taking from one another.[87]

85. Ezra Pound, *A B C of Reading,* Yale University Press, 1934, p. xii.

86. *Mozart's Letters* (Vol. II) to Herr Gottfried von Jacquin, Prague, Jan. 15, 1787. Oliver Ditson & Co., Boston.

87. *Editors' Note:* Earlier articles by Mr. Dodge are: "Negro Jazz," London *Dancing Times,* October, 1929, and "Harpsichords and Jazz Trumpets," *Hound & Horn,* July–September, 1934.

Hot Jazz: Notes on the Future

HRS Society Rag, January and February 1941.

Jazz, a new manifestation in music, having no precedent for a great many of its aspects, must seriously face the future. A folk art cannot continue forever the aliveness it feels in its infancy. For if there are no decadent influences attaching it on the one side, on the other it will settle back into a rut and function in this rut, century after century. When the original jazz players are playing we experience great art. The whole affair is on a high level including the player's intonation. But can that first high flush of improvisational greatness be maintained for long?

Jazz has had the advantage of disc recording at possibly its greatest period. The whole process of parturition, childhood, and adolescence is preserved on wax. That it has been preserved and will continue to be preserved goes without saying, but there is room for controversy as to how long this present run of improvisational creative activity can be maintained. We can almost see what will happen, if we will seriously consider the career of jazz to date. Let us say that in the 1920s, alongside the over-sweet orchestras, there were many jazz orchestras that were popular and, at the same time, quite consistently playing hot. Now, in the 1940s it would be more accurate to say that, although the sweet orchestras are a little hotter, at the same time the hot orchestras are very much sweeter! Both types are aiming at the same market, and when the final melding occurs, our original jazz feeling will have passed away. For jazz to maintain a vital reason for continuing, it must go ahead—not settle back as did the music of the Orient. But in order to go ahead, it must become more assimilable, more concentrated, more selective in its initial recording on discs. On some records we find solos that are far beyond the usual run of jazz. Occasionally we are lucky enough to find such solos on a record maintaining a high standard throughout, but as a rule the surrounding material is exceedingly mediocre. The improvement of this situation is a matter for the folk themselves, plus submission to critical guidance; only by guidance can the intonation and actual playing style, so prized by jazz lovers, be preserved on the discs and not lost in indiscriminatory arrangements.

Perhaps a first consideration for those interested in the development of jazz is a fixing of the place intonation takes in jazz—intonation regardless of the musical line it follows. Musical intonation and melodic line can not be fairly considered as one and the same thing. In some cases they are hard to separate offhand, but notation of the musical line will effectively do this for us if we are interested in making the test. Let us agree that very possibly instrumental intonation in jazz, as we now have it, can not improve upon itself—it is now at its zenith. Thus, we are saying that from the standpoint of intonation, jazz leaves nothing further to be desired. But is it not possible to enjoy a more compact form of what we already have recorded on discs? If there is only one worthwhile chorus on a record, is it not esoteric to simply play over that one chorus, avoiding the rest of the record? Is not music a time art and must we not go further than a mere intent listening for half a minute at a time? As musicologists we recognize the greatness in the melodic line of special portions of records already on the market, but is our sporadic listening completely satisfactory? Could we not look forward to listening at some length to recorded duplications of all the melodic solos we know and respect—transcribed with a feebler intonation, no doubt, than that of the original improviser, but allowing us a complete musical experience unmarred by trite choruses and tedious sections? In the long run the vital enjoyment of music resides more in the pleasure of living with a melodic line than in listening to its perfect rendition. When we get both at once, the experience is a poignant one, but if both, for some reason, cannot be present, is it not more satisfying to hear a good solo presented within a dilute intonation than a forceful playing style occupied with poor melodic line? When we add to this the consideration of a unique playing style, such as Bix Beiderbecke's, applied for the most part to cheap popular material so that enjoyment of his work is seriously disturbed, can we not further demand a deliberate assimilation of such an intonation and playing style and their application to a more significant ground melody? I feel that, if—unfortunately—our surrounding period merely accepts jazz melody as background for social dancing, more unfortunately still, a large number of jazz enthusiasts is carefully listening to intonation alone; that a very small minority is attentive to melodic line. Active band musicians hear this melodic line, but, not being musicologists, are unconscious of its importance to jazz. Let us remember that however vibrant a particular passage may be in so far as intonation is concerned, the future of jazz depends more upon what *is* said than *how* it is said; it depends more upon melodic line than upon intonation.

What progress has been made today towards keeping up the tradition of hot playing on discs has been made upon the demand of collectors, and actually brought about by organized centers of these collectors. Centers such as the Hot Record Society, the United Hot Clubs of America, the Hot Club of France, recording groups such as Blue Note, Solo Art, and others not only are responsible for the reissuing of old records, but are responsible for our entire wave of "jazz session" originals. All this pioneering activity has prompted the commercial houses to follow suit, to go through their old files for reissues and temporarily to organize the old hot players for new jam sessions. This is not jazz moving ahead under its own impetus. It is connoisseurship on a grand scale making the ball roll. Jazz in the 1920s was on its own; it was the hot and rare stuff responsible for jazz history. But only connoisseurship, improved and distilled through the jazz centers, provides what we have today. And only connoisseurship will take care of the future. If "sweet" times come about again, the commercial companies are not going to bother about maintaining and organizing a large public for hot music; they will become especially indifferent, for example, if no new generation of performers springs up, equal to that of the past. We may be past the era of jazz which could simply function by itself like all healthy folk arts in their infancy—the jazz which evolved as a phenomenon and maintained itself without organized motivation. It may be time now not only to concentrate on the preservation and continuation of a hot playing style, but to concentrate on the development of significant melodic line and its organization into a music of consistent playing quality. I believe it is the business of the jazz lover and critic to consider the future of jazz; that soon it will not be enough to record appreciatively the uneven moments of improvising jazz players. The old-timers will cease to exist. There is no reason to suppose the improvising spirit of true hot jazz will continue to work for the benefit of special recordings. When the spontaneous folk feeling has worn itself out, a people who now demand jazz will demand it no longer. We should begin to encourage the thoughtful organizing of already improvised material, the codifying of random ideas; a *notational* habit of thinking by jazz critics in the interest of a greater and continued development of jazz.

Now what is my understanding of the process of notation? I think the process involves something beyond the use of representative symbols for sight-reading performers. It is the means by which the improviser can pyramid the ideas impossible to carry along or develop in his head. Take an improvising pianist, for instance, like Meade Lux Lewis. The succession of ideas he has finally crystallized into *Honky Tonk Train* was simple enough for him to carry in his head. Memory took the place of notation—making notation superfluous in this particular instance. But memory cannot adequately cope with a wealth of ideas. The fact that few jazz players are now occupied with organizing their musical ideas, the fact that most of them are still improvising extemporaneously so that few records present any over-all consistent excellency, makes it imperative to point out the exceptional pieces such as *Honky Tonk* have come about in a way, that for all practical purpose of argument, is a method of notation—a method whereby the composer's memory takes the place of symbols jotted down on paper. Let it not be imagined for a moment that I expect Lewis to worry himself into conscious compositional activity. I wish merely to point out that, in a few cases where great folk improvisers receive recognition on the basis of three or four pieces—not blues or popular songs, *but pieces*—these pieces have come into existence by methods quite analogous to the writing habits of conscious composers. I wish to point out that a recording device is not taking, and cannot take the place of notation. It can only record, first, the unorganized activity of the folk and, second, if we are lucky, its step into the organized activity of a notating academy.

If the disc preserves the accidental best of yesterday and today, it is, for tomorrow, far too cumbersome a method of improvisation-and-playback if the jazz musician is planfully developing his creative ideas. And if jazz ideas are not developed when this generation dies off (leaving behind, true enough, a pile of records but no plan of writing), there will be nothing to impel future generations to go on with jazz. In fact, the urge to improvise will, in the nature of things, exhaust itself, and jazz will all too easily succumb to the popular song. A consideration of timesaving alone would prohibit any active composer from setting up a recording device in his room, improvising into it, playing back into it and then—through a laborious process of improvising, paring and cutting—committing the organized improvisation to memory for a final playback. No, the recording machine registers our playing style, our interpretive intonation; it is an invaluable replica of our musical actuality, but it does not take the place of notation in the creative life of the advancing musician. Moreover, if the use of memory is for all practical purposes a form of, or let us say a desire for, notation, any preconceived plan indicating which instrument will follow another in a string of variations shows an inclination for organizing. This soon necessitates the use of symbols, longhand if you will, simply to remind the director of that which he had planned. For the present, this is as far as the direct use of symbols need go in the development of jazz.

For the future of jazz, if significant recording activity can be maintained for any length of time, we must depend upon an academy to carry the art further along. In utilizing music's new accessory, the disc, such a "jazz academy" will not have to work under the handicap of past cultures. By being able to refer constantly to folk inspiration on the disc, this "academy" may make jazz become a greater music in certain aspects than any we have had so far in history. But it must be remembered that the recording of a complete, select, solidified, pared-down folk expression will take a long time and that only upon its successful accomplishment can this next step, that of instituting a new Occidental academy, be realized.

II

Jazz should take the next step forward—whatever the consequences. It cannot remain static; that is, whether it progresses or not, it is historically subject to change. If we cannot do much more than point out the pitfall of a careless *laissez-faire* attitude, if it is impossible to state categorically what the next move should be, nevertheless, we can urge jazz not to succumb to the pervading decadent thought and suggest ways it may entrench itself until it is ready to advance. This may help it to prepare itself for something ultimately well worth our effort—something which we can still call vital and important. After all, if we accept our whole heritage of Occidental music we must be prepared to expect decadence. So let us not deny jazz an extended development, a vital progressiveness, because of the fact that jazz will eventually succumb. Extensive recording of our early folk has already removed any valid reason for preserving a constant folk status similar to that of the Orient.

Let us consider the undeveloped status of Oriental music, a music which has never been subjected to a notational system equal to the task of developing it—a music where the players themselves, through their personal example, have been the real carrying agents from generation to generation. In spite of the fact that, similar to all dance music, complexities in execution are involved, it is this lack of notation more than anything else which has confined Oriental music to its tradition of undeveloped structural simplicity. It is a "way of playing music" which has been handed on rather than the relaying of a series of precise notes. This has left all Oriental music in a simple state of partial improvisation—the primitive notation somewhat resembling our own early pneumatic system. The pneumatic system consisted of an indicated *direction* of notes, either up or *down,* but it had no indication of actual pitch. According to *Grove's Dictionary,* Occidental pneumatic notation was developed into a system of considerable fullness and expressiveness, *but it only remained satisfactory* "as long as nothing more was required of it than that it should suggest to singers the proper execution of melodies already established in the ecclesiastical repertory." Obviously there is considerable difference between this quasi-notational activity, serving only to remind a performer-musician of something he has established recently (or is interested in preserving as a "ground melody" indication), and a developed notation.

If we grant that, owing to the fact that Oriental music was not able to develop and expand, it avoided succeeding decadence (the best of it was lost; nevertheless, a school of playing has been kept up, a passable semblance of the best), we must, on the other hand, admit jazz faces a different world. It faces decay. It cannot settle back as did the music of the Orient. Unless it goes ahead it will succumb to the surrounding atmosphere and only leave us a short history on discs—an amazing introduction to an absolutely new school of playing built upon the progress of our harmonically minded academy. It may be that jazz has been recorded in its greatest period of invention; on the other hand, perhaps jazz could support a long life in an academy. If the greatness of its early period is entirely preserved for us, why condemn the living remnant to an early decadence? Although, unlike the Orient, we have been able to preserve our best and perhaps greatest period, it is folly for us to count on maintaining a primitive status which cannot possibly exist within our prevailing decadence. Jazz should begin to think in notation terms and begin to give the future something akin to what it has been given by the past. For jazz, there was already in existence a highly organized, but simple, genre to develop within. Let jazz itself develop for the future that newness which it has brought to music; let it in turn provide the new impetus for the future to work on. It is the responsibility of the present not only to leave a significant residue (discs), but to make plans for the living future.

It is important for all jazz enthusiasts to know something about a music which has undergone notation—the immediate reaction and the ultimate result. In the history of Occidental music, the development of notation assumes an early importance. There is no doubt that the building of such a great structure as we now have in our music depended upon a written art-remainder; i.e. something more than the memory of an event to build upon. And this in spite of the evils following the art of a developed notation—evils perhaps not immediately prominent, but putting in an appearance soon enough. For, although it is through notational activity that Occidental musicians have been enabled to build the complex structure we have had for some time now, nevertheless this complex structure is responsible for our aftermath of melodic decline. But our melodic decline only comes as an over-ripening of past greatness. If the 20th-century popular song comes out of this over-ripening, and is one of its worst manifestations, nonetheless we have had the experience of great academic music and should try to understand and evaluate its protracted decay, its putrid final decomposition affecting jazz.

There are two significant elements in a highly developed music which, together, are beyond a simple folk state—*invention* and *form.* The act of *invention* consists, primarily, in the presentation of a continuous stream of a new material; *form* is, primarily, a consideration of the balanced repetition of *invention.* The editor of *Grove's Dictionary,* H. C. Colles, has said: "As long as musical sound consists solely of repetition, the monotone, it remains formless. On the other hand, when music goes to the other extreme and refuses to revert to any point, either rhythmic, melodic or harmonic, which recollection can identify, it is equally formless."[1] Obviously, *invention* is of far greater importance than *form.* We must look upon *invention* as the actuality from which *form* is achieved. Those who depend too entirely upon *form* lose sight of its function. According to Parry, it was the discovery of melody based on harmony, that ultimately nullified music's true purpose . . . "making composers look to form rather as ultimate and pre-eminent than as inevitable and subsidiary."[2] It is true that the harmonic balance *within* a tune involves *form.* Nevertheless, for present purposes, let us consider *form* in relation to repetition within an extended work as a whole, not the repetitions within the scope of a short tune. In *Solid Rock* (H.R.S. Original) the trumpet introduction, repeated as a coda, is a simple example of the larger form type. A string of variations, whether two or twenty, cannot be said to constitute a *form;* nor does its juxtaposition, common "ground" (tune), or time length combine to make a supervisory form. All we can discover is a certain prevailing balance, occasionally linking the variations through style, timbre, or mood.

There are limits beyond which the folk will not carry their art. If problems are not subconsciously solved, they stay unsolved; and the art, after the first improvisation, remains at a standstill. It is the academy which pushes ahead, which consciously explores every possible outlet for artistic expression. Ultimately, academic effort usually becomes incorrectly abstract. The flux of art, within any of its discoveries, does not remain the completely free expression natural to the original folk music. However, a succeeding folk, struck by this academic achievement, will engage in extensive academic borrowing. And, although such a folk approach to the academy may be naive, it can divert what it does not understand into something it can intuitively feel. Thus is manifested a cycle: early folk activity, in time turning into a more cerebral academy; the cerebral academic activity, in turn, working back to a new generation of folk; the new generation of quasi-folk functioning within the grafting. This new and final activity, although an apparent return to simplicity, is, nevertheless, a new and vital art expression.

Previously, I pointed out that there was no difference which could not be

1. *Grove's Dictionary*—Notation.

2. *Grove's Dictionary*—Form.

equated between the terms memory and notation; that memory and notation share similar objectives; that it is only in an advanced state of music that unaided memory limits the conscious artist. Also, that in order to advance, the conscious composer has to have at his command more than the means of memory, for there is no doubt the reduction of musical thought to paper entails a wholly different way of creation. It can play havoc with folk expression unequipped for such activity. In jazz, whether we like it or not, we already have a situation of professional arrangers and among them a few, very few, more apt at writing solos than are the average jazzmen at improvising. This situation is now in force through the demands imposed by orchestration problems, not through the arrangers' desire to compose. The conscious *composing* of such an arranger is premature and rarely approaches average improvising, the arranger having no aptitude nor equipment for the cold creation and handling of significant written material. Only long and considered research in their own early material will enable a folk to proceed with important independence.

Since the act of improvising depends upon factors over which the individual has no control, to improve upon improvisation, to make notation say an equal amount, the individual composer must absorb the *musical culture* behind his improvisation. Culture is not synonymous with improvisation—it is another activity. Culture involves an improved attitude towards improvisation. But, although this attitude is not necessary to the act of improvisation, only such an attitude can lead the folk artist to attain an understanding of his own music. Such an attitude involves constant and considered evaluation of his own improvisation. If, in the early stages of an art form, the many contributing factors which go to make up the improvisatory art can usually be found in an outstanding performer (a musician who will possess to a noticeable degree the qualities of both invention and performance), it is for the composer to distinguish the improvisatory line from performing-genius. He must live with the improvisation and then live up to it in his compositional activity.

Perhaps something will come out of jazz that will be as important for the functioning of the music of the future as was the discovery of harmony for our music. The major arts consist of more than one important element, and if significant elements are isolated, as for instance, *rhythm* was in Africa and *tone* in the old ecclesiastical chants, we find their divided art use very limited in scope as compared with their art use combined. A coming-together not only joins the individual resources but forms a new entity—a new entity which in its fused unity goes far beyond any mere pooling of resources. However, even in their segregated states, it must be borne in mind that quite unconsciously, or quite unexploredly, a kind of the unfamiliar element has always been used to enhance the familiar. For example, in Palestrina we find simple rhythms set within masses of tone; and in rhythmically sophisticated Africa, two drums of different timbre or pitch simultaneously employed as an elementary use of *tone* to enhance *rhythm.* The discovery of harmony, that is, the conscious use of it as a new element, made it possible for music to build its greatest structure. This structure, built upon the two older elements of *rhythm* and *tone* plus the new element of *harmony,* is evident in the cheapest contemporary tune.

It must be understood that when these elements first combine, the melding, at the outset, does *not* involve a more complicated art activity. Quite the opposite. Simple folk, in combining two or three elements so as to make one, make the new whole as simple to grasp as was any one of its former component parts.

As harmony was latent in the most primitive music (although it was not used as such), so, perhaps right now, there lies latent in jazz something which we need only to become actively conscious of to use deliberately. Out of the elements going to make up the whole of jazz, elements we are hardly conscious of, it is possible that one alone may be the mainstay of our future music. We may become poignantly aware of this element while jazz is still in its improvisational state; on the other hand, we may be able to extract it only after jazz has been considerably developed through notational activity. The ensemble attitude of the "jam session," in a primitive manner, has shown a way forward for the jazz orchestrators. Unfortunately, the commercial orchestrators have not been able to feel their way forward from this premise. Instead of working out from ensemble "jamming" towards a new clarity, they are occupying themselves with a cheap presentation of decadent academic procedure of symphonic composition.

In conclusion, whatever guidance we can extend, whatever innovations in orchestration practice we can suggest, *must* be of such a nature that the new practice can come through the best jazzmen of today. The jazzman's job is not done; the next step forward still depends upon, and can only be successfully accomplished out of, his subconscious and instinctive jazz sense. In spite of the continued enthusiasm of hot record collectors for the status quo, I believe we should consider the present standstill in the advance of jazz a disturbing situation. Jazz is still standing, but a success-lethargy seems to be undermining the progressive vitality it enjoyed in the 1920s. There is no reason why we should not enjoy a return to the good taste of the 1920s, but there is a great difference between the state of enjoyable *use* and that of self-indulgence. Let us enjoy this reminiscence but let us, at the same time, prepare for another valid step—if there is to be one.

Bubber

Bubber Miley died of tuberculosis in 1932. He held a unique position among the jazz musicians of the late twenties and in time I hope to write an extended story of his life, including a thorough review of his recordings. Let this short sketch outline his musical life and emphasize his importance as a jazz musician.

Bubber was born in Aiken, South Carolina, in the year 1903. At the age of six he was brought to New York. As a child he used to sing on the streets, sometimes bringing home as much as five dollars. By the time he was a schoolboy of 14 he was taking music lessons from a German professor—first on the trombone and then on the cornet. At 15 he joined the navy. Eighteen months later he was honorably discharged and joined a small band known as the Carolina Five. This band included Johnny Welch, soprano sax; Wesley,* piano; English, violin; and Cecil Benjamin, clarinet. They played gig dates and boat rides and moved through those places familiar to every small outfit in the early 1920s. They played at Purdy's and they played at Dupres' Cabaret at 53rd Street and finally landed, in the winter of 1922, at Connor's Cabaret on 135th Street off Lenox Avenue. They remained here well through the year 1923. Following this Bubber went on an extended tour through the South with a show called *The Sunny South.* When the tour closed he joined Mamie Smith's Jazz Hounds, along with Coleman Hawkins. From there he went into Duke Ellington's

*In a 1958 essay on Bubber Miley a blank space is left for the name of the piano player. This essay which appears in this volume (pp. 247–59) repeats some of the material offered here.

H.R.S. Society Rag, October 1941.

Roger Pryor Dodge with trumpeter James "Bubber" Miley, 1931, performing to "East St. Louis Toodle-Oo" by Duke Ellington. *Photograph by DeCamp Studio, New York.*

band with whom for the next four years or so, he toured, composed, made records and played steady engagements. The Ellington span roughly covers the early days of the Kentucky Club through the famous days of the Cotton Club in Harlem. In May 1929 he went to Paris with Noble Sissle for a month's engagement. When he returned to New York he joined Zutty Singleton's band at the Lafayette Theatre in Harlem, along with the late Charlie Green, trombonist. From here he went into Connie's Inn with Allie Ross's band and a floor show featuring Earl "Snake Hips" Tucker and Bessie Dudley.

At this time, in 1930, I had an act in Billy Rose's revue, *Sweet and Low,* in which we danced to the Ellington-Miley *East St. Louis Toodle-Oo.* In January 1931, I had the great luck to find Bubber dissatisfied at Connie's Inn and very willing to come with me. He played his own *East St. Louis Toodle-Oo,* spotted on a high stool on the stage, and stayed with me until the close of the show in late spring. He often played with Leo Reisman for recordings and broadcasts, and for one week, when working for me, doubled at the Paramount movie house, specially featured by Reisman's stage band. Dressed as a Paramount usher he would rise out of a front-row seat, play a hot *St. Louis Blues* as he stood in the aisle, and then streak his way out through a side exit to growl the *East St. Louis* for us. That summer, backed by Irving Mills, Bubber was enabled to build up his own orchestra. He secured the services of Gene Anderson, piano, and Zutty Singleton, drums. The band was placed in a show called the *Harlem Scandals.* They opened at the Lincoln Theatre in Philadelphia and subsequently came to New York and played at the Lafayette. But in Philadelphia, Bubber had been running a high fever and at the Lafayette he was an obviously ill man. It wasn't long after this that I got a letter from his mother, telling me he was sick. I went to see him at her house. He had dwindled to 76 pounds—a little shrivelled old man. It seems that he had had tuberculosis for some time. Later I got another letter telling me to come to Bellevue. Now he was James Wesley Miley and only his relatives remembered him. His mother told me he was to be taken to Welfare Island. I missed the visitors' hour when I went to see him there, and a few days later I heard he was dead. He was 29 years old.

My wife and I went to his funeral. It was held in a bare whitewashed parlor. Apparently no musicians were there although there was a large wreath of flowers from Duke Ellington. The mourners were out of his mother's life. Was this the funeral of one of the greatest artists of our time? The place Bubber had made for himself in music history was completely ignored. Not knowing who Bubber was, one would have thought it was a service for some good little colored boy. The congregation sang *Rock of Ages* and all through it we heard Bubber's horn, playing *Black and Tan.*

Bubber was co-author of many of Duke's most famous pieces. Miley told me that the inspiration for the *East St. Louis Toodle-Oo* came one night in Boston as he was returning home from work. He kept noticing the electric sign of the dry-cleaning store Lewandos. The name struck him as exceedingly funny and it ran through his head and fashioned itself into

> Oh Le-wan—dos
> Oh Le-wan—dos

Subsequently this piece became Duke Ellington's radio signature. The *Black and Tan Fantasy* was suggested to Bubber by his mother's constant humming of the *Holy City.*

Among the many great records made by the Duke with Bubber Miley are *Got Everything But You, Flaming Youth, Yellow Dog Blues, Jubilee Stomp,* etc. A disconcerting feature of the early Duke recordings made with Bubber is the fact that Bubber would usually take the first chorus with such charged intensity that the record as a whole would suffer through shooting its bolt at the offset. Once, to record the tune *Rockin' Chair,* Hoagy Carmichael got together a rare combination including Bix Beiderbecke, Bubber, Benny Goodman, Krupa, Bud

Freeman, Venuti, Tommy Dorsey, and Eddie Land and although as so often happens, they cut but fairly indifferent music for such a master band, nevertheless Bubber took three-quarters of the first chorus and did the best work on the side. Bix is hardly used and barely recognizable. Bubber's is the rhythmic, boyish, half-talking, half-singing countervoice to Carmichael's deep-toned pappy!

Miley made six sides under the label, Bubber Miley's Mileage Makers. They were exasperating recordings of commercial tunes. On one or two of them he did not even play, and the few solos he condescended to take were over before they began. To my knowledge, he made but two records with Leo Reisman; one, *What Is This Thing Called Love,* where he plays the first chorus straight and only employs his regular style as obbligato to the vocal, and a very good chorus on *Puttin' on the Ritz.* The latter resembles somewhat his chorus on *Diga Diga Do* made for Duke Ellington. It is not quite certain whether Bubber is playing on the old Perfect recording of *Down in the Mouth Blues* and *Lenox Avenue Shuffle* (composed by Ray-Miley and played by the Texas Blues Destroyers).

To Bubber Miley is attributed the first use of a rubber plunger as mute. Bubber would often tell of the time he went to the ten-cent store looking around for something new to mute his horn and how he suddenly came upon the rubber plungers used by plumbers. He said right there he took his trumpet out of its case and tried it, to his own and everybody else's high amusement.*

Bubber's melodic originality is hidden within his growl style. He fixed his notes into a beautiful melodic entity regardless of whether one likes this style of delivery or not. Listeners must judge growl music not only in terms of good or bad *actual* musical line. In the hands of Miley, Jabbo Smith, and Tricky Sam, growl music can be as musical, if not more so, than open brass playing. Equally, a great style minus significant music content can be a disconcerting experience. It is quite noticeable that facility on the open trumpet, even with such a great musician as Armstrong, leads into a florid style—an over-cascaded virtuosity. Whereas the mute, in competent hands, coaxes out a closely knit jazz with plenty of invention. The mute, besides introducing noises ordinarily foreign to the instrument, can inspire the player to subtle melodic invention. Perhaps the tightly squeezed-out notes demand more respectful attention than the easy array which roll out of an open trumpet.

Bubber rehearsed many numbers with me. Among other things he played the *King of the Zulus,* composed by Louis Armstrong, and played from notation his own forgotten improvisations on the *Yellow Dog Blues.* When I first showed

*The two paragraphs that follow here also appeared in a 1942 essay "Jazz in the Twenties." They have been omitted from that essay. P.D.

him notations of his solos, taken off records, he was quite confused—doubted he ever created them! But we soon discovered that when *reading* notes he used the correct valve, whereas when improvising he reached for them with his lip, sometimes reaching as much as a whole tone. I found through rehearsing with him that he was very conscious of what was important to jazz. He never had to warm up to play hot; he could play with immediate hot emphasis—even when his lips were still cold. He also *thought* in terms of musical invention and was never blandly satisfied with jungle intonation for its own sake. When he improvised a melodic turn that was inventive, he tried to remember it. Often before going on stage, he thought of new complicated little breaks to introduce. He was a musician packed with half-formed ideas for written composition. He was very slow. Unless he was supervised by a Duke Ellington who would see to it that any good idea was completed, he would leave it hanging in his mind or just play about with it in the dressing-room.

Miley was a player who, sometimes more, sometimes less, *set* his solos. In other words, after he had played a piece many times he was not entirely improvising from then on. What he did was to play a developed version of an earlier improvisation. Unless we are really familiar with an instrumentalist's attitude towards certain hot choruses it is over-confident to judge them as completely spontaneous improvisation. What is known as improvised music is not always *strictly* improvised. Moreover, when a solo is remembered from one time to the next, the tendency is for the solo to be much clearer in outline the second time. For if the kernel of the improvised idea is apparent in the early example, very often the surrounding matter does not stand out firmly on its own. In the recording field, a good example is Miley's solo on the *Black and Tan Fantasy* recorded by him on Brunswick, as compared with his later version recorded on Victor. On the Victor recording his sudden dynamic burst of notes following the first long sustained note covering four bars, has far more definiteness of form than the similar passage on Brunswick. All the way through the Victor record it seems he has a clearer view of his solo than he had on Brunswick. Every part of the Victor solo stands out in high relief by comparison. The same is true of the Victor *East St. Louis Toodle-Oo* as compared with earlier versions on Columbia and Vocalist. Bubber was a musician who could musically so crystallize an improvisation that the improvisation did not die after the impeccable white-heat delivery of its first presentation; his new outline had such backbone that it, in turn, could be used to take off from. In other words, another musician without playing the actual notes of the original improvisation, perhaps not even in the same style, could simply keep the general outline of Bubber's solo in the back of his head as you would a song, and improvise a new solo, not on the original tune or chordal foundation but on Bubber's already improvised solo.

Now, my greatest regret is that I did not have records made of all the material we rehearsed. Once, we did go to a very poor recording studio, and under unfavorable conditions Bubber was kind enough to make a new *Black and Tan* solo and a chorus or so of the *King of the Zulus.* But this is all we ever recorded. Nobody will ever know what he got out of *Sister Kate;* his countless variations on *East St. Louis Toodle-Oo* and *Black and Tan Fantasy;* his poignant preoccupation with the Largo from the *New World Symphony.*

Bubber's personal appearance was extremely natty. His manner was quiet but always on the verge of laughter. When walking down the street he possessed an endearing swagger and he could generally wheedle whatever he wanted from anybody. His speech was intense with the color of both his race and his profession, and he was about as careless with his money as he was with his horn—which was pretty careless. He left it everywhere. He sported a flashy Auburn car and had absolute credit at Big John's. Mention his name to any musician who played with him and you'll hear, "Now that boy. There was nobody like him. And I mean it."

Hot Jazz: Est-ce-du Bach

About a year ago there was great controversy over an incident involving a jazzed-up version of a Bach toccata. A lot of indignation went to print on behalf of the sacredness inherent in the works of Bach, and swingsters were warned not to distort our musical heritage. At the same time Olin Downes, a man of musical erudition whatever his attitude towards jazz, devoted a Sunday article to debunking the indignant music lovers' antipathetic points of view and to rationalizing the current practice of twisting non-contemporaneous music into jazz shape.

I'd like to authenticate Olin Downes's attitude and stress the vital influences of both the dance and secular behavior upon all great music, including jazz, without necessarily vouching for the jazzed-up toccata in question. My citations of a melody bantered about in the 16th and 17th centuries—first as a secular song and street-dancing accompaniment, next as a hymn for simple congregational singing, and still later, in the 18th century, as a significant part of a major work of art—should point up the fact that this revered religious concert music of Bach is no more nor less than dance music of a very rhythmic and secular character, born out of musical conditions very similar to those surrounding jazz.

Naturally enough, the dance suites of Bach are admitted to be actual dance forms which Bach turned out in his "lighter" mood. On the other hand, the chorales, fugues, cantatas, and Passions enjoy an extra reverence more literary than musically erudite. In Bach's Passions there is so much that points away from either holy origin or holy treatment that

H.R.S. Society Rag, November 1941.

one wonders why there is this insistence upon their being so different from his other works.

For instance, in 1601 we find Hans Leo Hassler publishing a book of *balleti* and *gailliards.* According to Schweitzer, twelve years later one of these melodies, the love song *My Heart Is Entangled by a Tender Girl,* appeared as a hymn tune called *With Love Does He Claim Me.* At a still later date this same tune reappeared in Bach's St. Matthew Passion as "O Skin All Bloody with Wound."[1] Actually, the solemn hymns were more often than not taken from such an unreligious source as the contemporary love song and in Bach's case, treatment of the religious arias and choruses is virtually dance in its style of writing. The aria "Repentance and Remorse" from the St. Matthew Passion is first of all a minuet—what the words make it is extraneous to the point. Spitta even called the aria "Have Pity" a sicilian.[2] In the same Passion, the final chorus, "We Fall Down with Tears," is a sarabande. Musically, if not philosophically, we know that practically every section of the St. Matthew Passion, except for the chorales which are either directly popular in source or modelled after secular songs, is a dance movement.

Again, Bach's church cantata "Sleepers Awake" is a glaring example of Bach simply swinging out into a bourrée. Even Maitland, editor of *Grove's Dictionary of Music,* says: ". . . the middle number of the cantata ("Sleepers Awake") starts with an independent bourrée-theme of the more inspiring kind, in the course of which the chorale is again sung in the tenor part."[3] We can understand all this better when we read what Parry, composer and Professor of Music at the University of Oxford, says under the heading Dance Rhythm in *Grove's Dictionary:* "Dance rhythm and dance gesture have exerted the most powerful influence on music from prehistoric times to the present day." He insists it is the combination of secular song with dancing which leads to definiteness in rhythm and that later, inevitably, these rhythms are used by the conscious composer, completely influencing his style and leading him away from "unrhythmic vagueness."

Coincidentally, a false reference has already sprung up around the early Negro spiritual—a reverence not extended towards the Negro instrumental genius now flourishing in the theatre. Spitta, using a similar approach in another era, observed that the endeavor to express personal emotion on the boards of a theatre does not differ in essence from the transcendental subjectivity of the hymn singers![4]

1. *J. S. Bach* by Schweitzer, p. 19, Vol. 1.
2. *J. S. Bach* by Spitta, p. 114, Vol. 2.
3. *Grove's Dictionary,* Chorale Arrangements.
4. *J. S. Bach* by Spitta, p. 479, Vol. 1.

One may conclude that if Bach's Mass and cantatas had been handed down minus their religious words and titles they would not be getting the religious-musical attention they now receive. Schweitzer reveals that even in the 16th century he has found such written comment as: "We could mislead all the purists of church music by putting before them an old secular motet with an accompanying sacred text."[5]

The tendency of musicianly taste in Bach's time was towards swing as much as possible in the dance suites and a satisfying rhythmic quality of dance in the fugues and church music. According to Forkel, Bach's first biographer, only Bach's "skillful treatment was necessary in order that each dance might exhibit its own distinctive character and swing." Also, ". . . even in Fugue . . . he was able to employ a rhythm as easy as it was striking . . . as natural as a simple Minuet."[6]

In 1758, however, criticism was launched against church organists in general: "Now are heard two-part variation and diminutions and playful passages, sometimes on the pedals, and sometimes in the upper part; then they kick with their feet, they ornament the tune and break it up, and hack it about until one does not know it again."[7] Nor did Bach escape reprimand by the elders of the church. One of the formal complaints charged him with "having hitherto been in the habit of making surprising variations in the chorales, and intermixing diverse strange sounds so that thereby the congregation was confounded." Thus we see that, even during an era fostering the most severe type of chorale, among the musicians themselves there was a bandying of melody. They used it where it could serve best, paying little or no attention to religious origin or religious treatment. Moreover, at the very apex of this whole contrapuntal style of music, both church and secular, we find a heightening of dance rhythm instead of a tendency away from it. Only after Johann Sebastian Bach's time do we see dance rhythm losing its grip upon the mind of the composer.

In conclusion, the whole situation revolves around a modern dogmatic reverence for religious music that is not religious, plus an attitude towards the composing habits of the old masters which they themselves never professed or demanded of their own forerunners. It revolves around the paradox of a Forkel, famed commentator in the academic camp, who praised Bach's music because it was like dance music and had "swing" to it and who commented with admiration on Bach's treatment of pedantic stuff such as fugues, because he made it sound like a minuet! And it also revolves around jazz lovers who are concerned solely with good jazz *treatment*, whether or not the original theme is

5. *J. S. Bach* by Schweitzer, p. 18, Vol. 1.

6. *J. S. Bach* by Forkel, p. 84.

7. *J. S. Bach* by Spitta, p. 312–315, Vol. 1.

a popular song or a blues or a classic theme. Personally, I believe jazz belongs with the blues. Nevertheless, I would far rather hear a soloist "get off" on a good classical theme than on the average popular song—providing the soloist is as much at home with the classic theme as he is with the popular song. Obviously, by this I do not mean "jazzing up" the classics as we see them in the stilted arrangements made by the larger swing orchestras, a saxophone section interpreting the music in a manner pleasing to neither the classic-minded nor jazz-minded. I mean a good soloist freely going to town. Then, although the outcry from the classic field might be louder still, at least we would have good jazz. Let us say that the derogatory term "jazzed-up" could be a phrase applied to what is usually done to a classic *piece,* while "getting off" on a classic *theme* could be a tribute both to jazz and to classicism.

Identifications

Every thorough observer has the opportunity to see new art aspects. It is the exchange of such "discoveries" that carries the critic along, rather than a solitary traveling within his own perception. Art is a mass activity and, involving fewer individuals, so is its appreciation.

There still exists a small minority of jazz lovers—a minority both interested in and sensitive to the art—who conscientiously maintain that the careful labeling of each piece is unessential to its appreciation. Familiarity with the exact title, opus number, and key signature is thought to be a special knowledge—a knowledge definitely extraneous to the fine appreciation of music. Especially is this felt in connection with the abstract opus number and key signature. But if this procedure is not an essential to the further grasp of comparative musical values, still, it would be difficult to contend that no process of identification is necessary.

There are two ways of identifying an artist: one is to recognize him through personal idiosyncrasy; the other to recognize him in the light of what we know he is capable of through our awareness of his artistic powers. Judgment by idiosyncrasy is very little more than a memory process, a hearing and remembering—a recognition of "facial" characteristics. Although this method is generally a simple matter of recognition—action as obvious as remembering a face—occasionally, however, such a method of identification can range into some difficult distinctions. On the other hand the identification of the artist through knowledge of his powers of production puts upon us the burden of preparing ourselves to judge and recognize these powers.

JAZZ, 1942. Also written for *JAZZ* in January 1945.

If we are given enough time, perhaps most of us can become aware of greatness, but to recognize greatness *immediately* needs preparation. Think of a great painter painting exclusively within a green palette; now of a minor imitating painting within a red. It would be an easy matter for anybody with normal vision to distinguish the difference between master and imitator. But a color-blind person would have to depend entirely upon his knowledge of the art. Actually, in the course of art, we never really experience such distinct separations. Conditions are always mixed. Even the weakness of the direct imitator will bring forward a slight difference in style. But it is only when this style difference has been definitely established that it becomes easy to distinguish between the works of the two. Similarly, it is hard to distinguish between identical twins. It takes much more thought than the identification of two averagely similar people.

Listening to an unknown improviser, we hear not only idiosyncratic original material but also idiosyncratic intonation, plus the presence (if it is there) of dynamic breadth of emotion. When we seriously try to identify this unknown player we help ourselves toward the solution by considering all these distinguishing factors—each mark helping or confirming the other.

The question as to whether or not such identifying knowledge is important, whether or not it helps us to appreciate the arm form within which the artist functions, demands attention. Is the time-taking activity of ascertaining the identity of what we hear of direct assistance to us in increasing our knowledge and enjoyment of the art in question, or is it mainly unwarranted?

Identification breeds *intensification* and in turn intensified specialization makes for more clarity in general. What once appeared to us as misty, intangible, upon becoming familiarly identifiable stands out in bold relief, capable of bearing a relation to itself and of being identified and compared with other material. Through studied familiarity we make a better contact within ourselves. The music becomes at home within us; it courses through our consciousness and permits us to make an evaluation. Indiscriminate attention to a whole period of music will not carry the listener as far as specific attention to that period at its best, although such an attention to the best must eventually *not* exclude consideration of the worst. The most fortunate position for the listener should permit a well-grounded survey and a specialized attention—a position which allows him to eye the art objectively and at the same time to feel it subjectively. Art experiences can be submitted to better evaluation after *intensification,* for *intensification* implies a proceeding *concentration;* and it is through concentration that we bring forward our best understanding. In general, art ignorance comes from unawareness born of poor initiative. Consciously, or unconsciously, we can prod ourselves into becoming aware of music values by inducing concentration through an activity like name-listing.

There are natural processes of labeling. Jazz records have popular names which we use as identifying agents. Although such labeling is taken for granted, nevertheless accurate recognition of various records by name is, in fact, a first step in concentration. A second step is the cataloguing of identities, the cataloguing of evident, similarities in style or of distinguishing individual traits. It is when the distinctions become more complicated, when the labeling becomes more detailed, that the casual listener, bewildered, develops a tendency to criticize such knowledge as being over-elaborate, develops a tendency to retire into the comment that *a thing is good when it is good no matter who does it or whatever it is called.*

On the other hand, intense concentration on one musical aspect alone can lead the uninformed into negligent misinformation. If a jazz-lover easily discounts Sidney Bechet, very likely he may be doing so because he has not heard the best Bechet. Those interested in more melodic content than a full tone alone provides might not investigate Bechet beyond a few recordings. However, if such a person consorted with Bechet admirers, sooner or later, through their concentration, he would meet up with a greater Bechet. Specialized pioneering in all departments of jazz being a practical impossibility, balance is induced at times by "fellow-traveling" with opposed concentrating influences.

It takes a long time to familiarize one's self within an art form, whether it is of an epoch, a nation, or one man. It is almost impossible to pioneer successfully in all the directions of a major art expression. But complete knowledge of an art can only come about through familiarity with its poor as well as its successful aspects. The observation of certain elements under different conditions brings these elements to us with more precise differentiation than observation limited to a single experience. We can conceivably subject ourselves to a complete art experience in our own period. It is our healthy resistance against confining ourselves exclusively to a long past period—our wide historic curiosity—which usually thwarts the possibility of complete detailed judgment of any past period of art production.

If the serious jazz listener tends to know only the subject matter within which he can particularize, he is almost forced into this specialized listening, because jazz is already too abundant for particularization in all of its phases. It is through such intensified attention that jazz reveals its nuances—nuances that otherwise would escape attention. Even so, such perception does not necessarily lead to the next step of appreciation; it is possible that out of all the perceived nuances only two or three may possess, in the final analysis, permanent validity.

The esoteric approach to jazz by those who do not really like it, those who simply admire certain elements familiar to their classical experience, cramps full appreciation of the art for them and reduces the validity of their criticism.

To travel significantly in any new aesthetic process we must not only make careful distinctions, comparing new material in the light of the old, but invest ourselves in a new emotion as we pass from the old experience to the new.

Tradition has narrowed down for us the few art elements commonly selected for critical consideration. Tradition allows the neophyte immediately to focus upon these important elements—a singling out he would not be likely to achieve otherwise. Thus, selected comparisons have become for us a traditional means of identification. In fact it is such a useful step to cultural advancement that the neophyte feels complacently sure of his identification powers, convinced of his own perspicacity. But it is easy to become confused. Do we not often come upon jazz enthusiasts who, aware of the similarity between McPartland and Beiderbecke, say of an unfamiliar good solo that it *lacks that something,* after they have been told that it is a McPartland solo—not a Beiderbecke? These are the same people who can be told that they are listening to a Beiderbecke solo, when they are listening to a mediocre McPartland and who never fail to say "there is nothing like good old Bix." although an attempt at serious art judgment has been made in both cases, it has been diverted at the source by misleading labeling.

Misrepresentation is commonly encountered within all music activity. Within the classic field, unless the art work in question has easy "facial" identification, the best informed mind will shuttle back and forth when confronted with a mass of unfamiliar, unlabeled art. In notated music the "facial" characteristics of the interpreter are not present to help us and our only avenue to the solution is through the idiosyncrasy present in the work as written. Thus, although Brahms and composers contemporaneous with him built upon the same familiar folk tunes, Brahms used a harmonic and rhythmic treatment peculiar to himself. In fact we may say that very often we detect Brahms through his orchestrational activity, not through his music matter. Likewise we find that work attached to, say, the name of Beethoven carries such import for us that we discover in his lesser works far more significance than we should otherwise find there. However, whether or not we unreservedly accept all the works of a great man, we must recognize that through a good portion of his works something powerful and significant must move to bring us to such a point that his name alone blinds us when we are unguarded. Moreover, imitation in musical composition, unlike that in painting, is not difficult, and the notion is quite feasible; of course, any success one achieves by style imitation in a short unpretentious piece does not necessarily establish the possibility of writing Beethoven's Tenth Symphony. Incorrect labeling can easily mislead the listener. Erroneous identification can frequently deceive the best informed.

The labeling-listing approach to art, however indirect, should be a part of culture psychology. The average mind is, normally, incompetent—it begets nothing and, without painstaking aid, takes in nothing. In other words, the mind cannot sit in judgment upon something new without preparation. It must submit itself to long training—training with guidance from the outside, or under guidance that it can give itself by a careful tabulation of sensations and the building of aesthetic apprehension thereon. Thus, through our defining memory we build a certainty, and this certainty we can check, recheck, and change with array after array of new sensations.

Sustained, good concentration is rather rare and seldom at hand for penetration into every deviation art may make. On the other hand, respect for a definite piece of music, or a certain player, sometimes stimulates the concentration-power so that the mind will assume too much; in trying to find what it expects it may find what is not there. But in the long run judgment balances itself—if we submit to prolonged reception-activity. The process of correcting an art attitude can be as mechanically simple as this. And so, although there is no special tribute due the man who distinguishes one thing from another by name only, nevertheless it is possible for such a man to progress immeasurably, because he has availed himself of this mechanical process.

Jazz in the Twenties

My musical education had been following the natural evolution of the serious layman finding his own way, that is, by way of Chopin and Beethoven slowly back to Bach and his predecessors. It was not until the winter of 1922–23 that I heard significant talk about jazz. I remember the heated arguments of musicians in "show business": was Paul Whiteman doing important things with popular music or was he not? I remember classical musicians who should have known better telling me that Whiteman was new and amazing—that he was getting tonal effects with a small orchestra that were better balanced than those of the symphonic orchestra.

Not thinking in the terms of a practical musician, it was naturally difficult to see what was the exact difference between Whiteman and pre-Whiteman dance music; difficult to gauge the exact value of his contribution to tonal arrangements. But even at that time, and I am proud of this, I refused to consider the Whiteman flurry anything more than the shop-talk of the profession. The *refinement* of Whiteman did not seem to make his music take any place in my classical music experience, and added nothing to the faint feeling of pleasure I got from Jim Europe recordings. Whiteman's rhythmic introductory bars were his one novel feature but such interest invariably fell into a tune handled in a silly flat fashion. I did not know what I wanted to hear. But I was looking for it. In 1923 I was really looking for jazz, but hadn't been lucky enough to hear the few good bands and recordings of the day.

It was on February 20, 1923, that I

Jazz, July 1942.

went to hear Alfred Casella. One of the compositions he played was Stravinsky's *Piano Rag Music.* I was taken off my feet. Here, for the first time, I heard what I wanted Whiteman to do. I bought Stravinsky's other jazz piece, *Ragtime.* I felt convinced this was the new music. And it wasn't until 1924 at Bea Parrot's apartment (now Mrs. William Soskin) that I heard my first hot jazz record.

It was Ted Lewis's *Aunt Hagar's Blues.* It seemed much simpler than Stravinsky's *Piano Rag Music,* but far more real. Whereas Stravinsky inclined me to look for nothing but startling rhythm, in *Aunt Hagar's Blues* I felt the impact of a music that was more like 18th-century music; it could grasp your attention by melodic significance and did not have to rely solely upon astounding rhythmical stunts. Miguel Covarrubias, who had just come up from Mexico, had also heard this record at Bea's. He told me of a much better one called *Blue Blues,* by the Mound City Blue Blowers. It was he also who told me he had listened to a record by a solo singer that was as good as both these instrumental records put together. This singer turned out to be Bessie Smith. So starting out with *Aunt Hagar's Blues, Blue Blues,* and, I think, Fletcher Henderson's *Strutter's Drag,* I was well on my way.

I did not go to Paul Whiteman's first Aeolian Concert, but went to the second hearing held at Carnegie Hall a few months later. This was in April 1924. The critical press struck me with amazement. The *Livery Stable Blues,* the only challenging music that had been played, was not even mentioned—played as it was in burlesque of honky-tonk music and only offered by Whiteman as a horrible example of what he and Gershwin were strictly avoiding—it was nevertheless evidence of a new music. But in New York, Whiteman's tonal effects and Gershwin's *Rhapsody in Blue* reigned supreme. About this same time England's foremost musical critic, Ernest Newman, reacted in precisely an opposite manner, and gave to the press an exaggerated and heated tirade against Gershwin and the jazz idea! That combined with maudlin New York eulogy led me to essay an exposition of jazz which I entitled *Jazz Contra Whiteman.* This was in the fall of 1925. I peddled it about to a few magazines with no success. The subject was too light for the serious magazines. Four years later (October 1929) under the title of "Negro Jazz" it was published in the London *Dancing Times.*

It was in the winter of 1924–25 that I first heard Fletcher Henderson at the Roseland Ballroom. Perhaps naturally enough, from merely listening to records, it never occurred to me that the whole vitality of jazz depended upon improvisation. In fact, such solo as that played on the full sax by Coleman Hawkins in *Strutter's Drag* seemed so perfect and clearly laid out, that I attributed it all to the conscious composing of Henderson. I thought the same thing of Charlie Green's trombone solo in the *Gouge of Armour Avenue,* also recorded by Henderson's orchestra. I told Henderson how much I liked them and asked when he had written them. "Oh," he said, "the hot choruses. I don't write them. No, they aren't written out. They're played ad lib." That night Fletcher Henderson featured a dance arrangement of the *Rhapsody in Blue.* I spoke about it. He said he thought Gershwin's music was outstanding. So I learned two things at once: that all the good music was improvised and that the scorn I felt for Gershwin and the enthusiasm I felt for the stuff Henderson and his men were playing, wasn't even completely shared by Henderson. At the time it was quite a jolt to find out that solos which seemed so inventive and comparable to the great written music of other periods were not consciously plotted and composed, but were simply played ad lib by players who thought that Gershwin was a great composer.

I now began to collect records—particularly what were then known as "race records." Early in record collecting I discovered that the *washboard* bands were, in a way, more satisfying than were the larger orchestras. The records of Ted Lewis had become increasingly commercial. From a true pioneer in hot music, Lewis became the tragedian of jazz under blue lights at the Palace Theatre! Even Fletcher Henderson had acquired the tendency of turning sweet and full-throated in his arrangement. But the washboard bands, with their few percussion instruments and single clarinet playing the melodic line from beginning to end, seemed consistently stimulating. I particularly like Jimmy O'Bryant and his washboard band playing *My Man Rocks Me.* Curiously enough, a few of Boyd Senter's records were along the washboard style of playing. His *St. Louis Blues* was inventive and his *Slippery Elm* was healthy. Boyd Senter, overlooking his lapses into corny intonation, made better records at that time than many of those put out by men of better taste.

The occasions when I managed to hear Bessie Smith in Harlem vaudeville were for the most part disappointing. In show business she generally sang quick-fast popular numbers. In these her voice hardly ever took on its fine quality. However, there were occasions when she would be featured singing the blues with simple piano accompaniment. Such moments would intensify all the beauty that is to be found on her best records. Constant playing of Bessie Smith records not only deepens ordinary enjoyment but sustains the seriousness of her music—an enlargement of aesthetic pleasure seldom encountered. The significance of the surpassing art of Bessie Smith has been overlooked by her own race to a much greater extent than the significance of the outstanding instrumentalists. Never theatrically pointed up by good stage management, Bessie Smith missed the mass acclaim of, say, a Louis Armstrong. Among white people the significance of her art has been shamefully overlooked in favor of Negro choirs and their diluted spiritual music. Bessie was part of a period that could manifest art straight and clean, but those who were supposedly trained to see gave no more than a superficial look. Once more one of the best examples of period art has slipped in and out unnoticed by the mature critics of the period. Without her recordings the late Bessie Smith would have left to our memory little more than general inclusion amongst outstanding Negro singers such as Ethel Waters, Florence Mills, and Clara Smith.

While in Chicago in 1927, I went to the Sunset Inn to hear the redoubtable Louis Armstrong. He was very slightly known then. He was delighted that someone had come simply to hear him. In Boston, a little later in the season, I went to another of those Whitemansque jazz concerts. This time is was under the baton of Leo Reisman. The most discouraging trash was played as usual: the same patronage of jazz; the same attempt to use it without respecting it. However, on the program appeared the name Johnny Dunn, then a stranger to me. He startled everybody at the jazz concert by playing jazz—the *East St. Louis Toodle-Oo.* At this time I hadn't tried to break through the anonymity of the recording jazz musicians. Many of the solos I admired and had transcribed and could play with one finger on the piano were still untagged with the composer's name. My dilemma was practically opposite that of the present-day jitterbug who approaches every player by his first name but shows a marked absence of musical curiosity.

It was in the spring of 1930 that I rehearsed an act that ultimately got into Billy Rose's first revue. I used the *East St. Louis Toodle-Oo,* training a young trumpet player (Clarence Powell, then a member of Claude Hopkins's orchestra), capable of strong outstanding tone. He played from notation of solos I had transcribed from the recording of the Ellington-Miley, *Toodle-Oo, Black and Tan Fantasy,* and Armstrong's *Potato Head Blues.* It was very interesting to hear him play Tricky Sam's trombone solo in the *Toodle-Oo* on the open trumpet. His success proved to me that a sympathetic reading of hot solos from notation, even on a different instrument from the original, lost nothing of the intrinsic beauty of the melodic line. Spontaneous "hot improvisation" need not be the sole characteristic of jazz. A good solo is always a good solo.

After the opening in Philadelphia, the usual cuts and changes were made. I was told I could not have such a large stage act (then myself, two boys, and a trumpet player) and was asked to cut down by using a cornetist from the pit. His name was Jimmy McPartland! It was not strange that his style of playing did not suit the *Toodle-Oo* which depends upon a dramatic jungle performer. Moreover, for a trumpet player to fill the theatre above a heavy exaggerated orchestration such as Rose had commissioned of Ferde Grofé, plenty of wind and force were needed as well as a hot intonation. McPartland had an intimate delightful style of his own and in the same revue he was outstanding, accompanying James Barton in *Sweet Annabelle Lee.* Accordingly, I went to see Bubber Miley. He was playing at Connie's Inn. He wanted to come with me very much. I had never dreamed of such luck. Actually it was only possible because Miley happened to be dissatisfied where he was. As Rose refused to pay for him I hired him out of my salary—that convinced Rose I knew nothing about business. But six months' dancing in front of Bubber Miley was an experience, extravagant or not, that I would not care to trade.

I have always felt that Miley was the greatest trumpeter in jazz history—in fact, the greatest musician of them all. I am aware that the wa-wa, or jungle style of playing, has been looked down upon by many swing enthusiasts. However, my introduction to this kind of playing was through Miley, and everything he did in this style was impeccable in taste. There is no doubt that undue stressing of jungle style becomes very boring, and without question a mediocre trumpeter had better play "open" than mangle the atmosphere with the wa-wa mute.* . . .

The act broke up. After a while Miley made some records under the label: Bubber Miley and His Mileage Makers. They were exasperating recordings of commercial tunes. On one or two of them he didn't even play and the few solos he condescended to take were over before they began. (Armstrong seems to be the only great instrumentalist who feels the artistic necessity of performing after becoming a band leader. Other soloists lend their names to the labels and let it go at that.) The last time I saw Miley was at the Lafayette Theatre in Harlem. Aside from an engaging solo on a piece called *Angel,* he seemed to have lost his original force and vitality. It wasn't long after this that I got a letter from his mother, telling me he was sick. I went to see him. He had dwindled to 76 pounds—a little shrivelled old man. This seemed hardly believable, but on entering his mother's cold-water flat I had asked this little huddled old man if I might see Bubber! It seems that he had had tuberculosis for some time. Later I got another letter telling me to come to Bellevue. Now he was James Miley and only his relatives remembered him. His mother told me he was to be taken to Welfare Island. I missed the visitors' hour when I went to see him there, and a few days later I was shocked to hear he was dead. He was 29 years old.

My wife and I went to his funeral. It was held in what looked more like a whitewashed shack than anything else. Apparently no musicians were there although there was a large wreath of flowers from Duke Ellington. The mourners were out of his mother's tenement life. Was this the funeral of one of the greatest artists of our time? The place Bubber had made for himself in music history was completely ignored. Not knowing who Bubber was one would have thought it was a service for some good little colored boy. The congregation sang *Rock of Ages* and all through it I could hear Bubber's horn playing *Black and Tan.* On the way home, walking through Central Park, I told my wife I felt as

*See pages 247–59 for material originally published in this essay.

though I had lost one of the best friends I ever had. Not that I ever shared his way of living—a high-geared night-into-day orchestra man—but that his music, which had meant Bubber to me, was gone forever. Now my greatest regret is that I didn't have records made of all the material we had rehearsed. Once we did go to a very poor recording studio and under unfavorable conditions, he was kind enough to make a new *Black and Tan* solo and a chorus or so of the *King of the Zulus* for me. But this is all we ever did. Nobody will ever know what he got out of *Sister Kate,* the countless variations on *East St. Louis Toodle-Oo* and *Black and Tan Fantasy,* his poignant preoccupation with the Largo from the *New World Symphony.* He never got into a rut of worn-out musical cliché. When playing, his greatest personal pleasure was to invent melodic variations on an intense but predominately soft trumpet. He didn't seem to need an *fff* attitude.

It was during 1931 Dudley Murphy put out a film called the *Black and Tan Fantasy.* Whether or not this short is important as a cinema, it is highly important as a document of Duke Ellington's Orchestra. That we have an aural and visual record of the original Ellington orchestra (excepting Bubber Miley) playing a masterpiece is something the future can thank Dudley Murphy for. Another film of his, the *St. Louis Blues,* built around Bessie Smith, is even more important. As a tribute to jazz it will become one of the rarest documents we have.

It seems that when the exponents of an intangible art form are still functioning and the means of recording their art are more or less available, they themselves are no less careless about posterity than their critics. The subsequent probability of a cultural blank for what was once a living art is never considered. When the period has passed and the great artists have been dead some time, then for the first time appears the serious critical notice of the deplorable recording lag! Or if the art happens to have been sparingly recorded for posterity the critic suddenly notes its rarity, and a trifle late in the game announces the recording as a posthumous treasure! Let us look forward to the day when 16mm sound projectors are more prevalent and the amateur can take care of the future, a function we might have suspected part of a critic's duty in a changing world.

Notwithstanding the point of view of many new jazz converts for whom jazz starts with 1936 swing, the period covering 1920 through 1930 represents not only the birth but flowering out of jazz. If we try to think of the history of jazz in terms of progression from cakewalk to ragtime jazz, there will always be the period of the 'teens more or less connecting ragtime with the jazz to come. But the Original Dixieland Jazz-Band, Sweetmeats Orchestra, Joseph Smith's Orchestra were actually not jazz as we know it. The Original Dixieland Jazz-Band and the early New Orleans combination, probably the closest to jazz, are more significant as a transition period than emotionally moving as early jazz. It is only from 1920 on that fragments of jazz appear which are in spite of their brevity the exact counterpart of "swing" solo choruses in 1939. Around 1923 appeared the first complete *solo* chorus—a style obtaining with us ever since. These early solo choruses were for the most part very rhythmic, and it was not until 1927–28 that the more florid, rippling solo came into existence—the result of increased technical facility. Fortunately with the advent of "swing," the jazz solo has gone back to the more rhythmic style of the mid-1920s. Early Negro jazz was lusty and ribald and only accessible to the musicians themselves at uncontrolled frantic moments. Such activity was obviously not suited to sentimental dancing and the dinner hour. The early solution was white sweet jazz. This premature compromise on the dance floor was definitely nourished by the equally premature compromise in the convert field. I say premature because jazz had not had time to establish the strict form so necessary to an art in order to withstand refined manhandling. Its only strength then, and its only real strength now, is the *subconscious* improvisation. The conscious orchestrators have made no correspondingly important contribution. We will notice that Duke Ellington's pieces, those of the late 1920s, are very loosely orchestrated; the whole affair a sort of arranged background for improvisation. There is still something staid about the so-called complete orchestration; but nothing even faintly comparable to the music not orchestrated. Prolonged uninventive fortissimos and "sweet jazz" are their two contributions. Even Fletcher Henderson was bogged down under his orchestrations. In no way is the musical significance of the improvisation adequately orchestrated, whether with swing or without swing—or with quasi Delius style! And yet the jazz world is always seriously admiring its dull orchestrations and casually dismissing its revolutionary melodic line.

So let us line up the 1920s. On the credit side, mark up the first appearance of actual jazz melodic fragments, the growing up of these fragments into full-length solos, and the hot, though somewhat florid, obbligato work. On the debit side goes the symphonic jazz orchestra, the bad taste of the first major jazz work (*Rhapsody in Blue*) and the commercialization of the hot virtuosi into sweet, full-toned, straight players. All in all, these first ten years established everything we have since used. And until the significance of jazz melody becomes ingrained in the mind of the arranger (later to be called composer, we hope) we shall have to continue to go through a period of self-culture, before we can expand, not simply expend, the precious material wrought by those first ten years.

Duke Ellington

Through the late 1920s and early 1930s, Duke Ellington maintained the unique policy of applying the improvisational powers of his musicians to his arrangements. Before Duke's time, the complete notated arrangement—occasionally incorporating the ad lib solo—had been principally featured by the large orchestral organizations. The Duke carried a loose arrangement method to the point of making one instrumentalist take a phrase out of the mouth of another and in turn have it taken from him, without interrupting the flow of a solo. The ensemble choruses and ensemble breaks were, naturally, written out. However, very often what was once an improvised solo turned up in later recordings of the piece as a duet or a trio or as work for a whole orchestra section. His orchestral parts thus whipped along with sparkle, drawing their strength from the original improvised source. This group method depended upon empty bars, here and there, for the talented improvisor to fill in, as opposed to written cues or full-length "ride" choruses. This spotted "freeing" of impromptu improvisation ability might be called sophisticated, tastefully directed jamming.

Jubilee Stomp is a good example of such spotting of improvisers to carry along the planned melodic whole. Here a simple, but satisfying, introduction leads into a furious sax chorus. Bubber Miley picks this up with a piercing note, leading into a rhythmic chorus. The style difference between the furious running rhythm of the sax chorus and trumpet solo with a slap-bass and orchestra behind it is highly imaginative. From here, after a short clar-

Jazz Magazine, January 1943.

YELLOW DOG

These choruses were arranged by Ellington. The Top, "Yellow Dog," was the introduction used in the recording. The Bottom, "Toodle-Oo," is the introduction to his famous theme song. Both of these were transposed from the record by the author.

inet solo, we run into a fine chorus by Tricky Sam. Bigard's break into the middle of the chorus has great power and, in this case, helps Tricky Sam instead of merely interrupting him. The whole record runs along this way, alive with piercing breaks and significant pick-ups.

The Duke had a startling sense of introductory theme-introduction as opposed to the tune itself. To *East St. Louis Toodle-Oo* he writes an eight-bar introduction in the style of an old passacaglia, although in 4/4 time. (Notation *Toodle-Oo.*) This is tastefully repeated as counterpoint behind the trumpet solos, and is finally used in the dramatic closing. Another great introduction theme is found in his arrangement of *Yellow Dog Blues.* (Notation *Yellow Dog.*) Not only is the theme and its compositional use good, but so also is the orchestration of it. Here, again, he closes his score with his introductory theme.

Outstanding introductions, loose arrangements, and orchestral *sound* are some three of the qualities responsible for Ellington's superior music. He was one of the first to bring into prominence the use of the slap-bass. He made it especially prominent in his recording, by placing it close to the mike. Moreover, at a time when big orchestra leaders were letting go the great heritage of jazz improvisation and relying solely on the complete arrangement, Duke Ellington enlivened the whole period. His music sounded more like jazz *composition* than popular tune arrangement. Whether or not he wrote it note by note is not important. In fact, even now, a loose arrangement plus a great band can supply more authentic music than any finished product of the most highly paid arranger. The Duke had the guiding hand that kept what was best in jazz and presented it in a more telling way than any other composer-arranger-leader of the period.

Wanda Landowska

At McMillin Theatre on January 16, Wanda Landowska gave her third New York recital, that is, third after many, many years' absence. This magazine is so strictly devoted to jazz that a review within the classic field may seem outside its announced policy. However, there are many jazz enthusiasts whose musical education began with the classics, and perhaps there are just as many jazz enthusiasts who are commencing to explore outside of the jazz field. To both groups I say go and hear Wanda Landowska. Listen to her records.

There has always been much discussion involving the point that classical music is one thing and jazz another. Of all classical music *since* the middle of the 18th century, this might be very truly said. On the other hand, I feel a real relationship between jazz and classical music before, and up to, the middle of the 18th century. This music, analogous to jazz, is the strong foundation the later classics built upon. True, it is a music which comes to us by way of notation, but stored up in this notation is the very best of music.

Up until the time of Wanda Landowska, early music had been receiving the same sort of interpretative treatment a jazz enthusiast dislikes when he hears the "long-hair" play jazz. Landowska revolutionized the interpretative attitude towards 17th- and 18th-century music. She brought back to it as much of its original "playing vitality" as is possible to recapture. She accomplished this after considerable study of textbooks written during the period and combined this information with her phenomenal insight into

Jazz, March 1943.

the music itself. In jazz, we also find "playing style" and intonation, but alongside these two great musical attributes, I feel, as in early classical music, that there is present the element which *can* be notated. I know I became acquainted with Joe Sullivan's *Gin Mill Blues* through its notation in *Down Beat.* I saw it was one of the greatest of piano blues. The greatness was all there. When I heard the record played by Sullivan I only heard what I would expect a great pianist to add, and it is this authentic interpretation that we find Wanda Landowska contributing to any of the written pieces that she plays. Because of what she has done, a new era of early music interpretation has been started on its way.

I repeat, go and hear the old dance suites, so filled with tunes and fugal contrapuntal devices. Hear them played by a harpsichordist who has a style of execution similar, no doubt, to the great harpsichordists of that other era—the harpsichordists whose rhythm at that time must have been as vital as the barrel-house and boogie-woogie piano of today.

Lu Watters Correspondence

ON THE RECORD

Roger Pryor Dodge

Lu Watters's records bring into sharp focus the substance of many heated arguments, on the air, in the fan mail, and in any parlor devoted to the discussion of "hot." While "purist" and "old fashioned" jazz are being flung from one side, the terms "commercial," "modern," and "stream-lined" are being flung from the other. The latter group holding that the true jazz was in the beginning.

There are a few points to consider in making a decision as to the correctness of Lu Watters's attitude, totally beside the question of whether or not they did the job well. There are considered approaches to music which do not even exist for the casual listener. Archaeological or cultural approaches will, in their pursuit, reveal aesthetic nuances not acquired readily in any other way. These are studies which when pursued for their own sakes expose the student to aesthetic nuances which are not apparent on a cursory hearing. Music in its own day does not need any of these "considered approaches." Sometimes it is even timeless enough to extend from one era to another. Thus we have the classics.

Therefore, music is to be listened to for all its sensual and aesthetic worth. Accordingly, music must be continuously inviting for us to listen to it for any length of time. If we are listening to records, one record should lead into another. We may originally intend to play one only, but sensual excitement, real craving to hear more music, leads us on; thus a

Jazz, August 1942.

complete enjoyment of music cannot lie in the knowledge of how great a lick here may be or in the isolated listening to eight bars there. Isolated high points in a new music are no more than the barometer of what we may anticipate, a pointing towards a future of fuller sustained stature. For those who know no other music than jazz, jazz, from beginning to end, not only suffices but satisfies. For such listeners, the infrequent high points of their music are the *extra* gratifying portions we may expect any music we thoroughly like to have. On the other hand, for those who impatiently wait for the high points only, there is no accidental lift, only sporadic confirmation of a great art form.

Living art grows with the people and matches them in stature. The people do not look for anything better or worse to come out of it—they take it for what it is. The average person either dances to dance music or sees others dance to it. Many intellectuals feel our contemporary life is mirrored in it. It does not portray our life, but by way of association they seem to attribute to it this power.

It is futile to assume that a "good ear" bounds *all* hearing. In the classic field if we listen seriously to Beethoven Brahms, Bach, and many other more or less constantly played composers of the last 250 years (perhaps a few excursions back to Frescobaldi) we cannot hope to receive more than a small part of what the 16th-century Palestrina has to give on these few occasions we are privileged to hear him. The specialized musicologist and the Papal choir may trace the mysteries of such music, but we cannot hope to really know it. In fact we come to the sad conclusion that no one can really *know* all music. A man is capable of just so much knowing, even when aware that what he may *never know* might have been a truly great experience.

How great was early jazz? Is the later jazz better or worse or only different? Even discounting improved modern recording methods, there is obviously a great improvement in ensemble playing. It is much clearer and far more inventive—throughout all the parts. At both fast and slow tempo, ensemble jamming has moved ahead for the better. In this sense, recapturing the past is to go backwards. Is the solo different? Yes. But we have not yet surpassed the past. There has been no single contribution to the solo field which can be called going ahead. Johnny Dodds's 1923 *Canal Street Blues* solo cannot be improved upon. Such a solo, removed from the dated surrounding orchestra activity, must still be taken as belonging to 1942.

Very early jazz does not invite continuous attention for more reasons than one. Early bad recording devices, a comparative lack of orchestral invention, the very real non-association between 1923 dance tempo and that of 1942, and above all the actual abundance of great jazz covering the last fifteen years, are some of the reasons. Moreover, a peculiar, cultivated attention is really necessary for a satisfactory approach to any past art. Accordingly, I would not advise the average jazz enthusiast, somewhat limited as to listening time, to take time away from his contemporary and near contemporary listening in order to catch a nuance affectionately hoarded by certain musicologists, musicians, and critics. The present has more to give. In fact, when we consider the shortcomings of previous to 1923 jazz why should we urge the average person, today for aesthetic reasons, to devote long listening time to such early recordings? If we do not turn back for Palestrina, why then, for pre-1923 jazz, which *as a whole* is admittedly less than that which has come afterwards? And why, above all, to a reconstruction of that past. Aside from the great solo artists, the old jam ensembles, as ensembles, were not as inventive as those of today and neither were the soloists as a rule. We must remember that it took a Dodds to create the *Canal Street Blues* solo.

Musicians have strange and mysterious ways of working. If they had not disregarded academic advice from the beginning we would never have had jazz. Let us only battle with those critics who pick the picayune things to praise, who single out some one thing to encourage at the expense of all other musical activity. Although Lillian Hardin and St. Cyr made up two-fifths of that great ensemble, Armstrong's Hot Five, they could not be said to point to what was to come out of jazz. If we want to recapture the *feeling* of such a band with a modern group, it is not only out of sentimental and nostalgic reminiscence that we try to restrain contemporary improved pianists, and guitarist for banjoist. Not improving on them is a "purist" fallacy. A return to the original Creole Jazz Band style without the original personnel or equivalent talent is a little sentimental. However, maybe our improved recording devices will encourage the average person into listening to a Lu Watters, and such listening in turn lead to ways of listening pleasure in actual early recorded jazz. Much appreciation comes about in just such a way. As for the musicians themselves, very often deliberate returns to the past will save an artist from decadence and bad influences of his day. It would not hurt Louis Armstrong to return. But Armstrong became angry at Panassié for intimating as much. There you are.

However, I cannot say that I find what Lu Watters's orchestra has done on the Jazz Man label satisfying. There seems to be no high-temperature desire to play. The playing is that of good musicians who have *gone through* the barrelhouse stage and with great satisfaction attained the more sophisticated manner. It is unhistorical sophisticated jazz of an unreal 1922. To play in the style of 1922 they hold themselves back. As this age abounds in good jazz it seems to me that these boys are holding back in order to approximate the past. Reconstruction by holding back is a poor substitute for the real thing. And we have the real thing. With a shelf full of good records I would not take one of Lu Watters's down to play many times. For example, in *Riverside Blues* a solo or two played in whole notes might be a relief after intricate work or a leading up to something that intends to build, but when these ad lib spaces, so inviting to all that is jazz, are never taken up, are consistently left as long-note interludes, we cannot avoid the suspicion that the players are inhibiting themselves.

The clarinetist Ellis Horns seems to play most of the solos. His solo on *Tiger Rag* has real gusto. Turk Murphy plays with a good old-fashioned heavy vibrato on *Come Back Sweet Papa.* Such a nonchalant solo can be delightful, but it has to be pointed up more for a record, or possibly the engineer should have seen to it that Murphy was placed nearer the mike.

The piano solo of Walter Rose on *Temptation Rag,* for modern ears, is repetitious. The kind of thing we play a couple of times for ourselves and only dig out afterwards to show somebody else how such rag music sounded. It seemed to mark the past more than the other sides.

As Bob Thiele points out, the Chicago style, as an isolated style, no longer exists. It seems a remark equally applicable to old styles such as New Orleans and Dixieland. Different conditions originally produced these different styles. Very quickly they influenced each other and lost their several identities. The New Orleans Album is not strictly New Orleans. It is New Orleans with an unintentional *plus,* and an inevitable *minus.* The going back, forward, or staying put is different in every case. What made the Chicago style was the fact that the Chicago boys were still learning. They accommodated their playing to their limitations. This "still learning" may very well have been the art basis of Chicago jazz. It is a good ingredient for a lot of very advanced art. Georg Brunis today may be cited as playing in "strict old Dixieland" but in reality it is Dixieland, plus what he has learned since. He is not trying to play as he did many years ago, but he is a living, vital exponent of Dixieland attitude. Lu Watters seems to have caught only a flat facsimile.

IN ANSWER TO ROGER PRYOR DODGE

Rob Rolantz

This article is in reply to the nonsensical tripe dished out by Roger Pryor Dodge in his review of the Lu Watters jazz band. First off, let me say that I have never heard any recording of the band in question. But strangely enough, that in no way bears on the subject. Mr. Dodge's discussion of Lu Watters's band raised my wrath as no one has done since George Frazier! And since I believe Mr. Dodge's discourse was infantile and incongruous, I am answering it.

It seems Lu Watters has a jazz band that plays New Orleans style. And it also seems that Mr. Dodge, ostensibly the record reviewer for *Jazz,* believes that any band playing early jazz has the wrong attitude, because that band is going backwards. If Mr. Dodge said only that, this reply need never have been written. But in explaining his "views," Mr. Dodge indulged in such banal statements, such half-cocked remarks, that the whole article left me gasping for breath. In plain words, does Mr. Dodge himself know what in heck he was talking about?

To prove Mr. Dodge's slight case of incoherence, I will quote from the second paragraph of his case against the Watters Jazz Band. "There are a few points to consider in making a decision as to the correctness of Lou Watters' attitude, totally beside the question of whether or not they did the job well. There are considered approaches to music which do not even exist for the casual listener. Archaeological or cultural approaches will, in their pursuit, reveal aesthetic nuances not acquired readily in any other way. These are studies which when pursued for their own sakes expose the student to aesthetic nuances which are not apparent on a cursory hearing. Music in its own day does not need any of these 'considered approaches.' Sometimes it is even timeless enough to extend from one era to another. Thus we have the classics." (Unquote.) Now if anyone reading that can tell me what it means he can have his choice of i.e., one of the spurs that jingle jangle, an off-key note of Helen O'Connell's, or a modulation from the sweeter music this side of heaven. (Aesthetic nuances, in case any of you musicians are interested, means "science of the beautiful different degrees of shading".)

Mr. Dodge goes on to state, after another few incoherent paragraphs that (and again I quote), "In fact we come to the sad conclusion that no one can really *know* music. A man is capable of just so much knowing, even when aware that what he may *never know* might have been a truly great experience." Quick Watson, the needle! Shades of Schopenhauer, Spinoza, and Whitman! With this profound observation, Mr. Roger Pryor Dodge now places himself upon Dr. Eliot's five-foot bookshelf!

I imagine that by now most of you readers must be wondering whom Mr. Dodge is kidding. (And you must also wonder what all these irrelevant utterances have to do with the Lu Watters band. Remember this is the way Mr. Dodge reviews a band.) But wait the best is yet to come. In one breath Mr. Dodge says that (pre-1923) early jazz is so awful we shouldn't listen to it, and then he contradicts himself by saying that our improved recording devices of today may encourage the average person to listen to Lu Watters, and that in turn will lead to ways of listening pleasure in actual early recorded jazz.

But the prize statement by Mr. Dodge is this one (and I quote for the last time). "Musicians have strange and mysterious ways of working. If they had not disregarded academic advice from the beginning we would never have had jazz." Oh Brother! Can't you just imagine Dr. Arturo Toscanini, at the turn of the century, calling together Louis Armstrong, King Oliver, and Kid Ory, and saying, "Gentlemen, your glissando is imperfect and that improvision business is out"!

Straight from the shoulders, don't you think our bands have enough trouble from critics without letting a person who can't even write coherently try to criticize a band. A critic's job is to find out what is good or bad, and why, and then to tell of ways or means to correct the bad points. But to let a critic write a lot of trash that doesn't even make sense, to let him use a lot of six-syllable words (which mean nothing) so he can impress everyone with his knowledge, isn't fair to the band or the magazine in whose pages the article appears.

Insofar as Mr. Dodge's theory is concerned, that a band playing New Orleans is going backwards, the whole idea is silly. Ain't the James and T. Dorsey string sections throwbacks to the old Whiteman-Waring days? Wasn't the old Crosby outfit a direct steal from the Dixieland bands of the early 1920s? Mr. Dodge's whole argument falls to this: the Lu Watters New Orleans Jazz Band is no good, because the New Orleans style that was played 25 years ago is no good! Any grammar school student can point out the fallacy in that line of reasoning.

New Orleans style is one of the most intriguing and interesting branches of jazz. To find a band that consents to play that way in these days, when chances are they won't be successful financially or in fame, is worth noting. And to have the band attacked by a muddleheaded critic, who writes like a freshman in English Composition, is enough to make one give up all hopes for good jazz. In spite of this criticism, Mr. Watters, keep up your work, jazz music needs more bands who are adventurous. And I hope Mr. Roger Pryor Dodge will muse upon his aesthetic nuances somewhere else than in *Jazz* magazine.

THE LU WATTERS CONTROVERSY CONTINUES

Reginald Maure

It's too bad that Rob Rolantz didn't quiet down a bit before he took a crack at Roger Dodge's review of Lu Watters's Jazz. Perhaps then he wouldn't have resorted to misstatement and misinterpretation. When criticism becomes acrimonious, it stops being criticism. It was the tone of Mr. Rolantz's answer to Mr. Dodge that sent me back to Dodge's review of the Lu Watters's records, and also the fact that I, for one, understood it. I hope Mr. Rolantz will forgive me for that.

Mr. Rolantz says several things that I should like to answer. Of course, it is impossible to quote at length, and unless the reader has both articles before him, it wouldn't mean much anyhow.

Mr. Rolantz objects to Dodge's "archaeological and cultural approaches . . . reveal aesthetic nuances" and defies anyone to tell him what this all means. To begin with, Dodge distinguishes the casual listener from the one who regards jazz as a serious music and who comes to it with a "considered approach." I will even grant Mr. Rolantz the benefit of this latter approach and ask him, when he listens to Mozart or Bach, whether he does not regard this music, as much as we both love it, with an attitude of antiquity. This is a "considered approach"—one which in our appreciation of that music is compelled to span some 200 years. It is impossible, as Dodge says, to "know" all music because we can not go back to Bach and hear it today as it sounded to those who heard it then. We hear it today with the sound of Beethoven, Mozart, and Brahms in our ears. The fresh initial impact of its newness in 1750 can not ring for us as it did then. We listen to it a little bit as we read ancient history. And what is more, unless we play Bach many, many times and catch the "nuances," the thousands of "nuances" which go to make up the 18th-century music, we won't even come near knowing it, any more than we will understand ancient history by a mere cursory reading.

What has this got to do with the Lu Watters records? This: Dodge feels that ensemble playing is better today than it was in early jazz. He sees no sense in going back or duplicating an outmoded style. And I think that he's right when he says this is going backward. It's asking the "average person" of 1943 to borrow his father's ears (circa 1923) to listen with. Now I'm sure that Mr. Rolantz wouldn't subsidize a society for the resurrection of ancient dance forms such as the gigue, sarabande, allemande, etc. Nor, I am sure, would he do as much for a group dedicated to resurrecting the bunny hug, Charleston, shimmy. Though he would not do this, he applauds the resurrection of the jazz music that accompanied these latter dance forms. He will even say as he does, "Mr. Watters, keep up your work, jazz music needs more bands who are adventurous." (*Venture,* to paraphrase Mr. Rolantz, in case any of you musicians are interested, means: the staking of a thing upon a contingency, that which is *unforeseen.*) One ventures forward, not backward.

Second, nowhere does Dodge say, as Mr. Rolantz would make him say, "Early jazz is so awful, we shouldn't listen to it." It's a matter of *fact* that Mr. Dodge was among the first, if not *the* first, to be expounding (in print, Mr. Rolantz) the jazz of that time. What Dodge did do was to give reasons why early jazz "does not invite continuous attention" for the "average listener" of today. And one of those reasons is that "a peculiar cultivated attention is really necessary for a satisfactory approach to any past art." "I would not advise the average jazz enthusiast to take time away from his contemporary . . . listening." And another reason is the fact, as Dodge says, that there is a "real non-association between 1923 dance tempo and that of 1942." The jazz enthusiasts of today, as of 1923, listen with their feet as well as with their ears, Mr. Rolantz.

And as to Mr. Rolantz's non sequitur on Dodge's "Musicians have a strange and mysterious way of working. If they had not disregarded academic advice from the beginning, we would never have had jazz." Mr. Rolantz's bile makes this mean that Dodge could assume Armstrong and Oliver playing under Toscanini.

Mr. Rolantz says that "it is a critic's job to find out what is good or bad and why, and then to tell of ways or means to correct the bad points." Only in a very general way is this true, Mr. Rolantz. Frankly, you can't tell an artist anything. He just does it and then you have it, and all you can do is evaluate it, but the chances are the artist won't even read it. He's too busy creating.

Perhaps Mr. Dodge has not made a clear distinction between reviewing and criticism (which is another thing). If he hasn't, it's because he's more deeply interested in music than Mr. Rolantz wishes to believe. Dr. Eliot's five-foot book-shelf, which Mr. Rolantz feels Dodge has placed himself on, can't be read like a Dick Tracy strip either. Try it sometime, Mr. Rolantz. You'll find all the big words there too, aesthetic, nuance, all of it. We don't think in monosyllables any more. And there's always some discrepancy between what a man thinks and what he can get down on paper. That's where our intelligence comes in.

LETTER ADDRESSED TO GENTLEMEN:

George F. Montgomery

It seems to me that the recent controversy about the Lu Watters Yerba Buena Band has omitted one point very important to your readers; namely, that the band's records are real New Orleans music at its best. With the dearth of records in the original style and spirit of New Orleans, I think it is unfortunate that when a modern band turns out some really fine music along those lines, your record critic, Mr. Dodge, has to throw cold water on their efforts merely because he does not believe the records are in modern dance tempo, and because he thinks that jazz music automatically improves with the passing of time.

Mr. Dodge reveals the inconsistency of his reasoning when he first condemns the records of the Watters band and also those of King Oliver's Creole Jazz Band because they are played in an outmoded style—original New Orleans—and then in your last issue recommends 18th-century music as played by Wanda Landowska on the harpsichord. I think Mr. Dodge paid the Watters band the highest possible compliment by comparing it to the Creole Jazz Band, and I think that a lot of your readers will agree.

It is a good thing you now have a record reviewer who can appreciate and point out the value of New Orleans music, as Mr. Dodge did in his fine review of the Bunk Johnson and Jelly Roll records on the Jazz Man label, or many of your readers would probably have passed up these classic records.

Enclosed find money order for one year's subscription beginning with your next issue, and for Volume I, Nos. 1 and 2, and let's hope that the unissued Watters records will not be too long in coming out.

CORRESPONDENCE ON LU WATTERS

Roger Pryor Dodge

In letters I have received, one of which was printed in the fourth issue of *Jazz,* I am pointedly taken to task because of my review of the Lu Watters records. The fact that I do not support Lu Watters's records finds disfavor. Bolstered by invective, my particular remarks on Lu Watters have been made to represent my general views on "New Orleans." Principally I have been attacked for trying to depreciate an honest effort. Needless to say, I do not question Watters's honesty of effort. But there are many misguided efforts in art activity which it is a critic's business to criticize, regardless of good or bad intention on the part of the artist. Moreover, when we consider that much modern art effort is instigated and nourished by a camp of critics leaning in the direction of that effort, it is even possible for such an organization as the Yerba Buena band to owe its existence to misplaced, or misinterpreted, confidence in a critic's dream. A whole trend of promising talent can be set astray because of such poor critical attitudes. Witness the acclaimed dead-end excursions in modern art, whether it be music, dance or painting.

Excessive praise of "those good old popular tunes" has always aroused my suspicion. It suggests that those who voice such remarks ask for nothing better as a jazz foundation. Without demanding from any period any more than it can give, nevertheless when there *is* good contemporary material, and plenty of it, I am not overjoyed by the constant re-appearance of such pieces as *High Society Rag* or *Panama. High Society* was a major vehicle for inspiration in one period and *Honeysuckle Rose* has turned out to be a major vehicle for another. Neither tune in itself demands much attention. (I trust a 1960 jazzman will not be told to go back to *Honeysuckle Rose* because some of the greatest solos came out of it during the 1930s.) But particularly distressing is the voiced satisfaction in the playing of such tunes straight. The structure of any tune has a binding force, but that does not mean that many tunes used both in the past and present do not demand considerable treatment.

What interests me most is explaining my reason for not going back to pre-1923 jazz with an attitude of "purist" integrity. A recording experiment such as Lu Watters's can and did arouse a tidal wave of comment, pro and con, among jazz critics. Some serious enthusiasts even went so far as to maintain that the jazz returned to was the outstanding jazz of all time. I have definite reasons for believing it is not.

It is a fallacy to think that music needs only to be heard once to be liked. We

Jazz, 1943.

cannot know a music without constant emotional recourse to it. Such a music was pre-1923 jazz. Let us think of listening to Chinese music. The average person will listen a few minutes and then give it up. On the other hand, let us live in China, for five years, and with their music constantly about us, we will commence to like some of it. Since all our Occidental music is based on scales and harmonies that are similar, we have an "in" when we hear a new piece. But this does not mean that what was once in the air and understood and liked by everyone holds significance for the next generation. The older generation heard their own living musicians in the flesh, not only on records. They did not have to listen attentively to become soaked in its style. But it is a purely cultural attitude if today, by listening attentively for some time, we begin to see beauties in a past music not so apparent on first hearing. By a cultural attitude I mean, going out of our way to expose ourselves to an art. In certain periods we cannot *escape* an awareness of an art, in others we must first have the desire of pursuit. Studying the history of jazz music, finding out who played here and who there, tracing the beginnings of certain elements is the archaeological approach. For those who pursue this activity rewards in musical beauty will also be apparent. But it must be remembered that it was the extra musical activity which was their initial point of focus.

What appears as contradictory in what I said about "listening" I thought I had made clear. My intent was to suggest to amateur and professional critics, and all those who have plenty of time to listen to music, that the more music of any period listened to, the more improved the musical perspective. On the other hand, I wanted to convey my candid opinion to all those who were not specializing in music appreciation, that is, to the casual amateur *listeners* with small time for listening, that the period Lu Watters chose faithfully to duplicate seemed to me the least rewarding in musical pleasure. For those with limited listening time I do not recommend pre-1923 jazz. It does not compare with that which came immediately afterwards. For the modern creative artist to write strictly in the polyphonic style of such an early composer as Palestrina would be futile, but then again, to take from it is another thing. Early Armstrong is on wax. To try to play as he *did* without sounding in any way like *anything* since, is also futile. Anything with real punch is going to have an element of something beyond 1923. Of course there can be isolated cases but I am speaking of the full life of a living musician. There have been musical hoaxes in the past. Paderewski and Kreisler perpetrated them. You can be sure the more they would have done so, the less inspired the result, and if inspired, the more apparent the hoax. We *are* living in a musically creative age, probably the best in the last two hundred years. Why then go back in the manner of an ancient instrument society? Such a society has its place, but only in an artistically dead world. Probably later, if jazz succumbs to commercial trends, it will be of merit. There is no doubt of the benefit in continually having recourse to the past, both creatively and reconstructively, but I do not think this pre-1923 jazz warrants it in a "purist" manner.

I never said that I did not like Dixieland or New Orleans, or that playing in those traditions was going backwards. I was attacking the purist attitude of disregarding all the best jazz since, and restricting ourselves to that style alone, including what was good as well as bad. I find that the trombone has gone ahead, and as much as I admire early New Orleans and Dixieland, I do not like a thickly coated "tail-gate." A little tail-gate is natural to the instrument but the corny tail-gate is obnoxious to my ears. Using the best of the past has always been my motto. I would like to see *more* of the past dug up and used, if to our modern ears it still seems to bear up. Plenty of it does and plenty of it does not.

The situation as regards the Bob Crosby and Muggsy Spanier revival of Dixieland band playing holds nothing in common with the Lu Watters experiment. Crosby uses "Dixieland" to suggest style source. No one familiar with the Original Dixieland recordings (circa 1920) would ever confuse the two for a moment. Bob Crosby's band plays within the familiar Dixieland tradition, but adds all the resources of the hot solo as developed by the great improvisers of a succeeding period.

The word "drive" has also been used in connection with these records. When I read the praise of orchestras *solely* in the terms of "drive," I feel that nothing critical has been said. I believe that, all in all, most of us are sympathetic in our feelings and I give them the benefit of the doubt by taking for granted that they mean to include other elements besides "drive." Then again they may mean nothing else but a "performance drive." Satisfaction in any one element is very often sufficient but we know that the jazz enthusiast demands "performance" above all. Of course there is "composition drive" as well as "performance drive," and the jazz enthusiasts' unconscious demands may be a balance between these two, favoring, naturally enough, the "performance."

Jazz, like all art, falls within the field of aesthetic consideration. It is no longer the special property of musicians and jitterbugs. If hot jazz is as good as most of us believe it is (unless we are simply confining ourselves to expressions of personal pleasure), it is worth establishing why it is so valuable. Consideration of the *valuable* in art inevitably leads to the use of words which have come to have more or less precise meaning for those interested in such speculation. It also forces in allied considerations which bear on the subject. It is easy enough to say what is good and bad in a band that is playing just what you like, but using such methods will get you nowhere when a whole policy seems a little askew. It was the same when I criticized Paul Whiteman back in the middle 1920s. To say that his band played badly was untrue; to say that they played too sweet and smooth could have been answered by Whiteman with—"So is

Mozart and all cultivated and advanced art." Probably most critics today do not know how the whole critical world was praising that sort of jazz at the time. The Lu Watters experiment is a definite policy taking nothing from the jazz of the best players since the early days. In the same issue of *Jazz* that my review appeared in, Gene Williams not only admits but apparently applauds the fact that Lu Watters "chose his musicians, not for their knack with 'hot solos' but for their ability to play the right ensemble part on their instruments." I agree that solo and ensemble playing may be two different arts, and that the latter may be the basis of the former, but I am quite certain as to where the cream of jazz invention exists. Gene Williams also says—"A good first man plays the melody, with or without minor variations. . . ." Yes, a first man can be a man *around* which the others may improvise, but such a stable position is pointless when he is the *only* man. Again we hear the desire for the old "pops," even played straight. The best of Armstrong was never straight. "Wild-cat" and "screw-ball" playing may, rightly enough, cause a reaction in favor of even playing a "pop" straight. However, we need not necessarily have to choose between two such extremes. There is nothing better than a long jam session to illustrate the limits of inventive powers. Very often we see a good player losing his inspiration and frantically trying to say something by over-reaching himself. In such cases it would have been better to go along *with* the tune instead of off on meaningless tangents.

In conclusion, I might say that these records seem to have introduced a new division in opinion. Whereas before, the division between hot and sweet jazz could be drawn along an ascending line, and we could more or less know where the other fellow stood, now we cannot by precedent be sure of how the different camps will divide up. A fiction has been introduced—a fiction of New Orleans and Dixieland. In fact it is not only a literal, but also a musical delusion. I do not think that this division expresses a desire for old-style assimilation, which would be natural enough for the older generation. I think it goes deeper into human gullibility. We are prone to idealize beyond all merit something which, although it had the seed of what was to come, did not in itself contain the maturity that was first fully realized at some later date. In its mature state the greatest popularity is usually manifest. (Of course, popularity in a *somewhat* restricted sense.) Some of us have a hankering to be a little purer than what is popular and for us there is nothing purer than that which we cannot, because of its age, further distil. There is this same feeling towards certain periods of classic music. Pure classic for some people becomes something handed down from heaven as a prototype for all that is to come afterwards. And so too there has been introduced into jazz a spurious criticism which no longer depends on "like and dislike."

Louis Armstrong

Jazz is a very young music. This becomes increasingly evident as we trace Armstrong moving all through its best history. His career stems back almost to the beginning of recorded jazz—not quite, but there he is alongside the King Olivers and the first of the truly significant batch of New Orleans players.

Louis Armstrong's musical vision seems to have no limits; there is no holding back in Armstrong. He may ride along, sometimes, in a way Charles Edward Smith considers ". . . suspiciously like quasi-philosophical jive . . ." but Louis never really sleeps on the job. His straight choruses have always been at least a little off-straight, with a well brass-coated tone.

Choruses such as those on *Potato Head Blues* must have sounded like wild-cat courting in the old days. Let us imagine a person without much jazz discernment hearing for the first time the *Potato Head* solo. Later, upon becoming well acquainted with the solo, such a person might find to his surprise that the trumpet, which once sounded like one minute of uncontrolled violence, is blowing actual notes in which every turn and phrase is filled to the utmost with invention and surprise. Weaving and flight all within the purest music fabric. Moreover, this person finds he might have gained a little insight for such trumpet playing if he had been lucky enough to hear *Cornet Chop Suey,* an even earlier Armstrong.

One of the greatest and longest solos, a solo in which his sustained thought never wavers, is to be found on Louis Armstrong's *Wild Man Blues,* the Okeh *Wild Man.* This is a Jelly Roll Morton thirty-

Jazz Magazine, December 1943.

two-bar composition, but it is not made up as a thirty-two-bar popular chorus; it more or less follows a line of groups of four bars, each four divided into two of melody and two of break. This blues is a "long thought" and Louis never lost sight of ways in which he might embellish this thought. He plays the theme off-straight with well-filled breaks and rhythmic twists and plenty of "accidentals" that cut the blues structure deeply and finely. Even the breaks flow out of the tune without appearing to be the extreme trumpet technical exhibitions that they really are. Each break demands attention but principally one should not overlook the third one, covering bars eleven and twelve. Not yet, but some day, this break may be recognized as containing elements of historic importance to music. The fourth break could be considered a study in virtuosity alone but the last four rhythmic notes give it the jazz tone. Near the end of the next section is noticeable a seven-note run, not on the arpeggio of the chord as Armstrong usually plays, but within a unit closely knit into the threads of the chromatic scale. Armstrong takes the first half of the record and Johnny Dodds takes the other, with Armstrong coming in for the brilliant conclusion.

An early record having compositional merit, although not an Armstrong composition, was the *King of the Zulus.* It is off the beaten track of the regularly

Trumpet Solo

WILD MAN BLUES

played tunes. Here, Armstrong's cornet holds the whole record together, one of the very earliest efforts at composition in jazz. A sense of composition tones the whole record with only Kid Ory, a little talk about chittlins and eight bars of St. Cyr to relieve the amazing Louis. Let us not say it is the best Louis, but perhaps it is the best all-round record of those days.

Armstrong made an outstanding contribution to the expansion of style in trumpet playing. If his style is sometimes overdone by other trumpet players, nevertheless, it is Armstrong who gave that instrument its new horizon. It is up to today's trumpeters not to sprawl all over it. By his rough and flexible trumpet manner of singing, by his extension of trumpet technique which has influenced so many musicians and by his great recordings, Louis Armstrong has enriched American jazz.

Jazz Critic Looks at Anthropologist

In the Record Changer, *Mr. Ernest Borneman, the anthropologist, contributes an interesting series of articles, "The Anthropologist Looks at Jazz." If Mr. Borneman's anthropological viewpoint is to promote rather than confuse an understanding of jazz, it must pass the test of musical criticism in which it provokes discussion. As a commentator in the literature of jazz, I offer my remarks in the hope of synthesizing what is valuable and sifting the critical grain from the chaff.* R. P. D.

Mr. Ernest Borneman's article, "The Anthropologist Looks at Jazz," contains a great deal of information that one does not encounter in the usual run of jazz literature. We are grateful to him: a specialist can always throw new light where we least expect it. In the early and simple stages of an art, analysis is not particularly difficult. It is not one hundredth part as difficult as when many elements have accrued in the growth of the art. It is in maturer stages, when the ingredients of the art have become entangled and fused, that the critic begins to make use of a measure of intuitive insight. Of course, an intuitive hypothesis, by itself, is not enough; but, unlike the scientist, we have not, in musical criticism, the benefit of acceleration of knowledge and understanding which comes by means of manipulating a controlled experiment in a laboratory. Instead, the critic must wait with resignation and some humility the results of his intuitive hypothesis. It is on this ground that I wish to meet with Mr. Borneman in friendly and constructive discussion, gratefully taking from anthro-

The Record Changer, October 1944.

pology what is really helpful and legitimate, frankly rejecting what may be alien to our purpose.

In the name of anthropology, Mr. Borneman, at the very start, dismisses any differences which exist between primitive and sophisticated music. In the name of anthropology, he wishes to make no distinction between the merits of one musician and another. If anthropology indeed dictates this ironing-out of distinctions, that is all very well. But why does Mr. Borneman proceed to discuss the *flowering* of jazz, the *decline* of jazz, and very particularly the *merits* of musicians? To satisfy its requirements, anthropology may take any suitable analytical standpoint. But Mr. Borneman is not always satisfied with remaining an anthropologist. He exists behind the scenes and re-enters in a brand new costume. He practices a little surprising and charming sleight of hand. In fact, he betrays our credibility and simplicity when, after announcing that he would view jazz as a specialist through the criteria of anthropology, he finishes his performance with rather conventional jazz criticism. We have sat respectfully before the scientist; we have listened; we have missed a transition point in the performance; we have finally heard only the familiar words—and we have been disappointed.

Let us analyze the nature of our disappointment. Mr. Borneman argues that "it would be inadequate to say that *St. Louis Blues* is a better piece of music than *Pistol Packin' Mamma,* or that Louis Armstrong is a better trumpeter than Clyde McCoy." Instead he wants to know *why* they came to exist and what *function* they perform. For greater convenience, let us change the two names—Armstrong and McCoy—to James P. Johnson and Art Tatum.* Does it not follow, then, that Mr. Borneman has chosen to label as "jazz" the school or idiom of Johnson, let us say, and that later he will surely examine the idiom of Tatum in order to show us its function and how it came about? At least the first few pages of his April article lead us to such a happy expectation—but, alas, nowhere does he follow this procedure. He speaks, indeed, of such things as he considers to have *value,* but like any of us outside the discipline of anthropology, he is very definite in divulging what he considers good and bad. In fact, for very good measure he offers advice on *what should be done!*—an attitude which, by any standards for understanding the social development of art, is putting the cart before the horse, sociologically speaking.

Mr. Borneman speaks of various elements as driving "American Negro music quite *inexorably towards one definite form* which would combine all surviving Africanisms with as much of the white man's as was accessible and acceptable to the Negro singers. This form is the blues" (italics mine). He offers this judg-

*McCoy has no standing, while Tatum has the greatest admiration of his profession and is also placed by Mr. Borneman in the same position that I place him.

ment categorically, as if the process had no alternative, yet we find in the West Indies and in South American countries a quite different music rising from the collaboration of the African and white man.

In the May article, Mr. Borneman speaks of four basic standards for jazz, all of which he calls fallacies. He speaks of the academic standard. He says a "new music requires a new aesthetic rationalization." But there is nothing new in jazz: counterpoint is always counterpoint, and invention is still invention. Even Mr. Borneman uses the word "counterpoint," certainly not a word used by jazz musicians, but a word descriptive of a certain style of classical music. The only difference is the material: the blues and the figurations of the instrumentalists. When it is counterpoint, we can compare it with the great contrapuntal period; and when it is invention, the invention follows the turns and patterns of all the inventive writings of our culture. The tunes and material *are* different, just as the tunes of one country are different from those of another. European tunes, however, contribute to the variety of tune form which we do find in classic music. After all, it is the tune (or general melodic content) which makes all the difference in the world. Appreciation of musical art is not just appreciation of the exquisite interweaving of melodies or the neat arrangement of notes within a rhythmic pattern found in a solo. The simple basic pattern must, in its conservative manner, hold the greatest import for us. It is only then that more advanced contrivances take on for us real meaning.

No, the source of confusion does not lie in the fact that jazz is different in its contrapuntal texture or in the similar use of figures for invention, but in the fact that the academic or classical man does not like this tune, the blues, and does not know the instrumental texture. He can be moved by the folk music of any country in Europe because classic music is built out of all the folk music of Europe, but the blues are new to him. It is not a case of the academic standard, but a case of jazz music coming to him by way of itself alone, with little to remind him of the music in which he was cultivated.

If we are musicians and can recognize musical elements such as counterpoint and invention, what do these things mean to us if this counterpoint and invention is *about* something having no meaning for us? The intricacies will remain obscure to us. To become familiar with them we must have an over-all attraction to the art, the pursuit of which will reveal the music. Its intricacies will then unfold and become clear. Moreover, I fear that when Mr. Borneman speaks of classic music, he seems to have in mind the symphonies of the 19th century. Why compare this late and highly developed music of our culture to a new music? Why expect jazz to resemble it or emulate it?

By slow development a baby becomes a man. There are many important stages it must go through before becoming that man. So it is with music. We cannot compare the various stages or expect one to conform with the other. The lifespan of art, however, is so long that most people consider the product of the various stages as different species. There are those who would like to see one emulate the other. There are those who frown at a natural growth.

Mr. Borneman says that "musical susceptibility, however profound, is no test of literate judgment. Jazz can be *played* without academic knowledge of music but it can never be *understood* without some musical literacy." I agree that the "putting of jazz into its natural perspective" does need a wider knowledge of music, but that jazz cannot be understood without this knowledge I do not agree. He says: "What sounds good to you is not necessarily good of its own *standard.*" (Italics mine.) What "sounds good" is by its *own standard* good. That is, if we are sensitive to the musical art in question. "Good" as part of musical history is something else. It seems to me that for any of us the test of "sounding good" is all we can go by. Not the "rational act" as Mr. Borneman says.

If a person is cultured in more than one species of music, he can compare the "sounding good" of one kind to another. Doing this he acquires perspective, but a further widening of knowledge does not become the "rational act." Mr. Borneman seems peeved at the seemingly naive enthusiasm of the jazz fans. Their passion for short-lived aspects of jazz is sometimes disconcerting and trying, but to dismiss their judgments is dangerous. It is about as logical as dismissing the African's likes or dislikes—the only basis of his creation and of his audience—because he lacks the historical view of music, and because he cannot, as Mr. Borneman would say, "rationalize." But if we keep acquiring these various cultural patterns, until the over-all aspects of one can be compared to the over-all aspects of another, I think we do reach a horizon which those in any one culture do not enjoy. There is a limit even to this, however, since the meaning of music, for any one of us, is not reached in a day. The more we skip about, the less we shall understand of each. We cannot cram in art.

Mr. Borneman says that jazz is not classical music, and again he uses the term "symphonic" as synonymous with classical. He also says that "the whole sense and purpose of jazz rests in this extraordinary ability of a group of musicians to improvise complex rhythmical and melodic counterpoint on a simple harmonic basis." I am afraid that this is how Mr. Borneman understands jazz. Nor is it at all surprising that for him the careful depositing of huge symphonies by "long hairs" is "classical music." This is the great fallacy into which all jazz critics fall. Any music that is to reach the proportions that our instrumental symphonies have attained must go through what jazz is going through today. In an article of mine published in *Hound and Horn* (Summer

1934) I quoted from a letter written by a certain André Maugars in 1639 upon the occasion of a visit to Rome. I will quote him again.*

> I will describe to you the most celebrated and most excellent concert which I have heard. . . . As to the instrumental music, it was composed of an organ, a large harpsichord, two or three arch-lutes, an Archiviole-da-Lyra and two or three violins. . . . Now a violin played alone to the organ, then another answered; another time all three played together different parts, then all the instruments went together. Now an arch-lute made a thousand divisions on ten or twelve notes each of five or six bars length, then the others did the same in a different way. I remember that a violin played in the true chromatic mode and although it seemed harsh to my ears at first, I nevertheless got used to this novelty and took extreme pleasure in it. But above all the great Frescobaldi exhibited thousands of inventions on his harpsichord, the organ always playing the ground. It is not without cause that the famous organist of St. Peter has acquired such a reputation in Europe, for although his published compositions are witnesses to his genius, yet to judge of his profound learning, you must hear him improvise.

I see *no* difference between this and what our jazz musicians are doing. We see the ability of Frescobaldi as improviser and Frescobaldi's writings compared by a man of that day. We can be sure that what was improvised on that occasion differed in no way from what Frescobaldi may have turned out on paper in his study. There need be no real difference between improvisation and notated music. After a time a difference does develop. Jazz did not wait for this time but has instituted scored music *before* making a practice of first committing her improvisations to paper. After all, notation is a means of preservation out of which has come an art dependent upon this medium. The art coming afterwards must be a consequence of notation's primary need, that of preservation; otherwise we get the hunting artist or arranger, hunting through the debris of European music for what he can find to syncopate and score.

I have tried to show in *HRS Society Rag* (January and February 1941) how jazz might slowly become notated. The musicians are not yet ready and the attempts by Henderson, to which Mr. Borneman refers constitute no argument that jazz is peculiarly an improvised music and that such orchestrations as we have are proof that it cannot be notated and orchestrated. Musical notation developed with the notation of previous music. Jazz finds itself with a notation too perfect for its present needs and a kind of orchestration alien to it but useful to its commercial establishment. To say that "the European tradition has developed all *alternatives* of scored music to such peaks of perfection that all jazz arrangers' attempts at originality of scored writings are doomed to look like

*Arnold Dolmetsch's "The Interpretations of the Music of the XVIIth and XVIIIth Centuries."

parodies of the real thing" (italics mine) is true of jazz arrangers. But it is by no means true that "all alternatives of scored music" have been reached. The development from improvisation to scored music may follow European music very closely, but whatever difference there is, that difference together with the present new material (blues and instrumental figurations) will be enough to launch a new scored music. Like improvised counterpoint and invention, it will compare to our classic music, but it will be a state of music in its own right, satisfying something that classic music does not.

Mr. Borneman feels that we are talking at cross-purposes and that if we recognized a thing for what it was we would not compare it with something of another kind. In speaking of jazz he says "it could not *possibly* be good by such standards as were evolved from the history of European music." (Italics mine.) Whatever new ingredient there is in jazz, and certainly there are many, the general manipulation, the general working out and those things that we *can* compare, put jazz in a position where, even if it did not have these other qualities, so dear to a listening public, and those special differences that Mr. Borneman has found, it would still compare "well" with early classic music. But jazz has seemingly more than early classic music had. I say "seemingly" because we cannot, as so many do, take the notated music as a faithful facsimile of all that transpired during the 17th and 18th centuries. They had their many "ways" of playing and significant playing style to which no notation could do justice. The musical literature of the period is witness to this fact. But in spite of notation's inadequacy, what came to us through it is great art. Jazz notated is comparable to this music and I have seen many musicians, though not liking jazz, still being astounded at the counterpoint and invention. I know we cannot compare something utterly new to something else but jazz, as different as it is, is not altogether far removed from the past.

Mr. Borneman says: "However well the Tin Pan Alley boys arrange their jazz pieces, the academically trained composer still keeps his headstart of five centuries and all attempts to beat him at his own game are therefore doomed to folly and failure." It is as though he had said that competition in orchestration is futile. Is not every jam session competing with 17th- and 18th-century counterpoint and is not every soloist competing with the inventive writing of those centuries? The trouble is that our orchestra men were too impatient and did not go *along* with jazz but wanted a quick job. They borrowed wholeheartedly and the results are our hybrid styles of orchestration which become dated overnight. Mr. Borneman says that "collective improvisation compares with solo improvisation as an exciting race compares with the dull clocking of individual competition." Am I to understand this to be the case through all music or only the case in jazz? Counterpoint is exciting, but excitement is not the end of art.

Mr. Borneman continues this subject in his June article. In speaking of duets he says: "In comparison with the tense counterpoint of these records, even the best of the solo records are disclosed in their mistaken ambition. Jazz is an orchestral music, not a solo art." No great solos in any musical art came from an art that was *only* a "solo art." One learns to play in an ensemble, learns to be creative, to acquire rhythmic sense and timing. We must first bear in mind that solos come after ensemble incubation and that it is not a "mistaken ambition" that prompts a soloist to take off. Unless, of course, we want to cross off all the great solo work of the 17th- and 18th-century contrapuntalists. Bach's great contrapuntal solo harpsichord fugues did not come about through the harpsichord. This session spoken of in my 17th-century quotation was the incubation to Bach's harpsichordal contrapuntal writing. Even the solo writing for the instruments capable of only one melodic line (such as the violin) had their genesis in such sessions as are spoken of in my quotation.

Mr. Borneman dismisses a great deal as not jazz. True he takes the best jazz, that which comes from the place of its birth, and calls only this jazz. What are we to call the rest—Bix and the like? To me they are a representation of another jazz dialect. When Mr. Borneman segregates Tatum, Goodman, Wilson and Bix into some unnamed category, we must first understand that there is a difference between hybrid jazz, vitiated jazz, and a jazz, though not New Orleans, yet born of dance music and speaking the rhythm of dance in its outline. Hybrid is the mixing of two or more breeds, both having been born out of previous digestion and going through no new digestion. Vitiated jazz is a jazz that has lost, through smoothness, flowing qualities of the instrument, and a reliance on the sensuous quality of its tone, any stamina it may have had. Classic musicians have satisfyingly done these very things but it has either been part of a composition or the underlying chordal structure has been highly significant. Unless it is done in good taste, it ends up in the endless ripples of either or both hands on the keyboard in the manner of a Liszt. It is a sort of music coming more from indulgence in virtuosity rather than from inner expression. Tatum is a musician *par excellence* of vitiated jazz. Any emulation by other instrumentalists only leads away from jazz, without the good taste we find in classic music.

There is very little that is hybrid in improvisation. Just as the earliest New Orleans jazz or pre-New Orleans jazz is the complete digestion of what otherwise would have created a hybrid music, so whatever we see of Debussy in the best of Bix was completely digested. Bix's piano is not the best of Bix and for Mr. Borneman to say "Bix Beiderbecke's little piano pieces which have been compared to Debussy would therefore tend to make us doubt rather than confirm his value as a jazz musician" is no logical dismissal when we know that

James P. Johnson also makes excursions to that whole-tone Frenchman. I might add that his digestion is not one-half so complete as was that of Bix, therefore bordering much more on the hybrid. However, for me, such excursions do not in the least detract from Johnson's greatness as a true jazz musician. Of course a digestion of bad influences is not beneficial and maybe the complete digestion by Bix and his followers of the latest academic musical distillations had more of a disastrous effect than the hybrid excursions of Johnson. Jazz it is, however, and therefore comparable to other forms of jazz.

I find great satisfaction when Mr. Borneman says: "All untrained singers, Africans as well as Occidentals, tend to sharpen the accented beats and to flatten the unaccented ones." Even if he may be wrong, such statements are basic, something we need in jazz analysis. It reminds me of Grimm's Law of Phonetics. When Mr. Borneman follows this with "strong beats shifted to weak ones by syncopation tend to be flattened in the process," I see reasons for certain changes. But his other statements, the statements of his that I have been quoting, I cannot agree with him upon. I know that there are many jazz enthusiasts who are in accord with Mr. Borneman's views in this matter of *jazz through improvisation* and *classic music through notation.* Mr. Borneman weakens the basic strength of the things that *can* be said with finality by putting equal stress on his later hypothesis. I can thank Mr. Borneman for putting his case quite directly and therefore in a position for my attack.

QUESTIONS AND ANSWERS

Ernest Borneman

I

Eight months ago, when we began to publish the first notes on the anthropology of jazz, we ventured upon a new field of jazz criticism. We hoped to provoke discussion and argument of a new and creative type, and when that discussion and argument came forth in numerous letters and articles, we gave voice to it in the present column of "Questions and Answers." Today we are glad to publish one of the most rewarding articles we have been fortunate enough to call forth—Mr. Roger Pryor Dodge's "Jazz Critic Looks at Anthropologist." The points of difference between Mr. Dodge and myself are diminutive compared to the all-important evidence that discussion on jazz need not disintegrate into personalities and that the *Record Changer* is rapidly becoming the foremost forum of discussion on jazz in the U.S.A. For this, we wish to offer our thanks to Mr. Dodge and to all others who have written in to give us facts, data, information and debate. Please continue to write in if you disagree with us or if you have any information which may help us to put our music into a wider perspective.

This is your column. We will publish all questions and data which are truly important to you. If you have something on your mind which you think should have been said long ago, let us know about it and we will give it due space. If it's jazz, and if it's true, "Q & A" will publish it.

II

Mr. Dodge thinks that there is some contradiction between my promise to view jazz as a social function and my conventional evaluation of individual jazz musicians. I think Mr. Dodge would have seen no such contradiction after the tenth installment of this series ("From Minstrelsy to Jazz") in which we are analyzing the social function of what Mr. Dodge calls "hybrid" and "vitiated" jazz. Yet I am grateful to Mr. Dodge for his interest, and I hope that the December installment will finally clear up the point.

If Mr. Dodge will meanwhile accept our definition of the anthropological platform, he will, I believe, also have to accept our evaluation of individual jazz musicians, even if the result of that evaluation should happen to coincide with what he calls "conventional jazz criticism." We defined our platform by affirming that "the best musician is the one who shows the least compromise with alien forms of music; who gives the widest development to the traditional framework; who shows the greatest variety within the unity of his chosen idiom." I think that all musicians praised in the "conventional" June installment will meet the test of this definition. If we define the social course of a civilization, we clearly have the right to define anyone who takes that course as a socially "good" or useful person, and we have not only the right but actually the logical obligation to define any deviation from that course as "bad" or useless; it seems to me that we are therefore quite justified in suggesting that the former should be supported and the latter discouraged. The horse is well hitched to the cart: it pulls its load.

III

Again, Mr. Dodge will see that his argument on the South American forms of Negro music has already been superseded in the sixth installment of the series which was set up in type when he wrote his criticism. The *blues* arose from North American civilization with the same degree of historical logic as the *samba* arose from that of South America. I would very much like to go into this Latin American development at greater length since it is much closer to my proper field of study than the North American phase of Negro music, but we are first and foremost a jazz magazine and there has already been much criticism of the fact that the "Anthropologist" column has so frequently trespassed on other territories adjoining those of jazz proper.

IV

Mr. Dodge appears to ignore the whole argument on African music if he compares European polyphony to jazz counterpoint. It is neither the theme nor the abstract pattern of counterpoint but the *basic tenet of music* which differentiates jazz from all but African forms of counterpoint. I am well familiar with Mr. Dodge's earlier work and I am as fond as he is of the great polyphonic tradition, but I am also aware that little of it has survived, even in the self-consciously contrapuntal twelve-tone school, and I think that my definition of "classical," i.e., academic music as "symphonic," is therefore quite justifiable.

The difference in basic tenets to which I referred above will become amply clear if Mr. Dodge remembers that the whole European tradition was a striving for *regularity*—of pitch, of time, of timbre and of vibrato—whereas the whole of West African music strove precisely for the negation of these elements. All Negro languages—and here I include more than those of West Africa—refrained from direct statement and aimed instead at circumlocution. The direct *naming* of the thing was taboo. Thus our European goal of honesty and clarity ("call a spade a spade") seems little short of oafishness and ill-breeding to the African. Our rational philosophy, which aims at reducing all phenomena to their basic principles, is utterly alien to African mentality. The very opposite is sought for; the veiling of all contents in ever-changing paraphrases is considered the criterion of intelligence and personality. Hence, language, too, is not a matter of consonants and vowels alone (for this could merely lead to precision) but also of intonation, and intonation, in West Africa, includes pitch, timbre, and rhythm among its defining marks. Music includes these same elements—not in a striving for regularity but in a striving to deny that regularity by teasing and eluding it. No note is ever attacked straight; the voice or instrument always approaches it from above or below, plays around the implied pitch without ever remaining on it for any length of time, and departs from it without ever having clarified its exact meaning. Similarly, the beat is rarely *stated;* it is *implied* or *suggested,* anticipated or retarded, tied or rubato. Finally, the timbre, especially that of the voice but also that of all wind instruments and of many string and percussion instruments, is varied by constantly changing vibrato and overtone effects. All this results in a musical mentality which is basically different from the European tradition even though "tune," "counterpoint," and "invention" are common to both musical traditions.

To sum up: Regularity of pitch, time, and timbre are common to both traditions, even though the one aims at *stating* them and the other one at *avoiding* them; both may make use of the same scales, harmonies, and instruments; the difference lies in the diametrically opposite goal of the whole performance.

The African musician and many of the best jazz musicians, white or colored, follow the African mentality without being aware of it. As creative artists, they have no need of rational awareness. The critic, however, whose process of appreciation is analytic rather than creative, must be quite consciously aware of this difference, or else he will miss the *sense* of the performance even though its *sound* may strike him as pleasant or unpleasant, according to his purely personal and quite unpredictable taste: Thus my statement that jazz can be *played* without rational knowledge of this process, but that it can never be *understood* without some rational insight into the jazz idiom's own rules.

I have no intention to dismiss what Mr. Dodge calls "the naive enthusiasm of the jazz fans." On the contrary, their fleeting enthusiasms and their peculiar standards of value are precisely indicative of what we defined above as "socially bad or useless" and as such they provide the most useful and informative material of anthropological research.

V

Mr. Dodge doubts my claim that collective improvisation is one of the defining marks of orchestral jazz. He refers us to his very intelligent suggestions on jazz notation. Clearly, there is a misunderstanding somewhere.

As far as twenty years back, Hornbostel developed a perfectly workable set of musical symbols to permit notation even of African music. No doubt that the same can be done with jazz. But what would be the object? If Mr. Dodge will accept the above analysis of the African tradition, he will see its implication in terms of notated music. Let me remind him that it was not lack of intelligence which prevented even the highest civilizations of Africa from committing their language to paper. It was the taboo on any form of abstraction and the deeply rooted faith in ambiguity as the criterion of man's freedom and spontaneity.

I have no doubt that a new American polyphonic tradition could be developed from the jazz idiom. The blues could be scored and twelve-part counterpoint in the Palestrina manner could be written around it. But what should we gain in the process? The inevitable loss of the improvized jazz idiom would be a high price to pay. Let Mr. Dodge not forget that the very phenomenon of African survival in America was based on the illiteracy of the Negro. Teach him to write, and he will write as you do. Teach him notation and he will notice that his native music was "wrong" according to the laws of scale, meter, harmony, and general purity. It was only through ignorance of our musical standards and our system of notation that the Negro was ever able to create jazz. If you rate our standards and our system of notation more highly than jazz itself, you should proceed to make a written language of the jazz idiom. But if you wish to preserve the idiom, you should not try to vitiate it by committing it to the rigidity of a written language.

Mr. Auf der Heide also criticized Mr. Roger Pryor Dodge's views expressed in his "Jazz Critic Looks at Anthropologist":

Dodge objects, as I do, to your evaluations. I agree for the most part with your conclusions as to what is good—what music has merits, and what stinks—but it would seem more appropriate if your efforts were focussed on establishing a grammar and syntax, in demonstrating the tools of anthropology so that your readers could use these as an assistance in making criticisms. I realize that criticism is one of mankind's favorite occupations—witness this present letter—but it would seem that critical operations should follow the mastery of the tools. In the same succession, an article on one of the knowledges should precede criticism of the object analyzed.

I disagree strenuously with Dodge's statements—"There is nothing new in jazz: counterpoint is always counterpoint, invention is still invention" and following. True, counterpoint has always been counterpoint, but there is a great difference between one man composing music for several voices in the quiet solitude of his room—taking months, even years on one composition; and the creation of seven men, with Afro-American backgrounds, each playing a different, yet vitally integrated part in the blues form.

He objects to your use of the word "counterpoint"—as a word not used by jazz musicians—but he does not consider that the jazz musician's vocabulary is less extensive than his musicianship. He states that "the [jazz] invention follows the turns and patterns of all the inventive writings of *our* culture." I'll wager I could play Mr. Dodge some records which would refute that statement. True, seven notes can only fall into a certain mathematically predetermined number of patterns, but the imitation patterns in jazz should be reckoned in geometric rather than mathematical proportions.

Dodge's tenet that it is the "tune or general melodic content" which is the important consideration; and that the "simple basic pattern," i.e. the solo, which "holds the greatest importance for us," is as profound as stating the alphabet, or the grammar of the English language is of greater importance to us than the works of Shakespeare or the King James version of the Bible. Of course the latter could not exist without the former, but who goes to the theatre to hear Gielgud perform the alphabet. The tools, i.e. grammar and tune, are important, but only as means to an end.

His next statement that the academic or classical man does not like the blues—presented, no doubt, as an excuse for the impenetrable cloud that obscures the aural horizon of most of them . . . and that "he can be moved by the folk music of any country in Europe because classical music is built out of the folk music of Europe" is most vulnerable. I have seen few classical or academic musicians who could or would listen to, or enjoy the folk music of Bulgaria, Jugo-Slavia, Sicily, Czecho-Slovakia, or even the Cante Jondo of Spain. Among my own acquaintances, I find people who like jazz but know little classical music, far more willing to listen to this kind of folk music and able to enjoy it than those who appreciate "good" music. The academic standard of classical European art music is very much like the old definition of a scientist: "he learns more and more about less and less, until finally he knows everything about nothing." The proponents of the academic European standard, and by this I include the academic American standard, are so tone deaf to all scalar systems but the tempered diatonic that they can hear no other music. The musical evaluation and criticism is so engrossed in the trivia of embroidering lace panties on the cadaver of European art music that it is unable to hear any kind of folk music . . . European or African. I agree with Mr. Dodge in this criticism of your statement that (jazz) can never be understood without some musical literacy. It does not seem to me that in all cases understanding of an art is necessarily a "rational act" or the result thereof. I objected to this statement earlier in this letter. Bill Colburn, one of the people who understands more about jazz than anyone I know, is not a rational person. His understanding of jazz seems to me to be a purely mystical process. This is the case with others of my acquaintance. Granted, this condition does not promote an academic standard, but it provides enthusiastic audiences, ripe for a revival of good jazz.

People are awfully wise guys when they aren't befuddled by the snobbery of academic standards. When they get a chance to hear the real music, I believe they will dispense with the stuff we call bad music—and they will do it for the reason, completely unscientific, that they like it.

Dodge sees no difference between the improvised music of Frescobaldi, and what our jazz musicians are doing today. I believe that Mr. Dodge has selected his quotation for his own convenience. Frescobaldi, in the preface of his "Tocatte" gives this as his first rule of playing: "First, this kind of performance must not be subject to strict time." . . . The dependence of jazz upon strict time is too well known to warrant repetition at this time. I have a suspicion that Dodge has in mind the "jazz" of Benny Goodman, and others like him.

Let's go back to the quotation. "But above all the great Frescobaldi exhibited thousands of inventions on his harpsichord, the organ always playing the ground." This, with a change in instrumentation, could well describe any of the Benny Goodman trio or quartet records. There is no difference. Neither is there a difference between what Frescobaldi did, what Liszt did, or what B.G. is doing. All three had an academic European background. Liszt's was flavored with romanticism and hints of Hungarian folk music. B.G.'s by impressionism, and hints of Afro-American folk music. It is precisely this featuring of solos over a ground bass, passing under the name "jazz" which has replaced a vital interesting music.

Dodge continues, saying, "Is not every jam session competing with 17th- and 18th-century counterpoint, and is not every soloist competing with the inventive writing of those centuries?" This seems to me the result of careless thought processes. I would call competition an act of deliberate intent based on a knowledge of previous or contemporary example. No (New Orleans) jazz band ever heard of the counterpoint of those centuries. A similar statement in another art field would be "Is not every Benin bronze competing with the art of Michael Angelo?"

Mr. Dodge replies:

Mr. Auf der Heide objects to my statement, "But there is nothing new in jazz: counterpoint is always counterpoint and invention still invention." He says that there is a great difference between a man sitting in solitude taking months or years to write and seven Afro-Americans improvising. Yes, there is a great difference in the procedure but when the result is the same then I feel we have a right to compare. One is polished, the other not. The business of the composer is to assure a well-fitting combination while the jazz people are not always so successful. I have not used the words "counterpoint" and "invention" in their widest possible meaning. I have used them with a connotation familiar to *our use* of them. I might even go further and say that I had in mind the use we make of them with reference to a particular style of music, namely 17th- and 18th-century.

There are certain rules for counterpoint, things you cannot do and things you should do. To an amazing degree the jazz people obey what has been felt to be good procedure in counterpoint. It may be held that any ensemble playing can come under the category of counterpoint. This is largely so, but I have heard other than Occidental folk music, which although played in ensemble, could not be judged by what we understand to be good counterpoint. In such cases we have no "in" to the music but must acquire new standards or patiently acquire a taste for its new scalar and melodic content. In our own music we have this same disparity between vertical and linear music. In jazz, however, we have a music based upon our own chordal progressions, scalar system, and a melodic content very closely linked with that of European music. It therefore needs no new standards but only a little acquaintanceship. For people to like it is another matter. I have shown jazz to musicians who did not like it, but who were nevertheless amazed at some of its counterpoint. They had this "in," however, and were forced to recognize that it was not an immature cacophony.

Mr. Auf der Heide implies that my quotation of Frescobaldi does not represent the period. He quotes Frescobaldi as saying that the toccata should not be played in strict time. Jazz has the same thing in some of its introductions. Witness Armstrong's introduction in *West End Blues.* I would say that it is bad practice in spite of the fact that there are great compositions in this manner. Mr. Auf der Heide shows little understanding of the period if he believes that there was not an *adherence* to the beat if not an *accenting* of it. We often see, in the writings of that period, advice as to how the music should be played, that it should not be played too strictly. What I interpret this as meaning is that they should be flexible as are our jazz musicians. Except for the toccata and the like, the music was dance music and certainly was played with at least a beat.

Mr. Auf der Heide says that with different instrumentation my quote of Frescobaldi could describe any of Benny Goodman's quartets or in fact what Liszt did. Yes I agree that the procedure of Benny Goodman is precisely the same. But why stop at B.G. I hear the same in many solos on Bunk's records, in fact on all the old records. Sometimes the other instruments retire to the background or all but a couple drop out completely. As for Liszt, I see no similarity whatsoever. In point of procedure Benny Goodman is the same as Bunk or Dodds and as I believe, Frescobaldi. The quality of the outcome is another matter. Liszt, on the other hand, is the perfect example of the romantic or academic treatment at its worst. What Liszt did is analogous to what a contemporary classic composer might do with *St. Louis Blues.* We would get a bang out of the habanera rhythm with filled out chords, hands spread wide apart, etc., etc. The great difference between pre-19th-century music (as well as our jazz) and the 19th-century procedure seems to escape Mr. Auf der Heide. I have already touched upon the question of solo and ensemble playing and have further expounded my views in the June *Record Changer.*

Dancing on Skates: Correspondence

Edwin Denby

Some remarks of mine on skating and its power of expression as dancing—a subject suggested by seeing Sonja Henie's Ice Revue recently—have led to interesting retorts from readers. The first of the letters below, written by a Marine is an answer to an opening-night notice of that revue. Private First Class Bromley takes the point of view that skating as Miss Henie does it doesn't need any further dance interest. The second letter, by a highly interesting writer on dance subjects, is in answer to a more theoretical *Sunday* piece. Mr. Dodge believes that skating by its nature can't be much more than Miss Henie makes it; that it can't become an art form of interest in itself but will always need a special star to put it over.

RESPONSE

Roger Pryor Dodge

Dear Dr. Denby:

Baltimore, Jan. 29

You speak of ice skating becoming a form of ballet, or let us say an art form with some of the interests that ballet holds for us. You also say that you feel that it "is more likely to develop from the comic style than from the graceful one."

When we consider skating, we are compelled to consider the finished presentation, embracing exhibitionism, drama if you will, plus whatever the thing itself holds for us. No art ever came out of exhibition of great technique. The advanced forms of art include it, but are not

Sunday, February 1944.

built upon it. Although shock and drama may heighten an evening's performance of ballet, they are far from being the backbone of that art, that something that makes it endearing to the real balletomane.

To watch any good dancer in a great school is a never-ending pleasure. But what we may see a few figure skaters do while watching the open rink at Rockefeller Plaza is all that skating has to offer. The Centre Theatre is simply the case of a unique skater or comedian on skates. A good comedian turns anything to value, whether it be the stage dancing of burlesque, the perfection of advanced acrobatics, or advanced figure skating. It is his attitude to heighten the interest of any activity. It is the ability he possesses as a unique performer that makes this possible. The art form itself, however, is not built upon such rare performers, but rather upon the phenomenal quality latent in the activity itself. Ballet dancing has this quality; skating has not.

The Dance-Basis of Jazz

The Record Changer, March and April 1945.

It seems quite strange that I should have to say that jazz is dance music—that it has a close bond with the dance, and that it is not mere coincidence that we find it enjoyable to dance to. To have to point this out to people who like dance music, who see it danced to, who on occasions dance to it themselves, is something I would never have thought needed talking about. But there are those who think that the rhythm of jazz comes naturally from the music itself and that people dance to it because it fits in with what they are doing.

The little controversy there has been on this subject stems from Hammond's remark after the Duke's concern that "Duke is dissatisfied with dance music as a medium for expression and is trying to achieve something of greater significance" and Feather's answer: ". . . who the hell wants to dance in Carnegie Hall?"

It is quite a problem to point out that the whole rhythmic structure of our *classic* music has its origins in dance music, and that dance music, in turn, is due to the dance. The excursions away from dance that classic music has taken and the romantic non-dance interpretations of music essentially dance in character have quite completely confused us as to the real origins of classic music.

An important critic who considers Ellington the greatest figure in jazz told me after hearing the Ellington concert and subsequently hearing him at the Hurricane that it seemed Ellington had to play for dancers in order to play well. Weaver in *Jazz* (December 1943) says: "But does jazz require dancing? It was first played for parades, and modern dancing grew up around jazz rather than vice versa. Jazz must have a definite beat, which dancing needs, but the fact that most musicians are employed for the purpose of supplying music for dancing should not mislead us into believing that the music is subordinate to the dancing or to the dancing public." I find that critics and public feel the same way. They do not concern themselves about the importance of dance to jazz but seem to take it for granted that jazz is what it is because of some quality innate in the music itself.

The jitterbugs with their not too impeccable taste in jazz prefer a jump rhythm consisting of riffs, and this might seem to be good proof that good jazz is not dance music. But the writers of the *Jazz Record Book* have a broader view when they say "that American hot jazz owes much to the dance, the mother of the arts, *should be self-evident*" (italics mine). Yes, it should be self-evident, but I am afraid that too short a viewpoint allows the present dance music to hold up to ridicule any long view we may have.

I may also quote Sir C. Hubert Parry, writer of the best articles in *Grove's Dictionary of Music.* Says Parry:

> Dance rhythm and dance gesture have exerted the most powerful influence on music from prehistoric times till the present day . . . the connection between popular songs and dancing led to a state of definiteness in the rhythm and periods of secular music . . . in fact dance rhythm may be securely asserted to have been the *immediate origin of all instrumental music.* [Italics mine.]

The reason for so many divergent ideas arises from the fact that there is no definite instance we can point to of just how the music derives from the dance. We cannot demonstrate how certain rhythms executed by dancers were taken up by musicians, or how these rhythms influenced the melody, etc. There may be some instances where we can point this out, but these cannot explain the whole of jazz music. However, it does seem to me somewhat pertinent that when jazz musicians do leave the dance, they leave rhythm. We see this in concerts and stage presentations.

The cause of most of the misunderstanding concerning the mutual interrelationship of dance and music is due to the long life an art can have away from its origins. Great art is no weed that immediately withers upon picking. We know that cultivated flowers will even grow after being picked if they are artificially nourished. The buds will even bloom. They do not immediately wither and die. And so it is with art, and in this case jazz. It can live away from its soil for a very long time.

Let us take the case of a jazz pianist who has been in a band for some seven years or more. We all know that a good piano-man must come from a band. The rhythm he learns there he could never acquire by playing at home, and what he has absorbed through his contact with the band will stay with him all his life. He may not get any better and he may make excursions into a less rhythmic jazz, but on the whole most of what he has will stay.

Let us further suppose that he has a pupil, a talented pupil but one who has never played in a band. The pupil studies hard and plays many hours a day, emulating in every way the style of his teacher. Though he will not play with the lack of rhythm that our classic piano students do, he will not quite have the rhythm of his teacher. Assuming these conditions, what will this do to him?

We must realize that certain activities produce certain effects on their participants. Certain attributes react upon those possessing them. Whereas the classical pianist has the whole realm of musical literature to work with and to activate him, the jazz pianist has practically nothing but some popular tunes or simple blues. He is entirely dependent upon his own resources, his own improvisational powers to satisfy himself.

Rhythm is one of the must satisfying elements. It completely drains the human system. Musical exhibitionism and self-expression fall within the pattern of art; they do not remain on the outside as artistically undigested emotions. So, a pianist with a good rhythmic background will either float away under its spell saying much of nothing, or be prompted to say something. Either alternative brings self-satisfaction.

The pianist without rhythm is not so completely satisfied. He is ill at ease, he is bored, and so he tries just a little harder than he would otherwise to say something. He becomes occupied with musical material which he is not able to digest rhythmically and in turn is carried further from rhythm. Expression is his only outlet. He pursues it further and further. Should this pupil have any students of his own, they would be in a less advantageous position than was he. Since they would not be studying with a band man, they would get their rhythms in a diluted fashion. These so-to-speak third-generation piano-men would have such a poor rhythmic background that they would go in wholeheartedly for music of no rhythm. Of course we can have no such simple laboratory experiment produced before our eyes, but I think it can be easily understood how rhythm would diminish in the process I have delineated.

Now if such an experiment could be carried out, the retreat from rhythm in the piano-man would be extremely quick. In the case of the jazz band removed from the dance hall (that is, playing for listeners) the divorce from rhythm would proceed at a less accelerated rate. Since the band has grown to maturity as a band, it can survive away from the dance hall for a long time without showing any marked effect. But when it leaves the dance, jazz loses two strong supports. One is the pulse of a mob moving in time, and this is far more solid than an orchestra playing; the other is the reduction in the number of musicians to a very few since they are now playing only for those who like to *listen* to jazz.

Just as we cannot point to precise elements as coming from dance and going to music, so we cannot point to precise elements that go from a band into a band piano-man. Nor is it these elements which make him a solid piano player, but rather being in the presence of the band itself. So too it is with the band. The band must be in the presence of the dancers themselves. Association plays a strong part here. To associate your playing with the feel of a human body moving in rhythm is to tie yourself down to a rhythmic pulse that the body *must* keep but which a musician *may* stray from.

When away from the dance over a long period of time, the general public, as well as the musicians, will think more of the thing listened to, first not demanding the rhythmic pulse and later disliking it. Though the jazz enthusiast is not at either stage as yet, this too will slowly come about, as has always happened with music in the past.

Only what is in demand will be catered to. If the musicians can either consciously or subconsciously give the people what they want, and at the same time do what they themselves want, all very well and good; but if they cannot, they will only do what the people want. In the playing of dance music the jazz musicians in the past did quite what they themselves wanted. And what they wanted was what the people wanted—rhythm and a tune they liked. The musicians gave them this. Since the playing of a tune over and over was monotonous, the people accepted the jazz passages (ensemble jamming) especially as in the old days the trumpet took the lead by playing straight. These are the public's tastes and demands, to which an art must cater or atrophy. The music was born through public taste and must live by it.

In a changing period a few highly sensitive musicians or a few listeners may demand a continuance of the best of the previous era. With a conscious attitude they may make a more presentable art, subduing certain basic necessities needed in the past and extending other elements or even substituting some in order to make it a more perfect art for the future. When music is removed from its basis, the dance, certain changes must be made. Bach did it in the 18th century with a music in public disfavor; today the small or private labels are doing it for jazz. In both cases we witness the quick crystallization of something that is almost in the act of disappearing and which has to be revamped, so to speak, for a public which, though not of the music's era, will consume it.

Some of the older musicians playing for dancers today are beginning to like the new stuff they play for them, a tendency which jazz enthusiasts bemoan; on the other hand, we have a few musicians who lament the fact that they have to play for the dancers instead of in jam sessions. Even the jam session is not the healthiest place for jazz music. Art Hodes, its most ardent participant, says: "A jam session today has come to mean an atmosphere of excitement, you feel it as you enter the room. It doesn't mean good jazz. You play for the crowd, and they demand effort, sweat. They want to be sent." That is to say, when inspiration leaves the player, and his important basis, the dance, no longer supports him, he becomes what is known as a screw-ball player. I must say that I prefer the jump style to the screw-ball style. When inspiration leaves a dance-man, his middle-of-the-road playing will be better than that of the uninspired jam-session-man gone frantic.

The musician need not worry too much about the taste of the jitterbugs. He could do a lot worse. At least the jitterbugs insist on a rhythm, the musician's most important asset. If he did not play for them he might be playing dinner music or presentation music, or giving more jazz concerts, with the result that he would become worthless for the present-day demands of the session. He may admit the influence of dance in the past but deny it any present import. Here again he is wrong, for the best changes will come from playing for dancers. Though he may regard as evil the present-day dance influence on jazz, it is but the lesser of many evils.

Of course changes must be made. But if made too fast they may hurt the older musicians who were born to a different playing style. This, I feel, is the only fault of present tendencies—the spoiling of old musicians. It is for the private and small labels to work faster and get more out of these veterans, and for collectors to furnish the market. For the collectors to carp too much and run King Oliver up to $90 is a sad state for a thinking public.

CONCLUSION

Although jazz is through and through dance music, history tells us that it must become something else. Art, like a great deal of organic matter, is born and nourished under conditions that undergo a change as it matures. To a certain degree, it must leave home. Rarely is great art the sole expression of one circumstance. Nor does it cater to one function. Nor is it the expression of one class of people. The trend is from a lowly birth to a more educated development—from the sugar-cane to 52nd Street via Rampart Street. The art is nourished by a multiplicity of functions: songs as *sung,* music as *danced to,* and finally music as *listened to.* Blues, as blues, have no further to go, whether sung in a cabin or by a score of Bessies. New Orleans had not much further to go whether Storyville's status quo remained or not. The rest is a working out, always under the dance, however. The jam session is the recognition of jazz as a listener's music. It is an auxiliary and the means of jazz's transition, if there is to be any, to notated music.

Unless a seed is planted in a hothouse it will not grow, *and* if kept there after it grows it will not fully mature. So it is with music. If it is born and nourished in the presence of dance it can only go so far, as the whole consciousness of the dancers and the players is subject to another art (the dance). Buds will develop that will never mature. It will never be completely an established listener's music. It can be packed with invention and germ significance, but on the whole it will remain more suitable to dance to than to listen to. While dancing we hear very little of its musical significance. It is precisely for this reason that these musical elements must further develop and spread out—become more of a listener's art. Its later significance, however, will depend upon the early buds maturing, not vanishing in the process.

I should say, in passing, that the hot solo and, especially, the jam ensemble require more attentive listening than does 19th-century classic music. The people, therefore, can only consume it as dance music. When it is not associated with their dance, they dismiss it. It becomes neither their dance music nor their listening music. It then becomes the property only of the collector-cult, whose listening does not make of it a listener's music. They are discoverers, they have found a beauty in an art that was not consciously intended they should observe.

The creation of art is in great part subconscious, but the function that the art is made to serve influences the intention of the artists and subsequently alters the art. Within certain functions reside elements that promote art creation. It is in this sense that I believe that dance makes its contribution to music. A listening public, however, certainly demands something that dancers do not, though what it demands may be detrimental to the art. I might even say that the demands of the listening public could never create an art. Its demands are not creatively inciting to the musicians. The listener, however, is the creator of this listening function. Whatever change an already developed art undergoes is due to the fact that his wants demand this change. The musicians strike a happy mean, a point between an art intended solely for one purpose (dance) but capable of serving another purpose (listening), and an art which can only serve this latter purpose. If a dance music contains the material for the greatest listening music then it is but a small change which makes it the greatest listening music. Any further subjection to the listener's attitude is harmful to the musicians as well as to the listening public. The whole question rests on the degree to which we think art should change and how it should change.

No supposition, however thoughtful and however theoretically presented as to what the changes should be, can give us as valid an assurance as can a cursory glance at musical history. The kindredness of Bach with dance music and the closeness of all of Haydn, Mozart, and a great deal of Beethoven to dance music tells us that if we are to stray from the dance it must be slow and never very far. If the forms of these classicists and romanticists are not of the simple dance tune type, we do see, however, that they rarely relax from rhythmic pronouncement. And when we have rhythmic pronouncement we can dance to it whether it be played in Carnegie Hall or not. It is far easier to dance to Beethoven at Carnegie Hall than to *Rhapsody in Blue* or *Black, Brown and Beige.* Some of Beethoven cannot be danced to, but for jazz to go this far in its present state is precarious. I am sure that the best music of a 19th-century Viennese waltz orchestra said more in waltz tempo than in the introduction to the waltz or in the music they must have played for the diners.

The one thing to remember is that dance music is born with the dance and that during its subsequent development it must keep the spirit of dance even if it does change its form. It need not remain the best dance music nor, on the other hand, need it rid itself of all dance semblance.

Metamorphosis is as rare a phenomenon in art as it is in life. In art it is usually found to be just a few changes, giving us the illusion of a total difference. We hear music that apparently has nothing to do with dance, but a little root tracing will reveal that it was once a dance tune.

I did not want to write an article making such close distinctions as I have. I know that they can very easily be taken as contradictions. The problem resides in the fact that we cannot criticize any specific period from a premise which only considers the beneficial and detrimental *tendencies.* It is the long subjection to bad influences which will eventually show ill effects. On the other hand we cannot guarantee great music by plopping a man down in a dance band. Although the dance is the ground upon which to sprinkle our musical seeds, there are periods when this ground is quite barren. An old dance-man may very well say more as an absentee than some young man playing away at "jump." It is just not good to be absent too long.

I have said that dance music must come from and remain with the dance and I have also said that this music must leave the dance—must become a listeners' music. Let me summarize this and restate it. Music born with the dance must develop as dance music. It must stay with the dance throughout its complete developmental and improvisational state. When played by the folk improvisers they must still be dance-men. A period may ensue in which they can play for the dance and at the same time improvise more of a listeners' music. This is a transitional stage.

We must remember that spot improvisation is the outlet of folk musicians and that such improvisations require the greatest knowledge of the instrument, perfect freedom of thought, a store of ideas, and a sure delivery through a good intonation. All of these elements are needed to insure a perfect path that will lead from the emotions of the player to the music heard. The changes from then on must be taken over by the composer who uses this dance material but slightly changes it, making it more suitable to listen to. In the past this obligation lay with the composer. The folk having lost its playing vitality, the composer had to compensate with better music—better from the standpoint of notated music. When the material is once created the composer can carry on. He does not have to have the attributes of the playing improviser. Today our live music must eventually do the same thing. Our records will be intact, and so whatever happens, we will not have lost what is on them.

When I say at one time that musicians must continually play for dancers and at another say that music must leave the dance, we must of course consider at what stage of development the music has arrived. It is the folk *musician* who must continually play for dancing, whereas it is the *music itself* which must leave these musicians to become a better listeners' music. This involves the composer and notation—the birth of the academy. It must also be remembered that when dance music leaves the dance it must never go very far. Dance music, rightly speaking, is that playing for the dancers. A slight change here or there, the introduction of a few non-rhythmic qualities, make it a new listenable music although still very easy to dance to.

Note: Since writing this article I have read Panassié's *The Real Jazz.* In it he states that jazz is dance music and best played when the musicians are in the presence of dancers. He says, "Today, for example, it is assumed that dancers must have musicians, but no one imagines that the musicians themselves need the dancers. . . . Yes, jazz is dance music and this is precisely its greatest attribute." Panassié also feels as I do about the importance of the tone of an instrument in inviting the musician to create when he says, "Sounds striking the ear of the musician who is about to play exercise an immediate and profound impression on him which warms and provokes his inspiration." I am sorry that it was too late to put Panassié's quote in the body of my article.

Jazz of This Quarter

It has been shown quite conclusively that the private label, encouraged by jazz enthusiasts, is quickly catching up in amount of worthy output. It is not only that the private labels have been busy doing what they can; there have also been other forces making it possible that a jazz of the past should still continue. I speak of John Hammond's search for Meade Lux Lewis and the hunt for Bunk Johnson—Louis Armstrong's teacher.

It is some time now since Meade Lux has had his real powers revealed, but Bunk is only just coming into his own. A close analysis of his previous records revealed all the elements that go to make a great trumpet player, yet they did not give complete satisfaction. These recent records on the American Music label supply what was missing. They gave me somewhat of a shock when I first heard them—the rendition of *St. Louis Blues* having the same restorative power, the same authenticity that one gets when 18th-century music is played on the harpsichord rather than on the piano.

In the second section starting with the line "St. Louis woman . . . ," Bunk comes through with an extremely inventive solo, never going too far, always letting you know that his solo is a variation on the tune. There is in this solo a strong resemblance to Bix, in outline of variation, tone, and manner of attack. It is that intimate weaving of Bix, though based on a greater blues tradition. In *Sobbin' Blues No. 2* on the Jazz Information label this resemblance is even more pronounced. I do not want to give the impression that he is a second Bix (Bunk has another style

View, May 1945.

William "Bunk" Johnson, Mrs. Johnson, and William Russell in Washington Square Park, New York, 1946, in a still from a film by Roger Pryor Dodge.

of playing, too), though strangely enough when he plays like Bix, Bix sounds diluted.

At the end of *St. Louis Blues* George Lewis plays a solo of two choruses that Larry Shields first did on records. Whose solo it originally was I do not know. I attributed it to Shields but I doubt whether Lewis got it from Shields. It was, no doubt, Negro property before Shields got it.

Tiger Rag is the finest recording of this piece that I know of. There is always a richness of tone and collective playing which is its sensuous base. This is so satisfying that invention becomes secondary though there is no lack of it. It is a quality of great folk music. *Tiger Rag* has always been a little pat in the observance of its traditional outline. Bunk's men break all this down. Bunk has a tragic sort of wailing solo at the end. It is a perfect example of tone dependent upon musical line and vice versa.

See See Rider and *When the Saints Go Marching In* are equally good in their pervading earthiness. *Burgundy Street Blues* is a clarinet solo with banjo and bass. Lewis plays an extremely full-toned solo. It is a little on the "singing side" but passionate in its simplicity. *Ice Cream* shows off Jim Robinson who has been in the background on the rest of the sides. The style is in march tempo. *Low Down Blues* is a little less inventive than the others but the quality, as always, is the same. Bunk plays a chorus at the end which he has recorded before. It is a trumpet blues like nothing else we hear. If only more trumpet players would take their inspiration from this chorus!

Four exceptional sides have been issued by the Crescent label. They feature Kid Ory's Band. The fault I find with them is that they do not have enough of Ory. His solo on *Blues for Jimmy* and his counter-melody in *South* are most beautiful. One chorus is not enough for such a fertile mind. Something happens when a soloist takes two choruses, that does not come through in the first. Omer Simeon plays the most solos, in fact is heard throughout the whole set. His clarinet plays an important role on these records. It was good to have Mutt Carey on trumpet, especially his solo on *Blues for Jimmy,* but he does not come through often enough.

On the whole these records of Ory's are like a good set that we may find in the 1920s. Whether made now or then, they are priceless as the heritage of the best jazz. However, they do not come up to present-day standards of direction. I have felt this about other similar attempts. On the other hand, the Bunk records are what they should be. Guidance could have done no more. To get them, however, Bill Russell had to record for six days straight, and in this manner every now and then one would click all the way through.

Commodore label has some good sides by Georg Brunis, Wild Bill Davison, and Pee Wee Russell. The best is the ten-inch *Tin Roof Blues* backed by *Royal Garden Blues.* The *Tin Roof Blues* takes it too easy in rendering the theme at the beginning and end. Before Brunis plays his traditional trombone solo, created by him some twenty-three years ago, he now plays a muted solo in a much higher register. The transition from one to another sets off movingly his old solo. In *Royal Garden Blues,* Wild Bill Davison's ensemble playing adds new life to a piece that has become a little pedestrian. The record has good solos. *Panama* is a little drawn-out, covering as it does twelve inches, but nevertheless marches along with ensembles and solos in fine fashion. Its worth is not apparent upon first hearing but the tune gains with continued playing.

Meade Lux Lewis is back recording again. He has two sides, *Chicago Flyer* and *Blues Whistle* on the Blue Note label, and an album of three records gotten out by Asch. There are some very good sides in the Asch album. *Boogie Tidal* is by far the best. It starts out simply in Meade's very relaxed style. Later he introduces a vigorous boogie and reaches a tremendous climax in a few well-planned choruses. The record is short but breathtaking. *Yancey's Pride* is sprinkled with Yancey phrases and some very interesting ones of Meade Lux. In *Lux's Boogie,* Meade uses a phrase from his introduction to develop within the piece itself. It is a little different from the usual in this respect.

On the whole I find that a ten-inch disc is too small for the present Meade Lux. He has become such a giant and has such a fertile imagination that before we know it the record is over. It seems that he builds his records with much more time in view than is given him to complete his thought. Somehow in *Boogie Tidal* it came off, but in the others he barely gets started. Meade Lux represents a highly developed folk which usually does *not* get down to fact in the first chorus.

Meade has at his command a new attitude that I noticed in his Blue Note harpsichord records back in 1941. There was in those records a complete freedom of imagination that I missed in his piano of that time. He now seems to be bursting with ideas of musical invention, polyrhythm, and dynamics.

In *Chicago Flyer,* his best side, he plays a wonderfully low fifth chorus. The plaintive theme as it is taken through the choral structure of the blues is music at its best. Then follow six choruses gradually rising until he reaches the top of the piano. We are then plunged into a chorus reminiscent of the low one except that we hear modern tone-clusters. From here he goes into a brutal chorus, and then into another to finish the record. In *Blues Whistle* Meade Lux gives us another whistling blues. Except for one chorus not really in whistling style, most of his whistling borders on twittering. It is interesting but does not wear well.

Blue Note has an excellent record in *Big City Blues* and *Steamin' and Bearin'. Steamin'* is well put together. There is an intense interest all the way through. This record, contrary to most, delays the ensemble beginning by a duet on guitar and bass. Harry Carney, a new man in the Blue Note galaxy, plays his four choruses in the sax school of the best Hawkins tradition. After Hall's fine choruses, especially the last, Benny Morton, another debut in Blue Note, plays two very fine choruses to be joined in by the ensemble for three more. In these last three he rides along, leading all the way through as would a trumpet except that he has more forward motion than most trumpet leads.

Benny Morton's theme in *Big City Blues* is more than reminiscent of a popular song currently making the rounds. Any jazz solo built on a tune takes on all the putrid nostalgia of the tune in question. The theme is the germ of a tune, the tune dependent upon how highly charged the theme may be. A striking bar or two very often can easily go into Mozart, Beethoven, Bach, or Irving Berlin, coming back then to Morton. Although the fragment of the tune he used was musically pregnant, only at some later date when the undesirable association with its tune has been forgotten will the solo live in its own right. After twice giving us the theme of this tune he extricates himself very beautifully and renders two very sensitive choruses. Carney's second chorus is extremely inventive and his musical line easy to follow.

Another two records on Blue Note are *Squeeze Me* and *Sugar Foot Stomp.* Kaminsky plays the traditional chorus on *Sugar Foot* with a slight variation and great insight. Both Edmond Hall and Vic Dickenson play their solos well. It is an excellent side. In *Squeeze Me* they capture the mood of the tune in its lazy drag motion.

I have reviewed in detail, for the *Jazz Record,* Art Hodes and His Chicagoans on Blue Note. We can safely say that without changing over into something quite contrary to the great past, we can still make records which are in that tradition by men not of that generation. The Commodore sides also fall into this category. It is a marvelous thing that the young musicians feel this way. I have read now and then of such younger musicians who are record-collectors and who have the critical attitude of the listening jazz enthusiasts. This is more than promising, making way as it does for the playing of music of authentic stamp with a desire at least of approximating rich meaningful tonal quality. These musicians are playing this way out of a feeling of appreciation, whereas the older musicians had no other really good choice. Today there are many schools to choose from. Many of them are exciting with little else to their credit. It takes, therefore, a discerning art intelligence to make the right choice. These Chicagoans have made it.

Categorical Terms in Jazz: Improvisation versus Arranged Jazz

To find the "word" seems to be the continuous quest of people who are trying to define a rather elusive element. A word appears and they find consolation in its use until everyone borrows it to name a very inferior offshoot of the very element we have held so significant. Immediately we say "No, that is not the same thing," but we can in no way make it very clear why it *is* different. We name our element X and then when X becomes the loose term for all sorts of activities, we run to meaningless adjectives—the true X, the righteous X, the original X, the real X, etc., etc. Whatever we use can just as well be taken over by the other side. Moreover, most of our word and adjective distinctions are a very cumbersome and much too obvious means of stating our differences.

And so it is with the word "jazz." Besides covering all of our popular and folk music, it has had many other meanings. A distinction was made in the use of "hot jazz." This was right enough when we made the separation between "sweet" and "hot" jazz. We could point to Whiteman as being the sweet and Henderson and Armstrong as hot. But since then there have come into prominence players whom we cannot call sweet and whom some critics refuse to call hot. This seems to me a little meaningless, unless we want to forget what we understood by the words "sweet," and "hot" so appropriate to our earlier distinction, and attribute to the word "hot" the connotation of one of our other adjectives such as "true," "righteous," etc.

When we find ourselves refusing to call Clyde McCoy and the present Harry James

The Record Changer, June 1945.

hot, the word loses all of its original meaning, since someone may say that they only assimilate hot but are not hot. In many instances, this may be so, but then again, though such kind of playing may be an honest expression of the player's feelings, yet in no way will it measure up to our standard of good jazz. All we are saying then is that assimilated versions are nothing but bad imitations with no feeling, while yet other expressions are, in our estimation, an inferior form of art but nevertheless an art expression.

If an art has meaning for some people then what argument have we to put forth that they do not experience it and who are we to say that they do not get as much kick out of their choice as we do out of ours? If a certain group feels our way and a certain group feels another, there is no firm basis that either can use to establish the superior quality of his side. Neither side can claim majority opinion. Throughout history we see a constant shift back and forth between a great majority favoring that which we have come to consider good and a great majority not favoring that which we have come to consider good. Time is probably the only factor which we can pin any faith on. But time is a very poor yardstick for anything of contemporary nature. However, if we believe in time and believe that the inferior will fall by the wayside, then all we can do is to call our object *jazz* and be satisfied with either calling it good or bad jazz, according to the way we feel about it.

There are two other words, "swing" and "commercial," that are used as describing the inferior brand of jazz. For a short while "swing" was used by a few critics as a new word for jazz with the thought no doubt that it was not so entangled in devious meanings as was jazz. But the word "swing" was pounced upon by the opposite camp with more vigor than was jazz, so that one camp had to fall back on the old word "jazz" again.

A lot of people do not like the word jazz. It has had so many derogatory connotations that they feel or hope some other word will come into existence. I feel its old meaning had made the two Z's take on for our ears a vulgar sense. A new word *may* turn up, but if it does not the word "jazz" will slowly rid itself of all these excrescences and as the inferior music falls away, in time the word will draw closer and closer to the music that will survive so that for everyone the object and its symbol will be one. The two will cling together, the word deriving its mental stimulus from the nourishment it receives from the music. The two will be as closely connected as is the word "fugue" with the fugues of Bach, and not as tiresome, heavy, and boresome as the word "fugue" is to those who only know it when associated with the fugal writings of some academic pedant of the 19th century.

The word "swing" has been used for many centuries to express the spirit of a certain tempo in dance music. Whereas jazz could be the embracing word for this whole phenomenon from blues to the most intricate manner of playing, the divisions could separate into slow blues, swing, and some word describing fast jazz (such as "fast stomp"). The dictionary meaning of the word "swing" does not cover slow blues or barrel-house piano any more than does hot cover all that is good in jazz. Swing has come to mean anything from arranged bands, the jump bands, to dance music of a moderate tempo.

The word "commercial" is extremely confusing when used to cover distinctions not so easily done away with by it. If a person uses it to differentiate between improvisational jamming on the one hand and the highly arranged bands on the other, then we clearly see what he is talking about. The same holds true with the word "arrangement." When we mean it as describing the *average* arrangement as opposed to an improvised solo or improvisational jamming of a small band then the meaning is clear. But when we use the word "commercial" with too much stress on the idea of making to sell and the use of the word "arranged" to mean uninspired music, I then think we get into a most confused state of thinking. Armstrong sells very good. Maybe he went a little commercial in order to do so. But what do we mean by "going commercial"? It is very possible that the change in Armstrong's playing made him sell well, or will it be said that in order to sell well he changed? If to sell well, one changed, then it is "going commercial," but if one changes with none but a musical motive and because of it sells well, I cannot see that we are justified in calling that person commercial.

Can we say that all commercial bands have "gone" commercial? Is it not possible that they like what they do and prefer it to other types of jazz? They fit into the category of commercial and they sell well but they have done nothing in the sense of diminishing their integrity in order to do so. For that matter it is very possible that Armstrong *did* bend just a little to insure his selling value. But in so doing he does not fall within the category of commercial but has in a sense "gone a little commercial."

On the other hand the band which we have no compunction in calling commercial may not have "gone commercial" at all. The arrangers and leaders may be expressing their truest expression, veering in no way from the music *they* like best. Certainly if there is a demand for a band in the commercial category then there must also be musicians who enjoy fulfilling this demand. Or would it be said that when a person studies music or is a musician he *ipso facto* has better taste? I do not think it follows. In other words for a band to succeed it must give the public what it wants, which is a simple music in which the planning is a studied variety and invention never in excess of the listeners' interest or knowledge, nor so presented as to stress a style not in perfect accord with the aesthetic tastes of its listeners. To hit this middle road takes the talents of a good arranger.

For a performer not to follow this course and yet succeed he must be an exception to the general run of good performers. He must have many elements extra to the musical one. He must have this musical acumen plus personality, a great technique, showmanship, a *sure* sense of improvised playing or some other attribute, such as singing, acting, or dancing. The greater number of these, the surer his success. These various extra-musical elements are so interwoven with the highly significant musical elements as to become one with them.

Armstrong has nearly all. He is an exception, however, and the normal course of events can never plan upon such a rarity. He is not only tops in the most musicianly requisites, but in all the others besides. He could, however, do with far less musical greatness and still be as successful but the reverse would reduce him to being *in* someone's orchestra or just a king of jam sessions and night clubs.

The use of "arranged music" is a right enough term when we think of the music we hear in this category. A categorical nomenclature in this case, as with commercial music, is perfectly usable and for that matter facilitates criticism and the talking about jazz music. When carried any further so as to include *anything* written, it is misleading. I have explained in length[1] the fallacy of establishing a great difference between improvisation and written music. In brief I said that all creation is improvisation. The improviser plays the complete composition at one run-through in a performance. The composer on paper improvises sections at a time, writing them down as they appear in his mind. He can go over and shape up his composition, cutting out the bad parts and improvising new sections to replace them. He can keep working on his composition until it satisfies him completely, whereas the improviser has no chance to go over his work, the improvised creation having to stand as it is, poor sections and all.

On the other hand great inspiration is fostered by the built-up emotion of the ideas plus the sound of the instrument playing these ideas at performance pitch. The ideas have an accumulative reaction which inspire the player. The tone of his instrument sets the pace of musical entrances and sustained prolongation. When a writer of music stops to write down he cuts short the vital flow of ideas. He must constantly start again and break off. The whole process becomes more cerebral and is attended by an artistic handicap, the conquest of which takes rare genius.

I picture two diametrically opposed ways of creation, but in reality or actual practice the opposition becomes less pronounced. Compositions are not written piecemeal but are largely improvised over and over, the composer remembering what he can. Composing is a process of writing out themes, using them in improvisation, going over and over until the thing flows to the best of the composer's ability. When it seems stilted it is at those points that the composer was probably writing bar by bar instead of through the long even flow of the improvisation.

Then again from the improviser's standpoint, we know well enough that every performance is not spot improvisation. A performance can be anything from spot improvisation to a piece which over many years the player has memorized. His

1. "Notes on the Future," *HRS Society Rag,* January–February 1941.

improvisations have slowly, one after another, caught on until the piece is set although not written down. So we can listen to anything from spot improvisation to a piece which, although once improvised, has become set, and be no more the wiser—we can see no difference. It would be an extra knowledge to know one from another, a knowledge such as having heard it before in actual performance or on a record.

This is in truth real composition, and when such compositions are written down they are the make-up of great art, the art born of improvisation. Great art is fixed improvisation. Because of the significant elements sticking, it becomes a memorized piece. It is only another step to write it down. This step, as insignificant as it may seem, is for the folk very difficult. What we have written down in the way of set improvisations is usually done from records by musicians with an academic training.

This has all been in regard to the notating of the improvisations of soloists. The arranger of the large band is another matter. When we consider a new art manifestation we should consider it as a homogeneous growth. Nothing is strictly homogeneous so far as any records we have of our past can show. There is a time, however, when a situation ripens and a sort of melting point is reached in which various artistic elements fuse. In fact it may be some new element or heightened feeling in general which fuses everything it touches. This is a most critical period and a period which actually creates the seed which is to blossom forth into great art. The significance of the consequent art, its ability to unfold, and its stamina to withstand foreign ingredients all depend upon the degree of artistic impact the original seed possessed. From this point on we can consider any development as homogeneous. If the beginning is significant and the development continually encouraging any future integrity will greatly depend upon how homogeneous it remains.

Let me quote Ernest Borneman (*Record Changer,* May 1944): "The best musician is the one who shows the least compromise with alien forms of music; who gives the widest development to the traditional framework; who shows the greatest variety within the unity of his chosen idiom." There is nothing homogeneous in the art of our present arranging. Before jazz had hardly acquired its first real magnitude, the magnitude it had in the 'teens of this century, the arrangement was flourishing.[2]

So jazz had to buck such entertainment as was given by such bands as Prince's

2. Whatever we may say against the arrangement, it is the arrangement that has supported the jazz musician and subsequently jazz. The people would have nothing to do with jazz in its raw state. A few hot spots will not support a nation of jazz. A few isolated players of the old school will not inoculate a new generation with the new dialect of musical expression unless this new expression is somehow, however diluted, a part of musical America.

band, the music that was the backbone of accepted dance music. The Whitemans, the Lopezs were influenced by the aura around jazz. The Hendersons and Noble Sissles went further, and used much more of the jazz idiom. But none of them used any arrangement that had a homogeneous growth. The arrangements of those days and of today are a complete borrowing of the classic procedure. The material became more jazz, and *was* jazz for that matter, but the putting together was classic. The large orchestra could do nothing else.

The homogeneous growth of jazz gives us two elements: the solo improvisation and the ensemble improvisation. The outline of the solo is not too difficult to follow. The ensemble, consisting of many outlines, is most difficult to follow and for the layman is not much more than a pleasant—though undistinguishable collection of sounds. The arranger has disregarded the ensemble tradition altogether but has used the solo "as is" in contrast to the arranged ensemble passages. It is used just as an intimate jazz touch to the body of the whole arrangement.

Looking at the whole procedure from a practical point of view the arrangers could have done nothing else. To arrange an improvised ensemble, although it could have been clearer to an audience and surer in its presentation, would still have been far beyond the average public understanding. Besides jazz had not arrived at any subtlety in ensemble, and the people do not want a constant blare continually given them. We notice in the past *and* today that the ensemble passages are always forte, whereas the solos may be in any dynamic register.

Solos do not have to be written for the poor players as anything which is spotted here and there need only be in solo style and nothing more. Besides, every musician can do it, whereas the reading of solos is most difficult. So our arranger takes fragments of jazz, riffs for the most part, and builds his arrangements with them. Haydn and all subsequent composers did the same. Although what they did is wholly satisfying and seems correct, any such use in the jazz arrangement seems a far cheaper music than the improvised ensembles. I am sure that if jazz could have existed in an insulated state, and still live, that out of the ensemble improvisations an entirely different arrangement would have appeared, an arrangement that would have cleared up the ensemble improvisation without letting down the aesthetic feel of the whole piece.

It seems that arrangers have not been satisfactorily able to have instruments play in unison. An arranged rhythmic section does not sound bad but when an orchestra of large proportions only uses this sort of an arrangement in order to back up a solo we have too much of a contrast between the rhythmic section and the solo. This contrast is exciting but we can see that there would be little purpose in having an orchestra of such proportions relegated to accompanying a soloist.

I find that the real let-down of jazz significance occurs when the orchestra tries to take over the exposition of melody. Whether it be an orchestration of a hot solo or the orchestration of a simple blues, the augmentation devitalizes anything it touches. Symphonic orchestration of classic music seems right while jazz orchestration does not. Whether the material is not ripe for orchestration or whether the arrangers are not composers or that jazz needs another *kind* of orchestration is hard to say. The material is as significant as classic material, in fact more so than the music of the romantic era.

One thing we know and that is that our classic music is a slow evolution in melody, musical form, and orchestration. They have all progressed from the simple to the complex. There was orchestration for a few instruments before there was for the many. There was a period of planning in which the instruments had a certain amount of liberty. Jazz orchestration is too complete. It either leaves a soloist absolutely free with an ad lib chorus or it allots every part to the men to the last note. Therefore we can speak of the present category of arrangement as a part of the jazz field but something that has been grafted, and without much success. There has been no homogeneous growth nor a complete digestion of the grafted material.

But to dismiss arrangements, per se, as not jazz is not musically realistic. We can speak of the weak parts as being so, but only if we are thinking of current arrangements. To build up a theory on the impossibility of arrangement as something outside of jazz is wrong. To say that jazz is improvisation is wrong. All music is improvisation and any music in its earlier stages is as much spot improvisation as is jazz. In other words arrangements at this time are in disfavor and naturally the name of anything that is in disfavor carries with it odium.

Attitude Towards Early and Late Jazz

I am often asked whether I think present-day jazz is as good as early jazz. Although I do not follow all the way with the extremists who throw out everything after, I do have a natural leaning towards the early jazz because I find it more vital. I do not mean to imply that I dislike all current jazz. Positive values in current jazz, values from the past, values that we *can* bring from the past and which we can extend, I do favor. However, it may be asked, if I call the early jazz more vital, why then do I not go all the way and prefer it to current jazz and acknowledge that it was the only jazz? In this article I intend to show why.

There are those who like anything that is earthy and primitive but jazz is not *just* earthy and primitive. If jazz were just a simple tune played with an earthy tone, when the earthy tone disappeared then it *would* be senseless to cling to this simple tune. But jazz was not only this. It was and is a music of extreme and highly developed figuration which in semblance is as complex in any single melodic line or intricate in its collective state as any music of which we know. The essential element in jazz, which I consider to be the melodic line, can and does have an existence and development of its own quite apart from the tone or texture qualities of early jazz. And it is where there is a longer flow of melodic thought that I find the keenest interest and would prefer it even when delivered by a blander tone.

The balance between the beginning, middle, and end sometimes comes about in a subconscious manner in the early stages of music. Although the construction of the solos and ensembles was hard-

The Record Changer, February 1946.

ly ever long-sighted, the virility and strength compensated for this short-sightedness. As music develops or grows older it has larger vision, there is a more consistent balance between the beginning, middle, and end of a solo or ensemble. There is, however, in the later music, a very slow deterioration going on at the same time. Because of the slowness of this deterioration music has time to establish a well-balanced composition with little of its original vitality lost and a great deal of compositional and consistent inventiveness gained.

Jazz might have followed this usual pattern if there had not been too many musically bad influences around it and had it not had to compete for an audience which was conditioned by the more suave compromises with Western music. This condition ruined most of the old players and created a bad musical atmosphere for the new musicians to be born in. All through this period of bad influences from 1930 on we can see the struggle of this first vital music trying to develop and at the same time hold on to its rich texture. We can see this conflict in current-day jazz musicians who one night strive to play within the great tradition of early jazz players and the next night succumb to the detrimental evils of sleek sweet jazz.

It is hard to estimate how much that is good will survive and develop with all the evil influences within the present-day jazz. Early jazz was not sweet nor was early European music. Just as the sweetness of Mozart was possible and acceptable at a certain point in the development of European music so may sweetness in jazz become possible and acceptable at a certain point in the development of jazz. Although we feel the change towards sweetness when entering Mozart's period we are nevertheless captivated immediately in spite of the sweetness and so we may be captivated with jazz even when it becomes sweet. We can sense a tendency towards sweetness and away from the earthy tone and brittle invention of early jazz. But when the best of the past has come to have the advantage of *development* and has become enriched it can satisfy us just as completely even though the texture of early jazz is missing.

As vital as an early music is, I do not think of it as listenable music. Except for a very few specialists its appeal was along other lines. The multitudes who liked it in its day were attracted to it by the sound of the instrument, its perfect function as dance music plus the excitement of the instruments in ensemble. Except when they played straight the liking was never musical. Any mass enthusiasm for a band "in the flesh" is never a strictly musical preference. Real preference is shown where the audience "goes for" the records. Although a large number of concertgoers have excellent sophisticated taste and a pretty high standard of acceptance of music written for an audience they cannot "keep up" with early jazz.

Even the most ardent enthusiasts are attracted to jazz for various reasons. Some are highly sensitive to intonation, others to rhythm and playing style.

They become completely sold to somebody's tone. Although they may respond to and appreciate a good line, it is imperative that it be delivered by the tone they like. It is a question whether they see the line unless delivered by the tone of their choice. I believe the greatest musical significance is to be found in the musical line. Although the early jazz did not have the musical line completely I do not say that the jazz since that time has been more successful. I only say that I believe it is more successful although verging on the cheap side constantly.

When under the influence of listening to a band in person some of the criticism made in this article may seem precious. I find that the atmosphere surrounding a band, whether a *son* orchestra in Cuba, a jazz band in New Orleans, or any authentic folk orchestra, is so all enveloping that we cannot analyze the nature of our total reaction or make critical distinctions.

Obviously early jazz music would not stand up if played on an instrument for which it was not intended and by performers foreign to its spirit. Music of the great contrapuntal period certainly stands upon the piano. This is a triumph of the melodic content of the music. It has great musical meaning divorced from the instrument that it was planned for. Naturally its significance is enhanced when put back upon its original instrument, the harpsichord. Even then we are not listening to the music as it was heard by its contemporaries. The whole manner of playing is missing and if we could recapture that then we would see how much richer the total effect would be. This music must have had a style that was rich in tone and phrasing. The question might narrow down to whether we prefer the highly sophisticated music of the 18th century that we do have to a music which, had we had recording then, came to us on records and revealed richness in tone, intonation, and phrasing but could not stand up well in notation. To decide this it becomes a matter of whether we value the sensuousness (tone) more than do we the aesthetic (melodic line).

Without a discussion of the significance of the aesthetic and the sensuous in music let us investigate which of these aspects we like the most in jazz. In the nature of the art the sensuous quality can exist by itself but the aesthetic presupposes some sort of sensuous surface. In reality, however, the sensuous is never isolated as we always have some tune, however simple and short, to carry it along. Therefore when thinking of the tone or sensuous quality we are never at a complete standstill, and when thinking of the aesthetic or melodic line we always have whatever sensuousness an instrument has in itself.

I find that music's sensuousness is a static sort of pleasure. Our musical appetite is quickly appeased at any one hearing although upon frequent repetition the sound does grow on us until it takes on a striking familiarity. Its pleasure is static because we have no incentive to follow it through, no expectancy, no satisfaction when it arrives, no leading into something else.

The aesthetic of a composition, when we become familiar with it, takes on a shape that unfolds in a very expected way. It is the fulfillment of anticipation which gives us our greatest delight. To be held in this position for some time makes for an engrossing experience. The longer we are held the more memorable the experience. Without the appeal of the sensuous tone there is no great immediacy, no memorable pleasure in the thing that passes, but only an interest in the unfolding of the line. When there is a rendition with little immediate attraction we shift the burden of interest to the aesthetic which, since it is not always inspired, is not always capable of compensating for the loss of tone quality. When this sensuous quality is accompanied by a paucity of aesthetic interest our attention tapers off very quickly.

In jazz the motor action of the strong beat and rhythmic suspension is physically so exhilarating that when combined with the tone quality a minimum of the aesthetic is needed. The overemphasis of motor action when there is nothing else makes for white-heat music only, which as listenable music is over-stimulating and creates a need for letting off an equal dynamic power in dancing.

The late inter-party controversy hinged on a discussion of where to focus our interest in the rise and decline of jazz. The earlier we situate ourselves in the jazz cycle the more commendable do we consider the qualities of "purity" and "simplicity" and the more reprehensible do we consider the attributes of later jazz which are frequently described as "overelaborate" and "Westernized." Those situated in the latter part of the jazz cycle, though offended by this name calling, seldom have weapons with which to fight back.

The questions of horizontal and vertical and of variation on the tune and variation on the harmony are far from the issue. These terms in no way cover what is felt by jazz fans nor are they correct terms for distinctions that are better "felt." In jazz there is no such thing as pure horizontal ensemble when we consider either the implied or expressed harmonic structure. There is no such thing as pure vertical music. Variation on the tune or on the harmony is abundantly evident in all types of jazz.

In closing one might say that the greatest of all possible distinctions within jazz is that which exists between the good old New Orleans jazz and the current 52nd Street variety.

The Psychology of the Hot Solo

To take content from a tune and still not explicitly state the tune in any part of its variation the improviser must have the tune safely buried in a substratum of consciousness. If it is buried too deep there is danger of losing the tune and a consequent falling back upon harmony alone. This often happens. But when the tune is buried properly, receding far enough into the mind of the improviser to allow freedom of thought, the improviser can sustain two voices running through his head at once. This might be called two-voiced thinking—one voice privately sustained in the head, the other publicly presented. It is not counterpoint where one voice is complementary to the other. It is a situation where the theme, matrix to the variation, is playing the part of an inner voice, the audible version being the newly created variation.

Naturally, such inner counterpoint is loosely conceived, and both voices, if made audible, do not necessarily complement each other. For the true improviser is not merely releasing a perfect obbligato for the tune in his head. He is creating a new derivation which will be heard alone, will have to stand by itself. If the orchestra happens to be playing the tune, it is only a background coincidence, not a countervoice. A variation which depends upon a background of orchestra playing the tune is an obbligato, not a new melodic entity, or independent creation.

An ideal state of improvisation is achieved by constant repetition of the tune until it has sunk into the subconscious. When it is successfully embedded, the musician is free to improvise, and the psychology of association may

*An earlier and shorter version of this article appeared in the *Atlantic Monthly,* July 1944.

Jazz Forum (England), May or June 1946.

work its full force. The first actual progression, or even the thought of it, immediately brings up in a flash the rest of the piece. In this way the improvisation becomes a *new tune*—its unity built out of the old tune. It is the comprehension of the place of harmony plus the maintenance of the intact unity of the tune that distinguishes a great improviser from the musician playing around the borders of a tune. For example, the act of circling about a succession of tuneful single notes with runs and rhythms does not vary the tune. It merely relieves a monotony of presentation. Although a weak device, it has been carried to comparative heights. A few creative pianists of recent times have made use of very little else, pianists like Earl Hines, Teddy Wilson, Art Tatum, or, in the classic field, Franz Liszt.

Another simple way of improvisation consists in forgetting the tune and improvising on the harmony as carried in the head or sustained by the orchestra. Here there is nothing the improviser does that will not be right enough, but lack of a cementing continuity of melodic thought keeps such improvisation from holding together. Unless the improviser can, on the moment, create a new unity of tune, we get much choppy activity—and little musical material.

To *imply* the tune at every stage of the improvised variation, to avoid obbligato, and to have variety and originality besides, is musical activity rarely accessible to anyone except the musician experienced as a professional folk player. Such a musician sees the tune in his mind practically as a whole, or, at least, each section. This image allows him perfect freedom. The musical line of his improvisation can disregard entirely the musical line of the tune. As an example let us consider the late Charles Christian's guitar solo on the popular-tune favorite *Honeysuckle Rose* by Benny Goodman's orchestra. There is nothing distinguished about the tune itself, but it is a perennial favorite at jam sessions.

Although it is possible throughout his variation to trace the melodic line of the original tune, Christian forces us to forget *Honeysuckle Rose* and listen to Charles Christian. The tune must have floated far back in his mind to permit him such an easy motion. One-finger piano-playing, at slow tempo, reveals the absolute melodic newness of his piece, although when played at tempo, the tune *Honeysuckle Rose* can be distinguished. We notice that Christian divides his solo into two groups of twelve bars with a closing eight-bar section, instead of the standard division of four series of eight-bar phrases. His first musical unity consists of a long line which nearly disappears in the ninth bar but goes on to finish firmly in the eleventh and twelfth. The thirteenth bar initiates a new musical line for him. This line includes a two-bar phrase, which, when it enters the release of the popular tune, takes on its character. Christian continues this reaching phrase and melds it into beautiful linear work of a continuous character to finish the middle section. The halting phrase in the thirtieth and thirty-first bars unmistakably but delicately commands the closing of the solo.

HONEYSUCKLE ROSE

Very often the improviser will want his new melodic line, his hot solo, to keep on going ahead to extend beyond the limits of a formal section of the original tune. So he obeys the demands of his melodic line, not the termination of the section, by extending his musical line beyond the normal limit of the section of the accepted tune. Having done this for a time, he is naturally way behind the course of the melody, but at any point he joins it with the greatest ease. Like a dog out for a walk with its master, running away and continually returning—returning to where the master *is,* not to where he left the master!

It is the adherence to a tune that gives a not-too-imaginative variation its significance. A *popular* tune weakens the quality of the variation upon it; a valid folk tune strengthens it. To *make use* of a tune's unity, not facilely follow its melodic outline, requires a faculty such as Mozart wrote about:

> . . . and I spread it out broader and clearer, and at last it gets almost finished in my head, even when it is a long piece, so that I can see the whole of it at a single glance in my mind, as if it were a beautiful painting or a handsome human being; in which way I do not hear it in my imagination at all as a succession—the way it must come later—but all at once, as it were. It is a rare

> feast! All the invention and making goes on in me as in a beautiful strong dream. But the best of all is the *hearing of it all at once.*[1]

On the notated page of music, art appears to be all on one level. For the average non-professional interpreter, and many professionals, a printed page presents a flat surface, that is, it presents no problems of what is basic or what is superimposed material. Without knowledge of harmony, or benefit of a facile technique, the music comes off the printed page flatly levelled; everything comes equal, harmonic progression, melody, superimposed harmony, the tune itself or the embellishments. It takes a player with both emotional and judicial capacity to see into the printed page of music. Moreover, even a player of outstanding technical and emotional ability will run after a vision of the written composition as a whole and, in so doing, leave out of his interpretation all the necessary justice due its smaller units. Especially will the matinee-idol conductor run after visions—and there is almost nothing to stop him.

On the other hand, with the folk (or near-folk) improvisers, their special creative process keeps every element naturally balanced and in its place. Theirs is a complete creative process, validly allowing them through emotion to change a good or bad piece into something else. It is in trying to recapture this creative surge that interpretive instrumentalists and conductors forget what the creator-musician never forgets, the actual values of the small unit. They lose it in seeking the excitement which lies behind it all. Virgil Thomson brings this out very clearly when speaking of Toscanini:

> . . . when one memorizes everything, one acquires a great awareness of music's run-through. One runs it through in the mind constantly; and one finds in that way a streamlined rendering that is wholly independent of detail and even of specific significance, a disembodied version that is all shape and no texture. Later, in rehearsal, one returns to the texture; and one takes care that it serve always as neutral surfacing for the shape.

And in another place:

> Poetry and nobility of expression are left for the last, to be put in as with an eyedropper or laid on like icing, if there is time. All this is good, because it makes music less esoteric. It is crude because it makes understanding an incidental matter. . . .

How well both the theatre and the "pops" concert hall understand this! The current theatre is designed for ready reception. Not much more cultural preparation is needed at our "pop" concerts. Thomson continues:

> Like Mendelssohn he [Toscanini] quite shamelessly whips up the tempo and sacrifices clarity and ignores a basic rhythm, just making the music, like his

1. Quoted in William James, *Psychology,* Vol. I, p. 255.

> baton, go round and round, if he finds the audience's attention tending to waver. No piece has to mean anything specific: every piece has to provoke from its hearers a spontaneous vote of acceptance. This is what I call the "wow technique."[2]

As in classic music, this same situation obtains when jazz or popular music is theatrically presented. If the music wavers, the rhythmic lapse is filled by the gusto of the conductor's baton! A lagging situation may have been saved but in so doing the more lasting satisfaction of a constant tempo has been irrevocably lost.

Obviously, there are two distinct attitudes towards music. One the creative attitude, the other the interpretive. Broadly speaking, these two attitudes represent two great developmental periods. The first period covers folk and advanced-folk creation—a period in which notation is a sideline, something not vitally important. Notation, in this first period, is *extra* to the period's musical experience. If possibly useful as an occasional reminder, it seems completely superfluous as a device for making music. Later, whether contemporaneously considered vital to creation or not, it is the success of adequate notation (usually achieved at the end of the first period) which determines the vitality of the succeeding period. The second period, for the creator, involves the immediate notation of every musical thought, the lining up of every fragment which can contribute to a composed work. His musical life depends upon print for presentation of his musical ideas. Moreover, in this second period, the conductor-interpreter emerges as a separate entity, resting his whole musical life upon the reading and interpreting of notated music.

The conductor-interpreter's musical self-expression seldom finds inspiration outside the printed page. A desire to see beyond the printed page, more often than not, leads to a distasteful boldness in the altering of tempo and dynamic indications (although the printed note values, since the interpreter cannot express himself creatively, remain obeyed to the last letter!). In direct opposition, a folk, improvising within its own period, will adjust its whole musical thought to a tempo! The music created is in perfect relation to this tempo. The pulse of an established tempo sets up a form into which is poured all folk musical thought and, when conditions are right, it is poured full and poured to fit. How different from the interpreter, who, although feeling the urge to create, is thwarted by lack of inventiveness and, since he must finger his way through the notes as written, will alternately rush and hold back, swell and subside, simply out of an impetuous desire to express himself. His emotion may come from as deeply within as that of the creative improviser, but the music is already set, and his only recourse is rhythmically to torture and twist it into a

2. Virgil Thomson, "The Toscanini Case," *New York Herald Tribune,* May 17, 1942.

new shape. This new shape may somewhat express his emotions, but the expression is an exercise in distortion rather than the substitution of a new identity. Because of his loss of understanding and of outlet in the art of improvisation, emotion only prompts him to mutilate that which he could never have written himself. Our most famous performers in the classic field speed through notated music without replacing their lack of creative ability with anything save their personal emotion, artless and generally unsuitable. It even seems as though they feel the *art* of music fetters them. The greatest music provides, and should provide, full enjoyment from measure to measure: it only needs enough *future,* or progressive activity, to keep moving. A great music is not a viaduct for anxiety sensations, a push rewarded by a climax.

Over a long period of time, two musical elements have crystalized themselves: harmonic progression and melodic progression. The folk musician, playing for dancers, improvises with both these elements well fixed in his subconscious mind. He acts upon them with varying intensities of conscious feeling. That is, he adds to this historical structure of definite harmonic and melodic progressions his personal emotional expressiveness and mechanical habits of musicianship. The phenomenon of a steady beat makes possible the perfect correlation of these diverse elements. The roots of musical improvisation reach far into the emotional, mechanical, and cerebral consciousness of man.

It is our good fortune, *now,* to be in the presence of the mystery of bona-fide creation—jazz improvisation. Later on, when the jazz period is over, our whole concern will be in reverting to records and the printed page, studying them, trying to find out how all the things there got there; unless, of course, those competent to formulate some idea will formulate it now, while jazz is still in its creative period and susceptible to immediate observation. Primitive and advanced spontaneous folk art creation, mystery that baffles critics and philosophers, is now accessible for critical and philosophical observation. Even those who wish simply to enjoy it should not take this music for granted. They, too, should realize that the creative process that produced Christian's solo has reappeared after a very long period of idleness. They should realize that once this present activity is done we may not be exposed to it again for an even longer time.

A Non-Aesthetic Basis for the Dance

Jazz Forum, September 1946.

John Dewey, speaking of primitive society, asks—"How is it that the everyday making of things grows into that form of making which is generally artistic?" Substituting *doing* for *making,* so that the question will have greater pertinence for the dance, we now ask: how is it that the everyday *doing* of things grows into that form of *doing* which is generally artistic? In answering this question I propose to establish a non-aesthetic basis for the dance—a basis of art achieved out of a community doing of things; a basis of *everyday doing* in response to everyday imperatives. Such a thesis must be established if dancing is to take its rightful place as a major art.

Let us divide the responsibility for art into three fundamental, mutually dependent parts, three parts which, it may be said, contribute to the creation of all great art forms, namely: non-aesthetic activity, mass participation, and subconscious aesthetic activity. Mass activity presupposes the need for the object, a need physical (concrete) or physiological (pertaining to the body function of living beings). That is, the folk *must* be occupied in something which is useful to them—something that will make them function in mass. Such activity can be the making of utensils or houses, the function of talking or the exercise of the motor activities involving play or stamping to rhythm.

If, at the very beginning of history, there is neither art activity nor any conception of art ambition, there is present, nonetheless, human ingenuity. There is the thought which produces the lever, provides shelter other than a cave, encourages the development of semi-articulate into articulate sound. But it is not until these needs are in actual process of being satisfied that the mass becomes well aware of participation in a new activity. It is only an instinctive simplification, or enlargement of ways and means to beautify the project already in hand that makes the mass aware of what was previously subconscious art activity. This awareness, firmly based upon its own non-aesthetic premise, can now be labelled *primitive aesthetic emotion.*

In other words, under favorable conditions, an art attitude is subconsciously developed in a people subsequent to their physical participation in an activity originally founded upon a non-aesthetic basis. It is the concern with various duties, the communicating of intention through language, the production of sound and movement involving pleasurable physiological release, which will, in the course of time, lead a people to become conscious of more than non-aesthetic need. When these various activities are well developed and the aesthetic feeling is contained in a tangible resultant the people discover a new dependable and quite unique pleasure, something that can be isolated and recognized as art emotion.

It is generally conceded that the basic element investing architecture throughout the ages with undisputed position as a fine art is its first function of giving shelter. It is not simple to follow the thought that, no matter what creative ideas are forwarded in any art medium, both their inspiration and check *must* issue from a primary and basic non-aesthetic foundation.

To understand how dancing is based on a non-aesthetic foundation, let us not think of the dance from the point of view of the onlooker. Let us create a sense of personal activity. That is, consider it with the sense of motivation familiar to any group engaged in an indigenous and contemporary folk-dance activity, whether it be minuet, mazurka, or fox-trot. This kind of dancing is dancing to satisfy a need. It is a craving which finds actual physical relief *only* in this way. In some historical periods it is a strong impulse, in others, weak, but *movement in rhythm,* if only a stamping of feet, satisfies a physiological imperative.[1]

There is beneath all aspects of dance the element of a steady beat. A steady beat or pulse is not an arbitrary or deliberate institution of the dance. There is good reason why we find it there. We cannot walk out of time. Anybody walking must keep time. A repeated movement which incorporates a good

1. This does not mean that by intellectually deciding to stamp the feet and do gymnastic exercises, an art will obediently emerge. Group participation in *one* person's mental or emotional activity, however temporarily absorbing, must not be confused with an art form arising out of the mass activity. Although there are many current systems in practice for moving in rhythm, we can observe that those systems decided upon by the individual or "little group" run definitely counter to that fundamental movement which emanates from the subconscious physiological urge of *any* folk.

portion of the body and is not either too fast nor too slow will fall into a manner of doing which will keep time. This is a phenomenon which is at the basis of all dance movement. It becomes an arbitrary act to keep *out of* time. Keeping time is both pleasurable and imperative. It gives form to variety and thereby becomes a vehicle within which expressive movement crystallizes, ultimately to become significant deportment.

We will notice that the two or three rhythms used by dancers are completely subject to a steady beat. There is no non-rhythmic separation. The pleasure in doing them is two-fold, the pleasure derived from a slight aesthetic inherent in a rhythmic step, plus the non-aesthetic pleasure derived from the constant drive of the over-all beat. Such dancing is a complete world to the dancer in which his thoughts are concerned solely with how he feels and the joy derived from treading out one or two rhythms to a steady beat. The place of deportment is a consequence of protracted indulgence, not an invitation to the dance.

A cursory review of any established dance academy readily reveals that the nature of its steps and movements are rhythmic. This fact exists all over the world and only in very advanced forms of the art do we see a retreat from rhythmic movement. This rhythmic movement, however, establishes great significance in an advanced folk period, though such significance is observed neither in the period preceding nor in the declining folk period subsequent to it. What is important, however, is that these rhythmic movements of the advanced folk type are of such a nature that they trace back to a period of pure motor indulgence. When the beat is the sole consideration, only then is it possible for the mass to participate. It is at such time that we find the genesis of significant movement.

At this time the consciousness of the fermenting aesthetic is practically non-existent. It is only later that an awareness institutes itself and much later that such aesthetic significance is handled as an identity apart from the pleasure of the motor action. Whether it be our over-developed ballet or any of the academized religious or secular dances of the Orient, the movements still carry with them an excitement due to motor action. In the academic period, such as ballet, steps and movements can be traced back to simpler ones. Yet when mastered to the point of flowing easily, these difficult steps will also give the participant an enjoyment purely physical in character. It is the dancers who can do difficult steps with the ease with which a child can skip, who not only receive but give the greatest pleasure. It must not be thought that motor action comes from jumping alone. Arm and body movements which repeat themselves as they do in the Orient give to the performer the fullest sense of motor action and feeling of rhythm.

There are other features to consider in relation to the conception of art emanating *from* a non-aesthetic basis. It is possible that the folk themselves consider this activity solely as art activity. It is possible that when the folk say they want to dance, they are looking forward to something that will give them aesthetic pleasure and not an exercise—a calisthenic. The folk, instead of thinking of their primary motor activity and accepting the aesthetic pleasure as a consequential corollary, think only of the aesthetic—the immediate aesthetic of which their dance-conditioned bodies are capable. On the other hand the folk onlookers themselves, as non-participants, only regard this dance activity as *art,* rather than as an exercise concomitantly beautiful.[2]

This attitude seems to be quite contrary to a people's attitude towards architecture. In architecture the use of the building is a primary consideration—its beauty is only secondary. This fact alone seems to suggest that these arts do not share a common non-aesthetic basis.

But is it necessary to contend that the crafts from which art emanates are all equally imperative to the human race? It is my contention that the art impulse becomes validly manifest *only* when it is founded upon a non-aesthetic basis. Art occupation is always carried on within a certain tradition, except possibly in the earliest historical periods. Those who function within any given craft also function within an aesthetic tradition because the aesthetic tradition is one with the traditional way of proceeding with the craft.

The outstanding difference between the non-aesthetic background of the architect as opposed to that of the dancer lies in the fact that the architect faces the task of solving obvious building problems whereas the dancer on the other hand, has no such obvious non-aesthetic problem to cope with. The dancer functions within whatever tradition he may have inherited, his pleasure deriving from motor action plus the traditional aesthetic. He makes no separation of this two-fold pleasure. According to his attitude, he will single out some one element that he feels represents the activity's attraction. No doubt, throughout the history of these two arts, architecture has always worked with the problem-to-solve and the dance has had no more than the primary inner compulsion to satisfy—a desire to learn and execute satisfying movement.

Society has easily overlooked the basic non-aesthetic sources of the dancer. It thinks of dance as aesthetic activity and although it seems to find no deficiency in a dance criticism not founded upon a non-aesthetic basis, yet it does not hesitate to acknowledge such a basis as a major consideration in the art of architecture. Civilization tends to restrict excessive exuberance and encourages personal reserve. Observation of child behavior reveals the child's constant recourse to highly energetic movement—movement functionally beneficial to his system. It is not uncommon to see any child stamp a temper away, keeping a rather strict tempo in doing so. In the child's stamping we can recognize the

2. The closest that the onlooker can come to motor enjoyment is through inner mimicry.

primitive physiological release so necessary to the later creation of rhythmic dance phrase. As we grow older society imposes restrictions on our conduct, and we are compelled to adopt *special means* to bring about a conciliation between observance of social propriety and our inherent necessity for physiological release. Curiously enough, society itself provides the way, through its facilities for sports, outdoor exercises, swimming, and social dancing.

In the exercise of traditional movements proper to the game, exercise, or dance we are participating in, the nature of the sport or dance limits the way in which we comport ourselves. The pleasure a football player gets in tackling is not only the accomplishment of downing a man on the opposite team. In a tackle he may throw himself at someone whom he is almost sure he cannot reach. He likes the exercise of the act. In tennis, on the other hand, except for a most extraordinary "try" in which he thinks he may be successful, the player never throws himself prone through kinetic impulse. Similarly, however exuberant we may be when we dance, we are restricted by the current dance decorum. Although the lindy hop can take care of any amount of exuberance on our part, we meticulously attune our exuberance to any other dance we may engage in.

Apart from establishing a non-aesthetic basis for the dance, this paper has been an attempt to prove that every art activity has its own propriety which is its art tradition. An effort has also been made to show how, through slow process, an art tradition is built and how on the basis of such tradition the public will accept art activity howsoever these actions may diverge from contemporary decorum.

In conclusion, it may be said that contemporary dance, whether it is the dance as personally participated in or the dance as a spectacle, is primarily thought of as art activity. On the other hand in admiring a building our first consideration is: this is something to live in. Although the contemporary dancer apparently thinks first of his delight in movement as an art activity and only secondarily receives the motor-action response, it is possible to imagine a past in which the reverse was true. Perhaps we can envision dance as a major art *only* when we have as a heritage an academy arising from and remaining close to motor action.

The Deceptive Nature of Sensuousness in Ensemble Playing

It will be my task to show (1) that collective improvisation, contrapuntal writing, and harmonic music in general do not constitute an advancement in art in the sense that the aesthetic becomes more complex, and (2) that such music carries with it in its very nature a pleasure merely sensuous in quality and similar to that experienced by a rich tone.

First we must consider the question of the simple aspects of sensuous quality in music and examine our attitude with reference to them. The tone of the instrument provides our most primary basis for musical sensuousness. No one has even been so careless as to state that the quality of the instrument is a determining factor in the art of music; yet, a composition might prove quite distasteful when executed on an instrument which, although perfect so far as pitch is concerned, has an objectionable tone. We can deduce from this that the perfect art is that in which we have a great composition executed upon a pleasing instrument. This does not suppose that these two entities enjoy equal importance. One is the expression of a great artist, the other the result of a great craftsman.

Although there are not many who will insist on attributing a very important rôle to the tone itself as part of the body of musical art, there are, on the other hand, those who consider a personal tone, a use of tones, the creation of musical line, rhythm, and the rest of it as one thing, no part of which can be considered alone. This brings us to the performer.

Between the instrument and the music itself there is much that goes to make up the great work of art. We have the inter-

The PL Yearbook of Jazz, 1946.

preter or performer who is a complex organism composed of many elements both sensuous and aesthetic. Depending upon the instrument, he can change its tone considerably. He can even produce a tone which can hardly be said to be the natural tone of the instrument. No matter how beautiful the tone he creates, it cannot be called anything but sensuous. He only adds to what may be termed the neutral tone of the instrument. He also may have various tones, as does a singer, which he uses at will and changes as he passes from one note to another. Such changes become his aesthetic contribution and can be carried to great heights. But his tone, however captivating, remains purely a sensuous contribution and though an important element, yet not to be compared with the aesthetic that goes to make up the art.

I could go into great detail with respect to the differences between what I consider the sensuous and aesthetic elements in the art of the improvising performer, but I am not concerned here with such distinctions. Quite simply, it is clear that a constant quality of tone can give an enjoyment that another tone will not. It is also obvious that variation of this tone, in so far as pitch is concerned, constitutes another enjoyment. Tone by itself, unvaried, provides a sensuous enjoyment; variation of note an aesthetic enjoyment, since the latter is conveyed through melodic design.

As soon as more than one musical line is introduced an auditory confusion is created. The addition of more voices (melodic lines) will bring us to a point where we will hear only one sound accompanied by ragged edges. If carried to extremes, the multiplication of voices will eliminate this raggedness only to create a complete confusion, but one in which we will perceive a texture of sound that in itself becomes sensuous. It is this sensuous sound derived from the texture created by many voices that we must analyze. Though it occupies a position apparently on the borderline between the sensuous and the aesthetic, we must also consider the state of flux that it truly remains in. By this I mean that it all depends upon whether there is clarity in the inter-relationship of voices or whether there is just an unbridled mixture of them.

As the voices are never in complete relief at all times, we always have a situation of partial confusion even in the most perfect part writing. We can assume then that in such cases we have a canopy of sensuous enjoyment with constant protrusion of the voices as our aesthetic element. The less apparent the path of the voices, the further are we removed from the aesthetic. We depend, then, upon the sensuous attraction for any musical meaning. If this were not so, such meaning would disappear in proportion to our losing sight of the voices.

We cannot receive very clearly more than one voice at a time. A great deal of the perfection of counterpoint is missed by the listener. I have in mind the perfection sought after in classic music with respect to the individuality of the voices. In ensemble playing we have the genesis of counterpoint. The quality of each instrument facilitated its distinction among the other instruments. In counterpoint on a keyboard instrument, the knowledge of whether a voice kept its own melodic line or not became the knowledge of the musicians rather than of the listener, and even then recourse to the written music was needed to certify this fact.

In the mingling of voices there is evidence that each voice at one time or another protrudes. Our attention, however, is so taken up with the voice in prominence that it becomes a task to keep aware of the path each voice is taking. In composing it is a purely pedantic procedure to obey these tenets, and although it may be good discipline for the composer, it does not communicate to the listener. Obviously, in ensemble playing there is much greater possibility of voice distinction than on a keyboard instrument. There is not, however, the complete clarity that one is given to suppose. What we get, instead, is a richness of tone and polyrhythm that tantalize, rather than a melodic clarity that reveals itself.

Counterpoint which has no directive soon becomes monotonous. Even a most intense business in the ensemble playing palls after a time. There must be a leading motif, something which has a singleness about it and in this single state describes a pattern that is simple. We get this in trumpet leads. They, however, play straight, so that our focus of attention is dominated by *what* they play. We can deduce two things from this. One, that there must be a certain singleness in art. There must be predominance. All else has its presence in relation to this singleness. Our other deduction is that the character of the piece is greatly conditioned by the lead (this predominance). For this reason the lead must have the greatest art significance in tune content.

To further elucidate my text let me quote John Ruskin's description of ornament:

> If the ornament does its duty—if it *is* ornament, and its points of shade and light tell in the general effect, we shall not be offended by finding that the sculptor in his fulness of fancy has chosen to give much more than these mere points of light, and has composed them of groups of figures. But if the ornament does not answer its purpose, if it has no distance, no truly decorative power; if generally seen it be a mere incrustation and meaningless roughness, we shall only be chagrined by finding when we look close, that the incrustation has cost years of labour, and has millions of figures and histories in it and would be better off being seen through a Stanhope lens. [*Seven Lamps of Architecture*]

The quality of a texture that is on the borderline between the sensuous and the aesthetic must be understood for what it is. It is not so much a case of trying to make extra close distinctions. It is more a pointing out of how we do feel about something which, as Ruskin says, is a little roughness when seen at a distance, and distinct composition on closer inspection.

Pursuing Ruskin's method, if we analyze musical texture, and find that each voice is charged with musical content, then we can rightfully be annoyed that it has not been disentangled from the mass. In this state it contributes *only* as a sensuous element whereas it should rightfully be brought out into the open as a robust aesthetic. Ruskin was not annoyed with ornament that had definite composition of light and shade and which upon inspection offered a meaningful detail. Likewise, in ensemble playing in which we are vividly conscious of one or two instruments at a time, we must not be too chagrined to find that after absorbing our attention, the voice or voices disappear into the general texture while others demand our attention; or that, though disappearing, they continue as potently as when absorbing our attention. In other words, in the case of ensemble playing we must consider the players themselves and grant them a musical line, which, although not always apparent to us, gives them a vehicle within which to function. We cannot expect good playing if we treat them as soloists for one or two bars and then relegate them to the secondary position of accompanists.

We cannot approach the subject of improvised ensemble playing with carping criticism. I am not taking a lofty attitude toward the folk, admiring them on the one hand for what they do, and on the other keeping them in their place with the intention of using them patronizingly later on as a spring-board for greater art activity or significance. It is rather that I want to single out what I consider to be valuable in their contribution to art and justly appraise the aesthetic when it does appear—not persuade myself that the greatest art enjoyment is attained when we are indulging ourselves in its sensuousness.

We are not melodically discerning when we listen to a great deal of counterpoint. Although acute listening singles out some of the voices, it is the fact that there are many voices which give us satisfaction. The pleasure is similar to that derived from the make-up of an instrument's tone. The tone of an instrument or the extra tone given by a player is not a matter of some indescribable quality in the tone of the instrument itself. The violin tone A can be approximated by three tuning forks pitched at A, its octave, and the E above. Some instruments are more complicated than others and in the hands of a player become even more so. The fact that we can break down the total qualities into simple or more basic patterns does not mean that the complex is intuited in its component parts. Even on the piano, where it is much easier to distinguish the make-up of various chords, we do not derive an aesthetic pleasure by being cognizant of complexity. The atmosphere in which we are so luxuriantly bathed almost precludes interest in complexity. We lose discernment because we are so completely taken in.

It is only ensemble playing that may permit aesthetic discrimination—that *may* allow a knowledge of what the voices are doing. We can single out one voice and follow it through. We may do this with any one of them. We can, therefore, become aware of the musical structure, so to speak. This awareness is not perceived as an aesthetic unity but as a result of voice-by-voice analysis. We like to have an art so replete with ideas that we keep discovering new beauties. There is a difference, however, between discovering what was there all the time and a deliberate process of blanking out the rest of the music in order to single out one voice—in order to discover whether or not it has some hidden beauty.

The interweaving of many voices does not make a composition artistically complex. There is no augmentation of the aesthetic. The constant addition of one voice after another presents a complex structure, but too much complexity clogs our apprehending powers. We may pursue two voices if they are justly woven, but anything more is bound to result in a sensuous pooling of all voices.

Ensemble playing can be said to be the crucible in which the hot solo is born. Extraordinary variation is not natural for an improvising soloist. If he is to play variations on a tune, the tune will be uppermost in his mind. He will not easily stray far from it. He will rather fashion his variations in such a way as to prevent the repetition of the tune from becoming too monotonous. Except for the lead and bass, most of the instruments have considerable freedom. The lead ringing in their ears inspires them to cast about for an expression of their fullest imagination. It is this which is responsible for the creation of new melodic lines. But for the greatest penetration on the part of the listeners, most of this creation resolves itself into a total effect of sensuousness.

The step that jazz has not taken is the use of the isolated hot solo as the basis for collective improvisation. Not every hot solo can be said to be a basis for collective playing, as in most cases it is only a variation of a tune already familiar, and does not acquire a real new tune identity. When the player does go far from the tune there is usually a paucity of tune content in his solo. It becomes more a vehicle for a soloist or an interlude within a composition. There are very few solos which have in their own right the quality of a tune, but they do exist and would make great foundations for ensemble expansion. They could be played practically straight by one instrument while the others again cast about for more undiscovered beauties.

The traditional trumpet solo in *Dipper Mouth Blues* is such a solo. So also is the trumpet solo in *East St. Louis Toodle-Oo.* The situation would be this: that whereas the beginning was a school of ensemble playing based on a simple tune, a further development would be the isolation of one of the voices (which the hot solo is today) and *its* reintroduction as a basic tune. It then would be exposed to further ensemble treatment and finally we would have a purity of tone and variation that could withstand the severest pruning and yet leave all the voices so situated that they would appeal to us aesthetically rather than sensuously.

The Place of Space and Time in the Dance

In a serious consideration of the dance as a major art we are forced to make some sort of definition pertaining to what *is* dance. Too many activities come under the heading of dance that are not dance. If these activities use the name of an art to which they are only remotely allied, they distort greatly any theoretical discussion of that art. Further, if they are mixed with the art in question, we find ourselves occupied with correlating a foreign material, a practice which is exceedingly disruptive in an art analysis. If these remotely allied activities are great art then they must have a non-aesthetic basis of their own which would have to be separately analyzed before any discussion of their relationship to the dance itself could take place. These other remotely allied activities such as expressive movement (Duncan adagio and modern movement) and choreography may be popular but unless they have the three fundamental elements mentioned in my article "The Non-Aesthetic Basis of Dance," I cannot see that they are anything but minor adjuncts to the real art of the dance. Certainly expressive movement and choreography as used today have in themselves no non-aesthetic basis.

An understanding of the *time* and *space* elements in the dance will greatly help towards a definition of "what is dance?" There will also be an investigation into the fallacious idea that the dance is composed in the same manner as music and a discussion of the adaptibility of music to extend in time and the inability of the dance to do so.

There is a simple categorical division in the arts according to whether they oc-

Jazz Forum, January 1947.

cupy *time* or *space.* Music to exist must occupy a length of time, therefore we call it a *time* art. A painting must occupy space to exist therefore we call it a *space* art. As dance is something we see occupying space we can call it a *space* art but when we consider that dancing is movement we find that to realize movement a length of time is involved so that we find dance is a *space-time* art.

For the painter a given space and how he arranges his lines, forms, and colors within it is his concern. What he does is in keeping with what he sees and it is only through his seeing that he manages to correlate whatever *is* there. It is this same manner of seeing that makes possible the enjoyment of this specific arrangement by an onlooker. The musician in much the same way is occupied with tones and their linear arrangement. It is through his ears that he is able to balance the arrangement and it is through the ears of the listener that the enjoyment of this specific work is received.

The dance as a *space-time* art presents a far more complicated problem. Not because it incorporates *space* and *time* but because there is a situation in which the dancer himself has a different attitude, in the early stages of his art, towards his own *space* consciousness than he does in the later stages of his art. If he were manipulating marionettes he would see them with his own eyes and would be external to them. In a sense he would share audience perspective. Such a situation, however, is not the case.

A dancer in the primitive stages of dance development is concerned only with his movement as he feels it. He feels the movement which is the *time* element. His consciousness of space has nothing to do with its outward patterns, the pattern which is of interest to the onlooker. It happens that what he feels *has* got pattern significance and thereby interest for the onlooker. Therefore, when using the categorical distinction of *time* and *space* we may safely say that the dancer in this connection is not concerned with *space* patterns but is only concerned with the movement or *time* element in the early stages of the dance.

Through the years of development of his art, a consciousness of these outward patterns occurs. The dancer's consciousness of this outward pattern as *space* element of the dance depends on the extent of development he has attained. As he acquires this pattern consciousness he will compose his movements with the conscious understanding of what his outward pattern *does* carry to the onlooker.

We can say then, that the dance for the onlooker is a *space-time* art, the whole aesthetic phenomenon resulting from an object in motion and readily perceived through the eyes. For the dancer we may say that it is the "felt" movement which is his concern and that out of this "felt" attitude slowly emerges a consciousness of what his pattern actually is. The full consciousness of the dance as a *space-time* art is a question of a growing consciousness within the dancer, a consciousness of what his outward pattern does look like. It is the "felt" dance that is his primary concern and which later becomes for him a pattern consciousness or a complete *space* and *time* consciousness.

It is important to the understanding of the place of the dance in society to understand the nature of "felt" dance as experienced by the primitive dancer and the growing consciousness of body pattern in the development of the dance. What we may do *with* it in the way of significant entertainment depends to what extent the dancer *has* accomplished a body consciousness. Whether or not we present the dancer "as is," wholly unoccupied with the visual aspect of his personally "felt" dance, or present him within the confines of a precise choreography depends on the stage of his development as a dancer.

The disparity between how the artist feels when he is creating and what he and others experience when viewing his art is unique in the dance. It is therefore important to know how little a folk dancer does think of how he looks and for us to know to what degree he *is* conscious when he does consider his outward appearance. The *space* consciousness must grow within the tradition and not be applied from the outside. We can *only* work through his already established consciousness and must not overstep its degree.

The complete dance cycle involves: first, the human imperative to movement-in-rhythm which satisfies the dancer's physiological need; second, his creation of the rhythmic attitude to movement and his enjoyment in its performance; third, his awakening to the significance of his body in *space* through watching others make the same movements. Finally, his conscious projection of short rhythmic movements and static pose. It is from elementary repetition of short primary movements originally springing from a physiological impulse that we arrive at dance form as a complete *space-time* art.

When considering the dance as a major art, it is important to remember that it is not a *time* art in the same sense as music. For those who believe there is no limitation to the development of dancing in terms of *time* art, it is important to know how it differs fundamentally from the *time* art of music. The essential difference between music and dancing is that music is *tone in rhythm,* while dancing is *movement in rhythm.* In music, both the tone and the rhythm are susceptible to mathematical analysis, while, in dancing, accurate position in *space* being undeterminable, only the basic rhythm is susceptible to analysis. That is, although we may seriously attempt to recapture a pose of the body, the actual repeated projection is never and can never be the same. An identical pose cannot be struck from a subconscious memory of the previous pose. An approximation is effected only by re-creating body stresses and pulls. If we wish to duplicate by voice the tone E flat, we can do so, either relatively or absolutely, owing to the phenomenon of memory of pitch. There is no such mechanical duplication or accurate memory process familiar or possible to dance art.

But embodied in a soloist's movements is the feeling of a whole people. The composition is fragmentary but the fragments can be charged with high art procedures. The many points of repose resulting from rhythmic stress, though only lasting a fraction of a second, are poses of dance significance. Such primary movement and its points of repose comprise the stuff which can crystallize into a major dance art.

It would seem to be a waste of time and energy to emphasize choreography since it can never capture the space actuality of the great "felt" dance. Unlike music in which even a child can play a great theme without detracting from its greatness, choreography can never re-create the greatness of truly great dancing. Music, which is purely a *time* art, may be so annotated as to embody the real character; whereas dance being both a *time* and *space* art may capture only the mathematical *time* element because it is measurable, it can never re-create the space element because there is no way of measuring exact movement in space.

Although this "felt" character of the art does not impair what is known as dance, it does, therefore, cancel all efforts towards a development of dance art similar to the *time* art development of music. That is, important development in dance composition or choreography comparable to symphonic development is definitely limited. Unlike music, dancing can never completely divest itself of the mannerisms inherent in a folk, or for that matter within a creative dancer. The art of music can advance in complexity and become unaccessible to the folk. It can strip itself of folk intonation, accent, and way of playing and still carry along into its new domain a vast residue of notable art material. True, the lost folk rendering deprives music of one of its most significant elements, but nevertheless, this element is not intrinsic to music's advance into a more complex and advanced compositional structure. Advanced compositional structure is not carried forward on traditional folk improvising habits alone. Music can divest itself of that folk material which cannot be carried along and must travel ahead only upon those folk principles which have lent it notational significance.

Let us consider a tune. A tune is a conventional thing. But so simple a thing as a folk tune represents true *progress* from beginning to end—a thing in which the whole can be recognized in any small section of its parts. The parts, however, never remain static but always lead *into* each other, the termination being the end of the composition. This is a structure all dancers should examine closely. A folk tune is a structure built upon a phrase, usually four bars, which is repeated and interwoven with other material so that beginning, middle, and conclusion are not only vital parts of the tune as a whole but are, in fact, *the* tune. Herbert Spencer in speaking of music says:

> The simple cadence embracing but a few notes, which in the chants of the savages is monotonously repeated, becomes, among civilized races, a long series

> of different musical phrases combined into one whole, and so complete is the integration that the melody cannot be broken off in the middle, nor shorn of its final note, without giving us a painful sense of incompleteness.[1]

In listening to a tune we experience a sensation that starts with the first progression and keeps us in a state of receptivity until the last note. This is because the component parts of a natural tune have their definite place.

In a new tune we look, simply, for melodic progression that is at once both familiar and strange. The familiarity will be the *style* of melody and the strangeness will be the substance that marks this particular departure. In other words, tunes of the same school follow a common basic formula, so that when we hear a new tune it is partially familiar during its whole course of progression; we know it in the light of the other tunes of that school. Our structure of a tune took hundreds of years to evolve, and it is this strength through slow evolution which has made a tune the heritage we can so significantly vary without recourse to revolution. But, though we can vary the tune itself, we can never take the closing phrase and arbitrarily move it to the beginning. Above all, in music, the same phrase may be found in any number of different tunes, but not only does this phrase *not* stand out as a unit by itself, but, on the contrary, becomes so much a part of each new tune that it passes unnoticed in itself.

In this sense, a small phrase in a dance composition never becomes a part of the whole but always remains merely a phrase in relation to other steps added to it. Every musical phrase carries its own internal necessity to stand in relationship to the whole but any dance phrase may be externally manipulated. If in a dance performed to music we do feel a beginning, a seemingly built-up middle, and an ending, this feeling of completeness in the dance as a whole is purely a musical feeling. The progress of the dance composition is merely the progress of the music to which it is danced. It must be constantly borne in mind that in dancing, although there are many different combinations of steps, each single step seldom takes more than two bars of music. A more extended figure than this is merely the arbitrary addition of other steps in two-bar combinations. This is definitely not comparable to the flow of a sixteen-bar tune, since between each group of two-bar steps there is no integral combination. True, when we are looking at the first bar we foresee the second, but, when this second bar is completed, our foresight ceases. We cannot possibly anticipate the arbitrary turn of mind of the dancer. A collection of his movements provokes no progressive anticipation, since the new steps, unlike the new melody, are not a *progressive* variation on a familiar form.

A dance *step* is the smallest unit in the dancer's art and has none of the completeness of a tune. When the dance is divorced from music there is no feeling of unity in any combination of steps that gives us the sense of *progressive*

1. Herbert Spencer, *First Principles of Philosophy.*

familiarity we get from any tune. When we hear any part of a tune we know what it is, and at what point we are in its progress; but, in reviewing any part of a dance, we see merely so many things strung together, all familiar in themselves but collectively and necessarily producing nothing that can be apprehended as a whole as a tune can be apprehended. We mistake the drama or emotion added by the dancer for an actuality of progress in dance composition. So-called dance composition is no more than dramatic juxtaposition of different movements and poses arbitrarily assembled according to the dramatic powers of the choreographer.

Whatever space art we may choose, statuary or two-dimensional representation, there is no sensorially felt interval. Therefore, no scale can develop. The variation is infinite whereas in music the variation, once we have a scale, is limited in pitch. There are bodily felt stresses and pulls which correlate up to a certain degree the space elements of the dance. They never acquire sufficient precision to enable them to establish a scale. Neither the interval within the pose of the body nor the line of movement taken by the body come within the measuring scope of a "scale."

To understand why there is no feeling for precise movement or even precise body posture we must look to an art that does have this phenomenon. In music we will find our answer. There is a scale in music on which the music is written and there is nothing of this kind in the dance. We will notice that music in its most primitive state has a scale while the dance only has movements allowing for great latitude in their execution. It is only later in the dance that certain positions are approximately fixed and arbitrarily adhered to.

Music has a scale for a very definite reason. The harmonic upper partials of a note are agreeable (consonant) according to their place within the scale of these upper partials.[2] For this reason the notes even in an arbitrarily devised scale would be bound to have different degrees of consonance. It is the ear's sensitivity to pitch that enables it to hear a degree of consonance which makes any interval become recognizable. Music is built on broken intervals.

It is through our sensitivity to pitch and to the degree of consonance in an *interval* that it becomes possible to create and recognize a positive *relationship* between musical sounds. Harmonic *relationship* is directly felt in a sensory way

2. The harmonic upper partials are the notes we hear above a note we may sound. If C is sounded the notes we hear are

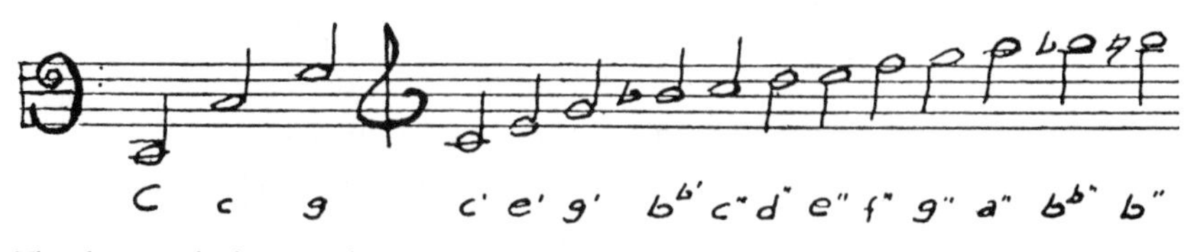

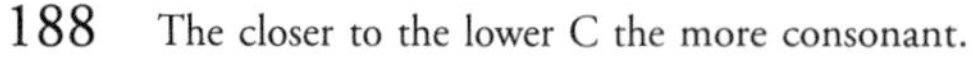

 The closer to the lower C the more consonant.

not in an aesthetic way until after the introduction of a second *interval.* In a space art two points make an *interval* but three points are necessary to create a *relationship.*[3] The *relationship* is felt aesthetically but is not experienced through the senses. We may see the *interval* in space and aesthetically feel the *relationship* between three points but no sense is impinged upon by the points in this *interval.* In music the aesthetic reaction derived from broken *intervals,* that is, a melodic line, is fortified and fixed by our sensitivity to the degree of *consonance* in these *intervals.*

It is owing to the lack of *actual progress* in *time* that the dance, as a *time* art, remains in an unorganized state throughout its popular existence. In music we say, "play a tune," and both folk amateur and trained musician alike will render a complete *time* art. They do not think of anything less, such as two or three notes or a phrase; for a tune is a tune—as precise as it is complete. But in dancing even the terminology is vague. We speak of "doing a step," and, although this is precise enough if we mean a one- or two-bar sequence, when we say "do a dance," this, to the folk dancer, turns out to be no more than an invitation to "do a step" or to do this step together with *any* number of other steps popular at the time. Combinations are then repeated either until the music stops or until the dancer is too tired to continue. Although, from the physiological standpoint, a folk dancer can dance with pleasure all evening (conversely, he will not receive any physiological reaction unless he dances for some minutes), from a *time* art point of view, his dancing is simply continuous exercise wherein actual dance unity is confined to isolated dance steps, or units, and the end of the music is the cue to stop.

Rhythm alone has a short time limit and a connecting of different rhythms is purely arbitrary since there can be no homogeneous or integral flow. The time element of music rests solely upon the fact that tones can be accurately measured into definite pitch and then further crystallized, by means of rhythm, into the sure and long sequence known as a tune. Dancing, not being definitely posited in space nor as in music setting up any inevitable vibrations in space cannot actually bridge one rhythm pattern to another. The two dancing bars are the closest, in fact the only equivalent of a sixteen-bar tune. It is obvious that two bars cannot provide enough projection opportunity to satisfy the onlooker, either in *time* or *space.* For the onlooker, the *time* element is a quibble and the space element is snatched from view.

3. For example, between A and B we can have no relationship, only a straight line

• C

A ——————————————————————— B

When we introduce the point C we establish a relationship because of the fact that now there exists a ratio between all three points which if enlarged is not altered.

In so far as dance music is made of short tunes to accompany the dancer's art it is not listenable music. Sixteen or thirty-two bars is not enough to satisfy the musician's ear. Occasionally the device of variation does render dance music listenable. (*See Note at end of essay.*)

Although large musical works are to a certain degree arbitrary, in that each composer must pattern his composition himself, they immediately take on, after a few hearings, a homogeneous completeness which embues them with a quality that makes the work so completely perfect that, although we know the composer and know that he must have worked over them bit by bit, nevertheless, it seems as though they were created by a magic wand, crystallized instantaneously into the perfect beings as we know them. There is an inherent quality in the tune itself which makes it possible for it to become integrated into a musical whole, whereas no dance step ever becomes homogeneous with other elements within a dance composition.

Except for convenient patterns, such as using a dance step so many times to do one thing and so many times to do another, as in some communal rite or pageant, the dance is a continuous affair similar to a walk and there is nothing in this continuous affair that can be regarded as coming under a unit control. The convenient pattern that I mentioned before is in no way an aesthetic unity as is the unit of an eight- or sixteen-bar tune. It is purely convenience.

For the dance, then, it is a continuous beating out of one or more basic rhythmic patterns. The dancer is no different from a walker who continually walks around and around in an open space using the steady pulse of walking to get about. So the dancer using a step or steps of a rhythmic pattern also goes about a given space. As this mode of movement has a great charge of the aesthetic, he need not cover so much space but can exercise himself in a small space due to the various combinations of the one or more rhythms. As his movements take on significance and through his body (the *space* element) can transmit to the visual world the subtle rhythms he is doing, an onlooker can watch with a keen aesthetic interest the continuous repetition of these few basic rhythms. Through this means we get a few basic rhythms and their bodily counterparts continuously presented in various combinations, various aesthetic nuances, and brought before the eye in various positions.

To establish any compositional integrity beyond that found in a few basic units is arbitrary and has no basis in the folk or primitive dance. The dance is definitely limited and any sequence of steps, which comprises the *time* length of the dance, is arbitrary and not a natural art development. And the isolation of any unit in the dance (such as two bars) for the purpose of presenting it as a total work of art is completely unsatisfactory as an art experience. Although for purposes of presentation and preservation choreography is necessary, we must always remember that the highest significance of the dance is in the natural state of continuous doing.

Note: As we can extend the tune structure to a much larger but looser structure, so might it be possible to extend the two dance bars to that of sixteen, and in this sixteen feel a state of completeness and unity. This sixteen, however, would only be similar to the extended musical composition in that it was a loose structure. It still is too short for complete satisfaction. Either variation of lyrics or accompanying dance detracts from the monotony of a repeated tune to which we may listen many times without boredom. No such repetition is acceptable in a dance of sixteen bars unless it is part of a game such as a Virginia reel or the like.

Alicia Markova and Alicia Alonso

Two outstanding exponents of great ballet dancing extant today may be found in the dancing of Alicia Markova and Alicia Alonso. Although these two dancers differ from one another they are both real dancers and not just well-trained performers. They both have a performer's rhythm, that quality which makes a sparkling dance and the lack of which renders lifeless the best performance. In studying this rhythm and trying to discover the elements of greatness in it we see that both dancers uniquely distort the basic rhythm of the dance by either lengthening or shortening certain steps within a step unit. They get this by acceleration and slowing according to their own interpretation rather than by following in a stereotyped manner the set rhythms of the dance. Or again they may strike a static pose for a fraction of a second in place of continuous legato movement. This rhythmic distortion is more effective in Markova because of her fast technical work. In her poses, because of her extraordinary sense of balance, a sustained rhythmic feeling is felt.

Markova is a representative of the traditional Russian technique—a technique of the classroom, and a technique which is both a repository and conveyor of all we have that is important to dance art. She is quite removed from the present-day attitude which stems from Fokine's influences. Her singularity is such that in evaluating her dancing we must consider two important elements. First, unlike her contemporaries she is an exponent of the great Russian technique; second, she is a great dancer.

The airy lightness that she possesses

The Ballet, March 1947.

when contained in a long tutu creates an illusion of thistle-down. Her technique supports this quality at all times, even during the execution of the most difficult steps. It would be hard to abstract "thistle-down" quality from the total dance to estimate its significance. In some of her dancing there would seem to be a certain lack of intention in her interpretation of choreography. Despite this fact she is a convincing dramatic pantomimist. There are times, however, when she achieves true and total mastery as a performer.

There is a remarkable style in the manner in which Markova handles her body. In order to understand the import of what she has done it is necessary to first appreciate the handicaps under which a ballet dancer works. In a training, such as ballet, only essentials can be translated from one generation to another. It is learning by rote. Markova inherits, as does every ballet dancer, the ballet archetype handed through the academy from one generation to the next. She does not, however, resort to over-humanizing this cold archetype. Nor is she merely a cold classroom dancer performing with meticulous care. She has borrowed from a past which did not come to us by rote but which was captured in graphic prints of the day. This past was a great period of the dance, a period when the pose and body deportment *was* part of the schooling. It is the period of Taglioni that Markova stems from. Taglioni was a famous model to the great artists of her day. The result is a gallery of dance position, a graphic repository of the way a great school of dancers "filled in" what has become the "gap" between classroom and stage. Markova, by her deep identification with the Taglioni print, has added to her other positive qualities a creative insight into the use of a ballet pose. We see in every movement and pose the greatest intention on her part when dancing in the style of Taglioni. We see a dancer with marked intention *doing something* with her body. Markova in her Taglioni ballet has given the classic pre-romantic ballet its greatest significance. It is a rare condition that the dancer best fitted for the task had the desire to fulfill it.

A desire and an ability to recapture the Taglioni period would have been no more valuable than a print come to life, had we not had a dancer with the rhythmic ability of Markova to make it more than mere successful research. Her dancing is a vital presentation which, like all great performance, has the quality of immediacy.

Markova has a performer's rhythm which *is* dancing. In the nature of the dance we do not get rhythmic refinements as we do in composed or written music. All of the steps have the simplest basic rhythm, a rhythm that is easy to acquire. In such a simple step as *glissade assemblé* we see Markova, in the galop finale of *Pas de Quartre,* give great meaning to the *assemblé,* accomplishing the position of the *assemblé* in the air with bent knees before she lands. This gives a fraction of an extra movement and adds to the rhythmic sequence thus making it more than a simple *and–one—two.*

Alicia Alonso is the better "all around" dancer. It is not rhythmic technique alone which makes Alonso great. We can see running through her whole body a unity of expression that correlates the torso, legs, arms, head, and even eyes and makes her something more than a choreographic pawn. Her interpretation of choreography is interesting and absorbing. A choreographer's work can "come out" pretty flat and pointless unless revivified by such a dancer. Alonso gives any movement a poignant meaning. She does more than just present mass and outline. For instance, if she is to turn her head to the left she will cast her eyes down modestly or stare in that direction. In her sudden stops, either the flow of movement continues from her body immediately after the pose is struck or a complete static envelopment is felt in contrast to the previous movement.

Alonso removes interpretive choreography from the limited inward-felt emotionalism of many good dancers and lifts it into the realm of external spacial relationship. She does more than try to put "feeling" into an arched neck or raised arm. She gives a rebirth to any material handed her. Alonso shows that she has intention in doing a choreographic part and vitalizes the thing as seen. She does more than go through all the conventional and unconventional movements with those sudden terminations in pose; she reveals an intention behind them, an intention that comes to the audience and creates a complete feeling of rapport with them. The variety in her poses shows a fertility of dance thought only possible through the inner urgency within her. It is a dancer such as Alonso whose interpretation implements the choreography to a far greater extent than a great musician implements the notated score.

Alonso never loses the awareness that what she is doing has style nor does she use ballet's stereotyped movements as though they were natural to her. Rather does she play with her classically trained body as though she were playing upon an instrument. She manages to keep a dual character: *using* her body as an instrument and at the same time never losing her human directness of intention. This keeps her on a level of great art.

Alicia Alonso is a great dancer although she is somewhat uneven. It is the complete giving of herself at one moment and a detached playing with ballet tradition the next which makes her so outstanding in using the ballet technique of the post-Marinsky period. And these qualities, so apparent in Alonso, are necessary to make ballet a vital and rewarding experience. Were these qualities always present in all dancers, ballet might become as important a part of our aesthetic experience as the picture gallery or concert hall.

Thus, both Markova and Alonso each in her own way bring to the ballet dance something vital, something much more than a competent portrayal of a part. Each one does more than maintain the level of a solo dancer and gives us an exciting and satisfying experience.

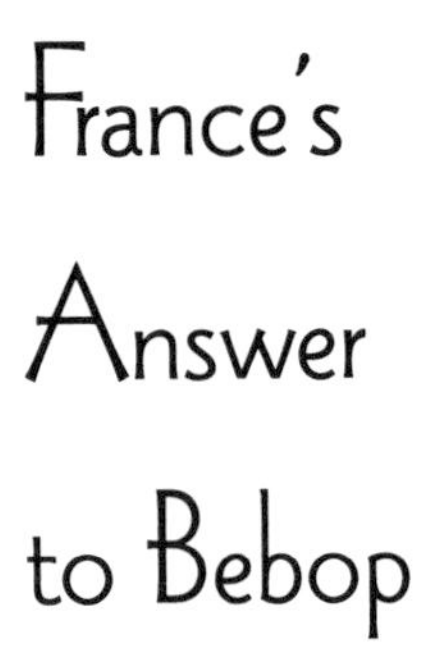
France's
Answer
to Bebop

Until it happened I never would have believed that there could be a revival of early jazz such as we are now experiencing in the "kid bands." I expressed this attitude when I was first given the Lu Watters records to review. At the time I had been going along with what I felt were the most significant aspects of a "developing jazz." I did not feel at the time that "going back" would make any contribution to jazz. For that matter I felt it was impossible to recapture the past when I said, "To try to play as he [early Armstrong] did without sounding in any way like anything since, is . . . futile. Anything with real punch is going to have an element of something beyond 1923." In speaking of Paderewski's and Kreisler's hoax when they attempted to reconstruct the past, I said: "You can be sure the more they would have done so, the less inspired the result, and if inspired the more apparent the hoax."

For some critics Lu Watters's first recordings represented the best jazz since Oliver. Whatever his artistic accomplishments he was certainly the forerunner of what has come to be the greatest thing since early jazz.

Except for the "comeback" of the early men in general and the records of the Bunk on the AM label in particular, jazz had certainly very definitely come to an impasse. Since the advent of this "comeback" I have found that I could not go along any further with what might be called a "developing jazz" and I have felt definitely that I had already gone along with it too far.

Since the early days of Watters a great many new bands have sprung up with

Playback, June 1949.

what may be called collector-players. Such bands as Bob Wilber and His Wildcats, Frisco Jazz Band, Century Stompers, Claude Luter, Claude Bolling, and Graeme Bell Jazz Band represent the solidifying of a new era in jazz. Because of the preservation of early jazz on records that actually reveal the quality of early jazz more than would ever be done by mere annotation, collector players have identified themselves with this great period.

Jazz went ahead too fast anyway. Early classic music did not hurry itself to its inevitable end in such manner. It lingered. The tempo in America, out of which jazz sprung, sped it ahead so fast that it burnt itself out. Except for those who believe in bebop, this is a fact.

I have always believed, and still do, that jazz can move into notation. This has nothing to do with arranged jazz as we know it. Jazz has deteriorated so fast since the 1930s, when it was first recognized as an art, that had notation occurred it would have been of little significance. With the appearance of this return to early jazz and the serious attitude of these players who feel the significance of the best of jazz, the way may be opening up for notation in the future.

Playing recorded solos, note by note, as Claude Luter and Pierre Merlin do, is a first indication of this. This deliberate and intentional re-creation of specific solos is a departure from the usual haphazard method of imitation, such as in *Dippermouth Blues* and *Tin Roof Blues.* Specific instances of this deliberate and intentional re-creation of solos may be found in Merlin's rendition of Tommy Ladnier's *Graveyard Dream Blues* solo and Luter's various Dodds solos.

Although Luter's band is capable of creative improvisation they are not afraid to become the interpreters of another's creation. It is no longer beneath a good musician's dignity to repeat a solo note by note.

But, it may be asked, why is notation important? Pure improvisation can only go so far. Guidance, alone, can save jazz from premature decline. Whether the greatest period is before guidance, or after, does not enter our discussion. This much can be said, however: Group improvisation made use of mass creativity. For jazz to survive, the individual artist will have to assume greater responsibilities. If he uses the fruits of this mass creativity he may so integrate it as to give a greater depth to the complete musical composition. Whereas before, an individual could only say his part in a solo or in contribution to an uncontrolled ensemble, now his powers are extended. A portion of the music of such a late contrapuntalist as Bach was sometimes not much more than integration. It is for these reasons that I believe notation is important and will aid in the resuscitation of early jazz and in the future development of the art.

Although music in primitive societies seems to be able to exhibit a constant change without deteriorating, the whole history of Western art has been to go through a cycle of birth, development, fruition, and decline. Unless young musicians can surpass New Orleans their music will decline, and unless they have recourse to guidance and notation they can never surpass New Orleans.

When I say "surpass" I refer only to the aspirations of the musicians and the expectancy of the audiences at the time. Just as King Oliver and Bunk felt that they surpassed a King Bolden, so did Armstrong feel he surpassed them. Every artist must have this uppermost in his thoughts. Mozart felt it and Beethoven after him. Even Chopin no doubt felt progress over a good deal of Beethoven. Though time may ultimately reveal that the link between successive artists was not improvement but deterioration, it is necessary to the artist and his contemporary society to aspire to surpass his predecessors. It is hard for me to feel that Luter et al. can surpass New Orleans as improvisational artists. They can, if they continue their policy of choice, and step it up as they go along, develop further.

Although I do not presume to say what will be next. I may give an example of what I mean. I should think a system of notation whether exact or free or both, somewhat in the Ellington manner (of course not necessarily in the Ellington style), could and might develop on early jazz. Applying choice both to solos and counter voices may be attained in all manner of means. Dissonance, so noticeable at time in the old players, could be used both in the melody itself and between voices. Conscious choice sometimes eliminates it. A consistent use of noticeable material without inventive overcrowding should be kept uppermost in the mind. For a good time to come use of material should stick close to the style of early jazz music as we hear it on records today.

The burden of sustaining jazz will naturally fall to fewer men. Their methods are unimportant. Musicians who would disagree with my contention that notation will be a means of developing and sustaining jazz are apparently unaware that even the repetition of a great solo or the leadership of a musician like Jelly Roll is an aspect of the process which leads to notation.

For the present we witness the solidification of the new era in jazz in Claude Luter. (We cannot predict how long this era will survive.) The great French clarinetist and his band play in the early style with such conviction and beauty that we are forced to accept the fact that men can go back and start again.

It is hard to think that the culmination of this has been provided by boys who were not reared in the city of New Orleans, hearing their mothers singing the blues before they could talk and hearing the best jazz (and little other) being born around them in a city that reeked pure jazz. They were born to hear a jazz absolutely diluted and lost and through a few records within their own rooms have been able to speak its language and make it their own. The thing is a fact now and with such a precedent there is no telling what will come next.

The playing of Claude Luter's band has an ensemble richness which dates back to early jazz when a freedom of parts existed more than it ever has since. Luter himself never stops playing and weaves his fine melodic thread all through the records. His tone is highly individual and gives great interest to his solos and is strongly reminiscent of Dodds.

Claude Luter is a player who has ideas, technique, and a driving power. Beyond these things he has what all great players have, the ability to ingratiate himself with his audience. After we have been astounded at what has come about in general and after a real enjoyment of a lot of it, Claude Luter brings back something else to jazz, namely, a real passion on our part for listening. Anything less than this does not make sense, musically speaking.

Nijinsky: An Appreciation

Note. In the original article there were eleven photographs: Spectre de la Rose, *a study made in Paris in 1912; Nijinsky the schoolboy in 1906, two years before his graduation from the Imperial School of the Maryinsky Theatre;* L'Après Midi d'un Faune; Le Dieu Bleu, *a study made in Paris in 1913;* Petrouchka, *a study made in Paris in 1912;* Giselle, *a study made in St. Petersburg with Tamar Karsavina in 1909;* Scheherazade, *a study made in Paris in 1912;* Narcisse, *a study made in Paris in 1911;* King Candaule, *a study made in St. Petersburg in 1908;* Carnaval, *a study made in Paris in 1913; Romola and Vaslav Nijinsky photographed in their automobile in New York City, 1917.*

I remember Nijinsky dancing. I was one of those fortunate enough to have seen him. I have not forgotten in 33 years that sight, and never shall. His passing on April 8 affected me as though it were something personal, as I am sure it must for all those who remember his glorious dancing.

A study of photographs of Nijinsky dancing and a perusal of the works written about him are all that are left when memory itself is gone. Let us look at some of the remarkable photographs of Nijinsky in dance. One must remember that these photos were taken 30 to 40 years ago, at a time when photographic technique was hardly at its present peak of perfection, and that to make an impression in a photo study in those days was a work of sheer genius, if not a miracle. The mere fact that so much of the remarkable presence, the fluidity, and dy-

Dance Magazine, June 1950.

namic movement come across in these dated photos is a testament to a genius that no poetry can describe.

His photos are mere hints at the remarkable quality possessed by Nijinsky for transformation into the role or thing danced, his penetration into the inner meaning of the role and the resultant transformation into the physical. His photos certainly grip the soul of the spectator as few photos of dancers, contemporary or historic, ever do. Some questions have been raised by Nijinsky's detractors about his technical abilities, so carping is the mind of man, so short of memory. Little I say can fortify his claim to genius, but I think his physical and technical abilities deserve some explanation.

Human beings do not vary in any great degree in their capacity to achieve physical feats beyond the power of each other. Many are the dancers who can leap as high as Nijinsky did or execute a *batterie* of entrechats as he did. It is solely in the *impression* of elevation and flashing *batterie* that he outranks any male dancer in living memory. His physical achievement *can* be measured; his beats *can* be counted, but the *illusion* he created was something that other powerful male dancers simply do not have, whatever their physical equipment.

Although Nijinsky was trained in the academic ballet, as it was known and understood by his teachers in the Russian Imperial School at the turn of the 19th century, there is no doubt in the mind of any who ever saw him that he was adept in the use of the torso as the most extraordinary of modern dancers today, far more so because he had at his command a perfectly ballet-trained body. With his perfect body, his fey intelligence, he was able to vitalize ballet in a way that no other dancer, even in his remarkable generation which produced so many great dancers, was able to do. I remember with absolute pleasure his ability to retard or hurry a beat at will, a feat which is a struggle for most dancers. The mechanics of dancing being for him such a "snap," he could devote limitless energy to the "play" with each step or movement.

As for his standing as a choreographer, there seems to be some disagreement among many of his contemporaries. Lincoln Kirstein, however, presents a strong case for Nijinsky's greatness not only as a choreographer, but also as an innovator, and he does this from the impartial standpoint of one who is not prejudiced by association with the Russian ballet of Diaghilev's times. Mr. Kirstein says, "In his four ballets (*Jeux, Sacre du Printemps, L'Après Midi d'un Faune,* and *Tyl Eulenspiegel*) something entered dancing which was not there before, which has never left it since. If, as Stravinsky, Ansermet, and Fokine tell us, there was little originality in his compositions, why was the quality which appeared in his work not manifested by them, either separately or in collaborations, before or afterwards?"

However, the relationship of performer to composer is not analogous to the relationship of dancer to choreographer. Upon the fallacy that such an analogy is possible rests an aspect of the Nijinsky controversy. Such fallacious thinking occurs too frequently and has plagued many an artist.

I, personally, see no need to justify him as a great choreographer in order to esteem him as a great dancer.

Nijinsky, in common with many artists, aspired to cerebral achievement. Kirstein says, "Nijinsky believed that the permanence of his reputation would not rest upon his fame as a dancer, or even as a choreographer. . . . But Nijinsky wrote a book, primarily a system of simplified choreographic notation, which he felt would make his name forever living to dancers centuries after him."

The book referred to by Mr. Kirstein is on the shelves of most public libraries and can be obtained easily for study. The pitiable circumstance about this book is not only that Nijinsky thought a dance notation system would immortalize his name before his merit as a dancer, but that this book was not written until the year of his madness, to wit: 1918.

Of all the Nijinsky ballets, none except the *Faun* is periodically re-created by this or that company. *Sacre du Printemps* represents a colossal musical and production job and is therefore discouraging to those who might wish to re-stage it; in fact, the only occasions since the eclipse of Nijinsky when this ballet was re-staged were for the last Diaghilev season in London in 1929 and in January 1930 at the Metropolitan Opera, when Léonide Massine re-staged the work, with none other than Martha Graham dancing the principal part of The Chosen.

Jeux, after thirty-seven years, is being re-choreographed by American choreographer William Dollar for the season of Ballet Theatre in New York, concurrently, almost sympathetically, one might conclude, with the passing of Nijinsky.

I saw Nijinsky dance *Tyl* in 1916. His other ballets reveal his departure from the academic framework of classic ballet. *Tyl Eulenspiegel* represents his return to the framework. For example, in one section he danced a great circle of the stage in 6/8 time, reminiscent of his circle in the beginning of *Spectre de la Rose.*

A little earlier in *Tyl,* he danced a small section with the same character he gave to *Carnaval,* skipping with hands on hips. His courtship dance with the chatelaines was a series of striding leaps and broad gestures of mocking reverence. His own part in *Tyl* demanded the utmost in physical equipment from a dancer, a fact that may restrain a contemporary choreographer from trying to re-stage *Tyl* as conceived by Nijinsky, even though there are still a number of people who remember the ballet well enough to do so.

Vaslav Nijinsky, poetic and dynamic dancer, is dead, but like the king who never dies, he is not dead to us. He graced and illuminated his brief time in the theatres of the world and gave us unforgettable beauty and inspiration. We can sorrow for his passing, but cry and sing that he came, even if only for a brief time.

Dance Magazine, August 1950.
Roger Pryor Dodge, author of the "appreciation" of Nijinsky, which appeared in stringently edited form in the June 1950 issue of Dance Magazine, *strenuously objects to the changes in "ideas" of his original text, the sentimentalization of his copy, and four errors of fact, which are joyfully set forth, in accordance with Mr. Dodge's desire to emend the following errors: (a) Dollar's version of* Jeux *was not the first version of the Debussy ballet seen since Nijinsky's made it, debut; he (Dodge) saw the Jean Borlin version in Paris in 1921; (b) London, 1929, was not the first time* Sacre du Printemps *was restaged since the original; Mr. Dodge saw Massine's version in Paris, 1921; (c) the choreographic notation which Nijinsky was averred to have done in 1918 was actually worked out in 1915 according to Mr. Dodge; (d) the book referred to in the Kirstein quotation is not "on the shelves of most libraries," as it was not printed. Presumably this book has no resemblance to the one published in 1918 on the choreographic notations of Nijinsky, which can be found on library shelves. Lack of space presents publication of the editors' more numerous objections to Mr. D's original text.*

Landmark: Landowska Completes the "48"

Johann Sebastian Bach's Forty-eight Preludes and Fugues reveal, or so it seems to me, an intrinsic rhythmic gaiety seldom found outside popular dance music. The generative quality of the dance was so manifestly in the air at that time that even complicated "listening" music such as this seems shot through with its sparkle. In order to appreciate Wanda Landowska's unique contribution in releasing this dance spirit, it is important to recall the interpretive concepts of the 19th and early 20th century which she had the insight to turn her back upon.

Wanda Landowska grew up in an era dominated by powerful and dramatic virtuoso pianists. This approach applied to the dance-derived music of Bach and his predecessors, went far towards depriving the music of most of its inner vitality. The era even produced the phenomenon of Busoni, Godowsky, and others, seriously rewriting Bach so as to give suitable scope to the prevailing grandiose virtuosity. All this, together with the use of the piano as substitute instrument, went to make the most perverse invasion of the 18th century ever put on by the academy.

Wanda Landowska deserves sole credit for her vital alerting of our age to the values of the harpsichord in the interpretation of Bach. She accomplished this through a genuine insight into the music. She accepted it as a dance-derived creation and acted upon it as such. This approach enabled her to bring out the inner vitality of the music and to give it performance pre-eminently suitable to its nature. It was a remarkable achievement for the time.

On a disk just issued, Mme. Landowska

High Fidelity, November 1954.

concludes her recording of the entire Well-Tempered Clavier (RCA Victor LM 1820). It occupies six records, the first of which was issued five years ago. These last eight preludes and fugues from the Second Book include one especially happy combination of Bach and Landowska, namely, the Prelude in G-sharp Minor. Here the full beauty of the music is revealed not only through her tasteful use of harpsichord registers but through her infectious rhythmic playing. The Fugue in B-flat Minor might also be singled out for mention, starting off as it does in a particularly strong and commanding manner.

Landowska has a miraculous gift for endowing a notated page of music with the measured feeling of dance. And this with no sacrifice of her overall artistry in subtle phrasing and just dynamics. If in the past she pursued this feeling for the dance with greater constancy than as of now, even so, we find her usual strong beat in the fast pieces and measured breadth in the slow. Contrary to general belief, strict time does not have to be broken in order to bring about interpretive expression or a less mechanically precise rendering. Such action may momentarily attract audience attention but it inevitably deprives the music of the great solace of tranquil measured time.

Curiously, now that the Wanda Landowska rebellion has shaped the contemporary approach to 17th- and 18th-century music, the music world finds itself involved in a new situation: that of the harpsichord "authenticists" who decry the harpsichord playing of the Landowska School as too vehement. For me, their historic theories lose force when put into practice. That is, their actual playing seems to reveal the music modestly, without vitally presenting it. The "authenticists" err in believing that the key to quality in interpretive style can be found in following detailed material from critical books of the past. Wanda Landowska, familiar with the same books, imparts a rhythmic playing style not derived from them. Such style is generally overlooked in musical controversies so inordinately preoccupied with academic considerations.

So, until such time as some future harpsichordist adheres more closely to the dance, not only in unaltered beat but in spirit, it seems to me Mme. Landowska still stands as the most important interpretive figure in the field. Her harpsichord mission has been so successful that the instrument is now a favorite of the Bach performers. The dance-conscious style she integrates into her keyboard playing is unique. Those occasional lapses into Romantic tension seem small in comparison with the death-knell she herself rang on this practice. All in all, Bach's 48 Preludes and Fugues for the Well-Tempered Clavier may never have sounded so well since the 18th century.

Jazz: Its Rise and Decline

The event of art transmission from one culture to another always seems to avoid a clear-cut termination and a fresh beginning. This produces what might seem an intolerably unstable and accident-prone method of nurturing new art. However, in this transmission lies the actual strength of the culture process. It permits successive art cultures slowly to emerge from the flux of one art condition into that of another. The better the time for such an exchange the more significant is the new emerging art. In the instance of jazz we can say an especially fortunate time-table aided its launching.

Since the evolution of jazz falls within this classical process, it is not surprising to find it originating in one cultural pattern and developing in another. This evolution in cultural process, easily observable in the main development of jazz, can be sighted and pursued down to its smallest cross-city exchanges. There is enough historical evidence to show that within every dominant civilization more than one culture force is responsible for an art. Where the one culture was only capable of producing some major or minor innovation, the other culture was set to develop it. In fact, research constantly reveals that an art activity which *comes about* in one environmental situation may well move into its development by way of another; the climate necessary to bring about an art innovation is not always the climate suitable for its development.

The open climate of New Orleans, with its Storyville, parades, dance halls, and general hubbub of urban life could never have created jazz had this art not had a previous life in the restricted rural

The Record Changer, March 1955.

surroundings of plantation life. Social interpretation of the music based solely on one aspect of the Negro's long life on this continent fails when called upon to explain the real process behind the music's creation and development. We may be sure the values of jazz as music no more report the dissipated social life of Storyville than it does the slave life of the plantation. Its progress reveals, as the history of all other classic music reveals, that, once emotional feeling is crystallized in melody, this melody assumes an artificial reality of its own and is musically susceptible to religious as well as secular expression.

A non-isolated emerging art creativity is always conditioned by the surrounding culture. Whether it will significantly rise or fall depends upon a congenial atmosphere of some sort to nurture its growth—congenial for the art but not necessarily the artist. In the case of jazz we imported a primitive people with their culture patterns and placed them in slavery within our culture. This prompted their primitiveness to move into new self-expression. A certain religious freedom of action, limited but present none the less, allowed them self-expression in musical utterance. By the time it was necessary for the future of jazz that they change their environment, the Negro life in New Orleans as differentiated from that on the plantation took on community proportions. Moreover, if in New Orleans the Negro had had no more personal freedom than that obtaining on the slave plantation—in spite of the fact that that very limitation may have been the historically necessary situation for the original impact—we can say that the New Orleans environment could have carried the art no further. However, by the turn of the century the Negro possessed unlimited availability for professional playing. Playing all night and every night developed him as an instrumentalist. He grew musically as he developed a new *way of playing*—a jazz way. This jazz *way of playing* not only significantly developed the American Negro's own melodic material, but it was so strong in itself and so completely self-reliant that it produced great music even when applied to the urban popular tunes. It could transform anything and everything it used into an entity resembling itself. Unfortunately, it was the surrounding white culture which finally halted the favorable development of jazz and forced it into its present decline.

Although American Negro jazz is related to other primitive musics, sharing some things with them, the great difference is that the American Negro, having absorbed our Western civilization, became a part of it and therefore subject to it. Jazz was born to Western culture at a time when that culture already had become musically decadent. It was more than difficult for the jazz musician to develop gradually his own genre uninfluenced by this surrounding deteriorative climate. Western Europe's own early art development had proceeded strong and vibrant in its own atmosphere, with no such decadence to hurt its musical growth.

A new art development in our Western culture situation seems to exhaust all its vitality in the process of *being.* Although the ideal primitive art situation encouraged change, it seemed to be able to hold to its basic substance and rely on what might be considered *surface* changes in its development. Such surface flux left the inner content significantly intact. Precisely the opposite characteristic obtains in the literate aspects of our present Western society. Here, change seems to invite a decay which goes straight to the core of the art. Perhaps the burning-out of jazz must be attributed to the debacle of Western culture.

It is very important to recognize that the Negro of the plantation period was actually grappling with song content itself when he created spirituals, work songs, and early blues. It was the vitality of this new song content which made the foundation for New Orleans jazz. It was his inevitable full exposure to the cloying character of contemporary Western melody which finally caught and consumed him and led to the present deterioration of American jazz.

However, in the beginning the Negro jazzman had the background of his own musical past to guide him and was able to ride along giving great invention to any prevailing popular melody. But his greatest success has always been when he violently attacked it. Curiously these musical attacks, however refreshing their approach to a stale situation, always seemed to retain their lingering sense of violence. On the other hand, the considerable violence perpetrated by the plantation Negroes on Christian hymnology subsided into the serenity of the blues. The fact is, improvising jazz musicians cannot be expected to bring about a continuous transformation of revolutionary significance equal to that of the blues melodic material itself. They can only vary their chosen themes, not transform them. And it should not be forgotten that an important part of the act of improvisation is the moment when the musician falls back upon the basic tune. Unfortunately, the constant reliance upon the popular tune unconsciously devitalized the Negro jazzman and conditioned him for his final capitulation to its eating decay.

This softening process started early. It began as soon as the jazz bands spread out from New Orleans[1] into locales where their success depended more and more upon catering to audiences unacquainted with the blues. The jazz musicians played what their new audiences preferred, and the audience choices were eagerly absorbed and associated with a superior way of life. It was this unfortunate but understandable respect the Negro held for his white audiences which led him to disparage the beautiful quality of his own musical accomplishment and link himself to the prevailing white decadence.

1. The term "New Orleans" can in a broad sense stand for *all* early jazz as well as that which is specifically known to have been situated in that Louisiana city.

Jazz is definitely a folk art. But it is quite a different music from that which is usually considered folk music. Because of the great division between today's folk art and today's academy, folk music has come to mean the simple tunes of a rural peasantry or under-privileged urban grouping. But in actuality it must be conceded that group music activity *outside* the academies[2] is in a large sense folk activity and its complicated output in more than one period has gone far beyond any such limited definition on our part. If it was once confined to the simple tunes bursting from a contained people, it is no less folk when similarly emitted from instrumentalists into the complicated jazz that we know. That jazz has in its variations the same invention, wealth of ideas, and contrapuntal fabric that we historically associate with our early academic music when it was at the same stage of structural development does not preclude the fact that it is folk. We must remember that the reading and writing of jazz are attainments applied to an already highly developed and untutored music, and even in this latter period the best jazz still functions without these attainments.

II

The present decadence of jazz, with us in spite of great developments in player ability and versatility, owes much to the fact that throughout this last stretch there has been unavoidable and constant contact with the classical music of the 19th century and the modern classical music of the 20th. The popular tune, located in the whole-tone scale and approaching atonality, has been an unhealthy deviation for jazz and for the jazz musician, as we have indicated. Although it has truly brought about a new jazz era, it is a jazz era occupied with an almost systematic exclusion of most of its past. I am convinced it is this denial of its own roots, pervading the *whole* material of progressive jazz, which deprives it of any future. If jazz is to be extended through bop and/or its associated cool and progressive techniques, we find jazz in the process of a losing battle.

If it was the early vocal blues and related virile song activity which contributed to New Orleans playing style its most important element, it is the unfortunate choice of the popular tune as basic material which has qualified progressive playing style. Moreover it is not simply a matter of the bop or progressive musician being *incidentally* concerned with popular songs. I believe their basic *attitude* toward these songs—their attitude which enthusiastically endorses the whole gamut of torch song and torch singing—is aesthetically debilitating in its channelling. It is the embrace of the ubiquitous mike-singers with their

2. Popular tunes, although at times closely related to folk music, are quite different in conception and final attainment.

over-intimate mannerism, their cozy "singer" delivery, which to my mind has deprived the progressive musician of all pretense to a "solid" lyrical song basis. It is easily the most serious defect of bop and its latest progressive manifestations. Although this essentially shallow foundation of their music is somewhat concealed by instrumentation, invention, and outright technique, it unwholesomely pervades and seeps through the whole output. It is the basic defect of modern jazz.

Perhaps further clarification of this fundamental dichotomy between the New Orleans jazz era and that of today can be best illustrated by setting up a comparison between two famous singers: Bessie Smith and Sarah Vaughan. If there was the extensive presence of the blues in anything and everything that Bessie sang, in the Vaughan singing of either a blues or popular ballad there is always, without fail, the residual flavor of the derived torch song. Fortunately or unfortunately, the pervasive entity of any singing style always goes straight to the core of the music associated with it. In the present instance in spite of extended forays into chromatic scales and atonal aspects of melodic variation, the sad fact remains that bop and progressive improvisers cannot help but reveal their vitiated song basis. Sophisticated harmonic activity, no matter how advanced, cannot counteract such manifest melodic decadence. In fact the complicated texture of modern harmony is itself an inviting hazard to greater degrees of decadence not only for jazz but the academy as well. Chopin, Brahms, and Debussy progressively enlarged the simple harmonic horizon of Beethoven, but in so doing released a concomitant "sweetness" impossible to the Beethoven context. Theirs was an experience-advance which may or may not reduce their critical rating. However, in the case of jazz it is a death march. An illustration is aptly provided by the basic triad tones of bugle usage. Although they are limited in expressiveness their melodic variety never declines to those saccharine levels possible to melody drawn from a wider range. (I am inclined to believe the more our tonal possibilities are extended the greater the need for self-discipline.)

In any event the creative output of a musician who is part of an entire declining music is inevitably colored by it—no matter how outstanding the talent. For example, an inventive mind will always come up with intriguing involutions providing their own kind of interest, but if the general style of period from which they came is unhealthy, sufficient interest in the music is lacking for the listener no matter how ingenious it is. Moreover, it might be said that music *dependent* upon ingenuity alone is doomed from the start. For ultimate significance of an invention lies in its style, not its originality.

Historically the progressive player is the immediate descendant of an enervated style of jazz—the tail-end of Dixieland. This associate atmosphere almost at once precluded a virile playing style. For bop and progressive musicians, similar not only to the early jazz players but to the folk of all eras, do not spring from areas of cultivated taste. In their case it is a pity, since to survive this shoddy period superior taste is really needed. It is this cultural lack which forces the bop and progressive musicians into the position of being tolerable only when they create the extraordinary item. Unfortunately the large bulk of their output is a mish-mash out of this rewardless struggle in semi-sophistication. However, this is the morass that has engulfed all jazz players trying to strain ahead, as they naturally must, since the end of the 1920s. In this connection the progressive jazz aficionado, in most cases as culturally deprived as the musician, defends any extreme deviation in progressive jazz as evidences of an advancing music. For them, some form of progressive jazz is clearly the next step in jazz. But going ahead merely by adding an ever-increasing harmonic *charge* to our already existing song form is not the only means of progress. Historically, Western music progressed by means of song-form extension before harmonic charge was resorted to. An extension of the song form itself would tend to interdict improvisation and of necessity introduce notation, apparently a direction its aficionados do not relish. So far, no doubt the best creative efforts in bop and progressive jazz have been derived from improvisation, but is it not a fact that progressive musicians are for the most part intellectually familiar with their work, actively engaging in conscious musical contrivance entailing notation? Is there not quite a trend toward advanced academic study? If so, obviously the musicians themselves have no aversion to the act of notation.

The awareness of the historic background of jazz which led New Orleans aficionados to decry *all* use of notation is not necessarily an example for progressive aficionados to follow. A thoughtful use of notation might encourage them to venture beyond their present exploitation of our limited popular song form and help them to avoid their tiresome over-charging of it. Moreover, a breakthrough on the restrictions of the song form might reduce the present wallowing in harmonic charge to minimal importance. Fortunately, for the extraordinary achievement of Occidental music, it is only lately that the West has taken to leaning so exclusively upon extended harmony. In fact it is more than naive for progressives to base an adverse criticism of old jazz upon its lack of harmonic change. Melodic material established on no more than three harmonic changes has long been the basis of extraordinary art achievement.[3]

When the progressive player uses complicated chords or scalar passages foreign to the tonality of our previous jazz, such use does not necessarily imply the presence of distinguished creative activity. When new effects—that is, new

3. Long sections of Beethoven alternate between two harmonies only while Wagner wrote a prelude on one harmony, the tonic chord.

to jazz in that they may be, say, chromatic, Oriental, whole tone, or pre-diatonic—are applied to our popular tunes or fragments of tunes, such novel application tends to suggest extraordinary freedom of movement—perhaps because our popular tunes fall within the diatonic system. However, the pattern of this new approach, which at first hearing may have caused one to assume the presence of a creative imagination, soon reveals itself as a repetitious identification of material—a cliché as familiar to its own environment as the displaced tonality. This does not mean a real union might not be effected but it does mean that *real* creative activity on the part of the improvising player is as necessary as ever and will have to emerge. The progressive musician cannot rest upon a removed exotic effect forever. His so-called progressive tonality can only continue to appear highly charged to those still marvelling at the absence of diatonic progression! For those of us in possession of the pattern of the new jazz tonality it is now up to the bop, cool, and progressive enthusiasts to define its worth on a more mature level of established music criteria. In any event, perhaps it should be conceded that something infinitely prosaic in itself may be introduced into another musical system and give off for a time an exciting and original effect.

Now, however, as the progressive improvisers commence bar by bar to make fuller use of the modernism around them, they seem to be losing grasp of the piece as a whole. Perhaps it is because modern music is a music of great consciousness made available to composers mainly through the facilities of notation. I doubt if progressive musicians, quite aside from their present lack of discriminating taste, can ever appreciably advance their music through improvisation—that is, a modern improvisation comparable in interest to the more absorbing aspects of modern academic music. Maybe chromatic melody and modern harmony need the opportunity for reconsideration inherent in the process of notation and are per se beyond the scope of significant improvisation.[4]

To sum up bop and its derivatives, let us say that in spite of their own complicated development they function in essence as a music on a much lower level of musical significance than either early Dixieland or New Orleans. Further, let us say that this music has developed too far to justify any expectation of future improvisational activity, and that all matters of taste aside, notated compositional expansion alone can enable it to use more tellingly its present cliché within chromatic freedom. In fact, let us say flatly that there is no future in preparation for jazz through bop, or through any one of those developments known as cool and progressive.

4. The lack of scope should be understood as referring to the capacity for improvisation exhibited by average musicians (folk or otherwise) and not to the capacity of a musical giant such as Bach or Beethoven.

The opinions on progressive jazz that I have set forth are held, I believe, by most of those who favor New Orleans jazz. However, if we still feel that progressive jazz is an issue, this is not so much because of the fact of its existence as it is for the infiltration of its critical tenets into certain contemporary evaluations of jazz. It is those staunch defenders of this music who, through their articulate enthusiasm, have made it necessary for our protracted consideration of it. As a group to contend with, what are their credentials? What are their criteria and what is their history?

Jazz changed in 1928, driving a wedge deep into jazz criticism and appreciation. The split, one of many, was permanent, and two classifications, pre-1928 and post-1928, are indicated by the metamorphosis. The tenor of post-1928, whatever the developments, extraordinary players, arrangements, ensemble procedures, singer attainments, and harmonic advances, is in a category inferior to and entirely different from the previous jazz. The progressive adherents align themselves whole-heartedly with the entirety of post-1928.

There are two groups in the post-1928ers. One feels that New Orleans jazz with which it has had considerable (if boring) experience is outmoded. The other, appreciatively raised on a mixture of New Orleans and Dixieland, was brushed along by the tide of ceaseless change; the latter accepts pre-1928 as desirable in its chronological place and at the same time endorses the current product as justifying the course of post-1928 development. As the music changes, worsening from the start—ultimately reaching the dubious present—groups of adherents broke away from the parent body. Had there been no progressive jazz splinter the post-1928ers would have remained immersed in the creations of the 1930s. There may be exceptions who have not voiced their opinions but on the whole it is the post-1928 supporters, whether they feel attached or cool toward New Orleans, who get their most intense experience from some manifestation or other of post-1928 music. These elements represent a powerful body of opinion, holding jazz in its present course.

It may be that in the *best* of progressive jazz there exists something which those steeped in New Orleans cannot see—a matter of not being able to look beyond one's generation. Nevertheless, in the attraction to the progressive we are witnessing nothing more than the involvement of the coterie, a readiness to plant flowers for cutting as well as blooms of a more continuous growth. Progressive jazz offers change-for-change's-sake, the coterie's requirement. It is difficult to find evidence to modify this statement for the reason that such opinion is neither discriminatory nor appreciative, remaining purely *progressive.* We have here, in other words, the possibility of a new creation with no valid collective recognition. The New Orleans followers are mainly blind to it while those partial to it are not critically well founded, to say the least. They have no way to enlist our support.

Unfortunately, if there is no future for jazz in progressive jazz, there is hardly more in a continuation of Dixieland. The vitality of Dixieland today merely refers to the fact that except for a few newcomers the players are still the same men who developed this style and grew up with it. Understandably they have had no urge to carry jazz any further, being diatonically and technically satisfied with their own high performance. However, it is obvious it is only a matter of years before these players cease to exist. When they pass away so will the vitality of Dixieland. The vitality has persisted simply because non-eclectic individuals set within a rewarding experience of their own do not change after they have reached a certain period in their lives.

In the functioning of any culture we always find the older artists in an art-lag of their own, independent of their culture's progress; the instinct to rapid change is the concern of the young men. This same phenomenon prevailed when the earliest New Orleans musicians did not shift over to the style of the incoming Dixielanders. Now in their turn the Dixielanders, ripened in the styles of the late 1920s and early 1930s, do not move into progressive or revivalist attitudes. Thus each new jazz style has been implemented by the oncoming group. The fact that active Dixieland has been with us so long cannot therefore mislead us into believing that it *can* hang on. It is sheer delusion to believe that a *living* Dixieland, in contra-distinction to a *revived,* can continue to be a part of our present-day musical experience. Where the talented material in the late 1920s went naturally into Dixieland, today potential material for an expanding Dixieland will blend naturally into progressive-bop or neo-classic revivalism. Let us face the fact that when the last great players have passed away, Dixieland as we know it will have passed also.

As for the situation facing the revivalists, curiously the very act of *revivalism* itself implies perhaps the newest attitude within jazz—that is, an attitude founded upon critical and eclectic choices within the past based neither upon popular nor modern trends and contradicting both the persistence of the present-day Dixielanders and the modishness of the moderns.

Actually it was the enthusiastic appreciative insights of the record collectors which seem to have inspired the first young musicians into active recapture of the past. This was a new audience which had cultivated listening practices geared to rejection as well as acceptance. Although major divisions of opinion soon split the ranks wide, nevertheless out of this situation sprang the new jazzman: the revivalist. Heretofore the perennial jazzman had been immersed in developing what he fancied to be the forward moving jazz of his own period. He exploited the outlets open to him with little thought of retaining past values. The revivalists, on the other hand, took on a collector's attitude. They turned back to redefine those special jazz qualities they felt other jazz musicians had carelessly lost in their march forward. Therefore a situation has been reached where teen-agers both here and abroad under the far-reaching influence of the collector have recaptured old New Orleans jazz music and can play it for us in the flesh!

Once again, can this situation continue for long? I hardly think so. The revivalists cannot help moving ahead from the point at which they pinned down their retreat. The restlessness of Western culture which historically forced early jazz to move into its middle jazz period will likewise force the future revivalists to move ahead in their reviving. History will repeat itself—and, if anything, "advance" more quickly. However, in spite of no foreseeable protracted future for improvising revivalism we still are indebted to the present revivalists for the opportunity to hear their fine music. Theirs is of a generally high standard because they can achieve significant music without dependence upon virtuoso personalization of the past, for early New Orleans jazz had a strong overall style which was self-sustaining with or without virtuosi. It is this strength persisting even when *great* soloists are lacking, this simple playing-off-straight on a good blues which was always highly gratifying and is comparatively gratifying when seized upon by the revivalist. On the other hand attempts on their part to revive early Ellington reveal how thin such a music is when deprived of the great soloists who made it.

In any event, the revivalists are hemmed in on all sides. For if to do what is natural for a creative artist, i.e., to reach out and explore, will ruin their recaptured quality, sooner or later the bad effects of *holding back* will have an even worse effect. Let us say revivalism can only be a step toward a future—not *the* future of jazz; that the thing we knew and *had* as jazz is now finished. The effort of revivalism when all is said and done, is merely effecting a past music on a lower vitality level.

In this connection it is curious how a large proportion of the progressive group relegates not only revivalism but New Orleans jazz itself to the lowest position in the hierarchy of jazz art. The very latest jazz aficionados seem unable to get out of early jazz even the lesser pleasure that the modern classical music lover gets out of any number of past styles which he may deem of secondary importance. Early jazz is to them, at best, no more than an historical link in the evolutionary line from the horseless carriage to the latest model. Such an approach smells of fashion. Fashionable excitement-levels can be very misleading if there is no critical musical approach beyond that of contemporary associative affinity.

We believe that the best of jazz need not have stopped when it did. Furthermore, in spite of its magnificent attainment, jazz will not, as it should, become a part of our musical life like the classical music of various periods.

The extraordinary style-difference between jazz and European music, and the persistent change still taking place in progressive manifestations, will eventually cut us off from the records made in the early days of jazz. Our interest, instead of being in close affinity with the music, will become similar to our removed attraction to ethnic records, a pride in a dormant rather than a living heritage.

The boppists, cools, and progressives are surely stimulating a dissolution within the vagaries of a non-jazz world. The revivalists, on the other hand, have made a start in the right direction, but we can see that without new attitudes they face a dead end similar to that of the laissez-faire Dixielanders.

A Listener's Hierarchy in Jazz: Historical Precedents for the Future

The Record Changer, September 1955.

I have indicated elsewhere the seriously deteriorated condition of jazz. I am compelled to this reiteration because jazz is an extremely important music. There may be a way out for it, taking the line of a New Orleans extension. By this I do not mean the current line of the revivalists, laudable as that is.

Significant musical events in the past suggest that it is possible to salvage jazz. Usually I consider it a waste of time to indicate the future of an art, but I see a plan for jazz—and I believe it is cogent, suggested by developments in certain past musics. We have seen many arts evolve and we have seen what usually happens to them at maturation. Jazz can be of enormous importance beyond its own time. For this reason I suggest that all historical procedures that indicate salvage are valuable.

The continuation of the bop-cool-progressive wave now in progress may make superfluous any suggestion on my part in relation to change. But a change is vitally necessary. I intend to show (1) that jazz records will gradually lose favor if a live jazz of equal or nearly equal importance does not come about, and (2) the act of revivalism is indicative of the fact that *all* thoughtful jazz musicians do not want to join the latest developments; they may be open to proposals for an innovation comparable to improvisational revivalism. The saving channel of growth that I see requires careful definition and consideration.

If one feels, as a number of critics do, that jazz has been on the decline since the 1920s, and that the current interest in progressive manifestations fails to support the contentions held for it, are we witnessing an inescapable arc of art inception, growth, and decline? I think not and I will endeavor to point out why. When I say this I have in mind a jazz which I believe would be—for the future generations, at least—a more suitable musical experience than that which we now have on records. I might say that suitability for the future need not be interpreted as predicating a necessarily greater music. Great musical folk expressions may reach, in essence, a height not attained by a music more feasible to create and more suitable to a future audience.

First we must discuss the extraordinary power of the musical recording. It has created a revolution in musical awareness. Without it our knowledge and consideration of jazz would be rudimentary. Is it, though, entirely advantageous? We must admit that through it we enter a period of great eclecticism as far as the knowledge of our musical heritage is concerned. The inroads different companies have made into ethnic cultures have given the public for the first time an insight into the musics of other people. Classic music within our own culture is now available to all, whereas, not long ago, it was economically impossible to concertize for the general public. Nevertheless, all of this musical awareness, good as it is, must be differentiated from the disposition of the music that we choose to specialize in and practice ourselves, the music of our preference.

It must be understood that I am not against musical erudition. But erudition must not tend to displace a more intimate acquaintance. A music that remains attractively alien or exotic lacks such intimacy. Any music loses its alien and exotic flavor when once it becomes intimate by a daily passage through our consciousness.

We must differentiate between our life as record listeners and a musical life outside of recordings: attending live performances, amateur playing, the professional life of a musician. In classic music we have this life. This is the music of the schools. It will not be jazz music. In other words, jazz as it is will not have seeped into our musical consciousness as has Western music into Euro-American culture. Because music is a time art we are limited in the time we can devote to a miscellany of idioms. I must say that if we confine ourselves by way of records to an ever-extended horizon we will never acquire great intimacy with a style we could call our own, for music above every other art is close to personal participation.

By this I mean that any music with which we are well acquainted we hum, whistle, or actually play, together with hearing it in performance. In this way it penetrates our musical consciousness, rewarding us with a musical experience not realized by a cursory knowledge of many idioms.

Our great heritage of jazz on disks must exist *together* with a living music otherwise its full import will not enter our consciousness. In the past it was part of all living or played music. We must remember that jazz records are today played by an audience whose members have had a life of particularization not attained by today's neophyte in the field. Besides, the experience of hearing the late Bunk Johnson at the Stuyvesant Casino is barely a decade away while the major feed from that music is being tapped by the Dixielanders and the Revivalists; Johnson, of course, endures complete in the George Lewis Band. The day will come when these vault originals and re-issues will stand alien and exotic for us as does the ethnic record. They will have as much meaning for China, India, Europe, or Africa as for us. It is because jazz is so indelibly fixed on records that we will only be able to admire it removed from the possibility of effective participation. It is where admiration and involvement meet that we receive the greatest musical reward.

Early record collectors had an incentive that will never exist again so far as jazz is concerned. The "collector" spirit is a great contributing incentive to a complete knowledge and appreciation of a thing. The digesting of musical matter, one item at a time, together with the excitement of discovery lays a thorough foundation that is hardly gained by the easy buying over the counter of old platters dubbed onto LP disks. Of course I'm thinking of a listener in the future who can easily buy these LPs, but who otherwise is surrounded by a style of music heard in dinner places, dance places, or on broadcasts having no resemblance to these disks, as compared with one who grew to know the best disk jazz at the same time he was hearing the same or closely related styles in the flesh. To expect a person to become rapidly acquainted with a style of music he has never heard before (for which he has no personal performance guide) is asking a good deal. Unless he is one of the future jazz zealots, his appreciation will rarely get beyond our present attitude toward ethnic records.

For the future, then, something should transpire which will bring jazz closer to us. I speak of the great music public whose indoctrination must be effected in an entirely different manner from such indoctrination responsible for the high level of artistic appreciation attained by most record collectors in the past. If jazz on disks has deeply penetrated our consciousness through the extra-musical avenues of approach (dancing, dining-listening, collecting, etc.), these accessories are diminishing and we must find a new means.

A significant perennial popular-folk music within the framework of a highly cultivated urban life is futile. Western history has shown that, except in the rarest instances (for example, Flamenco-Spanish music), a popular-folk music either disintegrates because of catering to popular demands or settles back into a placid life of working in motions of a once great music.[1] The great periods must be capped by a process of codification and musical geniuses must come

1. Oriental music is in a different category we cannot go into here.

forth who can work in the medium of codification with the eclat of the previous improvisers. Without a few great names the 19th century would be almost barren. The revivalists, today's great pre-1928 stylists, play a losing role if out of their ranks do not appear men of talent, taste, and eventually genius, who not only help their own particular bands but who through the notational (and eventually compositional) means used by past musicians extend their influence to bands of lesser talent.

By now, I think, it is clear that the revivalists are not succeeding. They do not seem to sustain what at times they have attained, and show more and more what they really are—musicians without roots. Revivalism is part of the ever-recurring criticism we find artists directing toward the prevalent art standards around them. Within this group we may note the practitioners of modern art in all media and the pre-Raphaelite painters who are eclectic by nature, functioning without cultural roots. Each artist chooses from other sources, and singly establishes an art the validity of which depends on the degree of genius behind it. Although the revivalists are in the same category they function through the means of ensemble cohesion and not as individuals. We might say that it is inconceivable that groups of musicians without roots can keep as high a level as did the early New Orleansians whose roots branched out in every direction.

The New Orleans bands were miracles of homogeneity: in fact we may say that every period of jazz has this quality of homogeneity. However, it was only the unique achievement in New Orleans that had any real significance for music. But significant or not, such homogeneity is a rare, almost impossible achievement when the locale of the band is not saturated with the style it is working in. All the more praise for the revivalists. But music cannot continue under a rare achievement dependent upon a group of men working in difficulty toward a developmental unity. It can, however, continue through the efforts of single musicians, as we have seen in classic music where the musicians' output, especially during the 19th century, had little in common with the surrounding music. Now the integrated creativity of New Orleans is a thing of the past. However enjoyable the playing of the revivalists, they cannot be expected to do much more than simulate the early style. The products of a borrowed New Orleans improvisational psyche are ephemeral and in no way solve the problem of the necessity for a living music. When they first went to work on the problem there was excitement in the very act of accomplishing it, but now that that part of it is over, group interest wanes or is bound to become diverted in other directions. It is the individual, the composer, who, firm in genius, crystallizes the whole period.

While I have always envisioned a progress in jazz that was more conscious than that through the means of improvisation, I *now* find that this becomes a necessity for future culture when we consider the impasse of a knowledge drawn only through records. I would have liked to see jazz become America's classic music. As highly developed as it is, it cannot stand as a serious American music in the sense that it has permeated the activities of our culture as have the musical practices of European music. It should be a music that we play as well as listen to. This is a shame, for from it could have flowed a music as classic as any developed by Western culture. Although it came from a segment of our culture, it has re-entered it in a highly dilute state, elevating the popular quotient of our lighter music with a cool and offensive sophistication. Any use of it by academic composers has been sporadic, never seriously influencing modern music.

My conception of composition in jazz[2] has led me to consider many examples of what I have gradually come to see as very poor jazz indeed. Whereas composition cannot make of a poor music a great music, certainly it gives scope to any music to which it may be applied—an important ingredient for a time art received only through listening. In music schools they could be teaching jazz classics, but we still find them immediately attaching the young to Haydn, Mozart, Beethoven, and Chopin. Before we outline in more detail the possibility and necessity of notation let us reiterate through history that we are not proposing anything new. Growth from non-notation to extended notational composition may be called standard in Western culture.

The tradition of jazz development is not inherently different from the tradition of classic music in its early phases. But the deepest fallacy in musical perception shared by both jazz and classic critics alike is that the development of jazz music is different from that of classical. Perhaps it is only equalled by the fallacious corollary contention that jazz *must* develop in a *different* way than all other musics! The unfortunate result of this shared point of view has led the jazz musicians to believe they are path-cutters in creativity through improvisation, that they are essentially within a different musical process than that of early notated classic music. They overlook the fact that classical music has only been created solely by way of notation for the last hundred years or so. If jazz *is* different—and it surely is—it is *not* because it is improvised.

Historically the whole structure and content of composed Western music is built on the *once* improvised dance music of the 16th century.[3] This impro-

2. "Notes on the Future," *HRS Society Rag.*

3. Sir Hubert Parry, entry for DANCE RHYTHM, *Grove's Dictionary of Music:* "The connection between popular songs and dancing led to a state of definiteness in the rhythm and periods of secular music . . ." and ". . . dance rhythm may be securely asserted to have been the immediate origin of all instrumental music."

Curt Sachs, *Rhythm and Tempo* (New York, 1953), 281: "Eager to replace the roving shapeless monody by steadfast forms, they [16th- and 17th-century musicians] had no models in the

vised material engrossed generations of early European musicians long before the individual composer seized certain aspects of it and crystallized it into the written material now handed down to us as the only residue of that era. Improvised dance music was certainly the forerunner of 17th- and 18th-century music suites and in a large sense the basis of the symphonic extensions of the 19th and 20th centuries. From a historical standpoint, therefore, improvisational activity has long been basic to Western music, far from adverse to finding its way into notation. However, through a kind of journalistic pleasure, critics and jazz musicians like to think of the two as mutually antagonistic.

Jazz followed historical precedent up to the moment of notation and here, for one reason or another, it stopped—possibly because composition as practiced by the Fletcher Hendersons was not a significant reordering *within the values* of improvised jazz. Obviously the 16th-century composer, for example William Byrd, in contradistinction to Henderson, was intensifying his material, not obviating it. So far, important composition has been beyond the talents or tastes of any jazz practitioner who has attempted it—they have merely created *another* music inferior in kind. For all practical purposes written jazz could sound as much in the New Orleans vein as early New Orleans itself. It need not *appear* different but it must have certain requirements that the other jazz may or may not have had. Whether or not it is satisfactory on the same contemporary level is ultimately immaterial.

William Byrd is important to us in this respect because he did not create *another* music inferior in kind, but instead organized what was around him through his writing. The importance of his own work for him, whatever its comparative value to the lost music out of which it came (and quite aside from the question of Byrd's awareness of this value), is in his love for what he wrote and could set down. For the subsequent music world this approach, so generally held by the composers of that time, had untold importance whatever may have been its import to the Byrds of that day. An age working towards the development of a new folk growth is at a double disadvantage. Its efforts are certain to be rather staid when compared with folk improvisation and rather simple when compared with the reigning academy. This was true of 1600 and would I think be true today. It is present-day aspirations towards detached and lofty monumental compositions which derogates such an effectively straightforward approach to jazz. Scott Joplin's ragtime sheet music is practically the

flowing polphony . . . of the early century. . . . The dances of the time, clear-cut . . . were the only point of departure."

Thompson International Cyclopedia of Music and Musicians, Dance Music: "Independent dance-music for instruments must have existed throughout the Middle Ages, but practically none of it has been preserved, since it was handed down from generation to generation as an improvised art."

same as his pianola rolls. In crystallizing in written form what was currently improvised, he did not substitute a new music. However, jazz *arrangements* are invariably different from jazz improvisations. Oddly enough those who recognize jazz only in its improvised state easily accept the written-out ragtime as a bona-fide ragtime!

There is no denying that the creative mind can show far more energy during improvisation than otherwise. The act of improvisation not only releases a profusion of musical ideas but provides significant rhythmic elements within the melodic contour. However, the fact that out of improvised jazz there sprang the only unique innovation in playing style now practiced in the Western world plus its most unique and significant music material does not in itself suggest that these elements can not be created, extended, and fixed beyond improvisation.

In acquiring playing style, proficient performers can be extremely moving playing music which at one time was the sole property of the improvisers. Claude Luter renders a Dodds clarinet solo in such a pleasing way that the listener receives great satisfaction whether the superiority of Dodds is present or not. Certainly the recapture of a former playing style is not too difficult for those musicians who are not at too great a remove from it. In fact the revivalists have shown that a past playing style can be used in executing old material exactly or similar new material.

The detached creation of the material itself, however, has not gone well, providing a problem both vexatious and vital. The rhythmic melodic line so easily brought about in jazz improvisation has never been equally available, bar by bar, to the conscious jazz composer. Perhaps he did not want this rhythmic line. Nevertheless, the sound composing jazz talent, when the time is ripe, can overcome this handicap as satisfactorily as its 17th-century equivalent. The music that runs through the inwardly turned mind needs only the accompanying ability to notate it, and there is no doubt there is an additional advantage in the opportunity for reconsidered editing. Of course, "written jazz," to date, has not been evolved in such a situation. A sickly modern academic culture has been patiently applied to the most decadent aspects of jazz and the indulgence does not seem to be on the wane. As long as the early virile New Orleans school of improvisation could function there were good grounds for dismissing any approach to the creation of jazz other than that of improvisation. But thirty years or more have passed. After the great New Orleans period (and in the 20th century as opposed to the 16th) only one generation may be sufficient to shift and smooth a virile music toward premature senility.

The composer's achievement can be a permanent blueprint for live performances, needing only an orchestra of the revivalist type for its interpretation. But the composer should have his own way, permitted some access to innovation, for his working habits and incentives are entirely different from those of a performing band. Although he may create a music only a little different from the band, sooner or later his results will differ decidedly from those of the band. Given appreciation of the music and sufficient taste to avoid the known undesirables in composition, there is no earthly reason why the composer, together with the revivalist bands as interpreters, cannot furnish us with a strong and vital music fairly comparable in practically every way with the original New Orleansian.

Because of the completeness of our present notational system we forget that notation was originally and primarily an activating agent for composers who needed only a few indicative symbols expressive of musical practice. It facilitated the retention and communication of conscious consideration. Such activity does not dispel improvisation in a single vast drive. The serious improviser-composer therefore might abjure *complete* notation but not *all* notation as such. He would be in a position to proceed by way of *building* on the style of his choice without necessarily *changing* it. The improvisation of the folk is subconscious and it is their major expression, but conscious expression presumes the ability to decide when to return to improvisation and when to employ the consciously placed specific material.

Jazz people do not deny the validity of the world of notation but their complete differentiation between it and improvisation would lead one to believe that notation emerged from nowhere. They seem incapable of *crediting* the fact that the controlling power of notation in the developmental extensions grew out of the first phase when notation served mainly to preserve and make accessible to other players a piece generated by improvisation. It was created as an aid for a music moving in the natural and basic ways of improvisation and was not instituted to develop anything but improvisation. Early improvisers developed a sense of the desirability and necessity of a communicating record in their practice. Notation began to widen under their hands according to a sequence of situations evolving under developmental extension.

The problem of recording rises again. How far can the influence of the music on record of an early and vital period take the place of development by composer musicians? Recorded music (as I have tried to show) cannot sink far into our culture. It can equip us and take us to an important stage midway in our composing venture, but not beyond. In order to realize a continuing culture from jazz we are forced to the means taken by the past. The past solved the very grave problem in its early music by notation and the initiation of composition. If, since then, we have evolved the recording, we have not, in so doing, necessarily obviated historical precedent. We have a superb adjunct which enables us to preserve a precodification and precompositional period but it often distracts us from the grave developmental problems of the music.

We now must take into consideration what it is that the listener wants. We must deal with the question of the difference between a music for listening and, in this case, one for dancing. If playing for the dancer produces certain musical traits, is it not true that the dance demands certain things that the listener does not? While the dancer may get more than he requires it is possible that the listener may get less. Jazz was not conceived with the aid of that notion of a thorough listening so important to us now. In its rightful context it was a musical expression conducive to the activity of the dance. What we find is a music of the greatest vitality, a dance-created music serving a listening audience for whom fortunately it was not intended. If it had been aimed at listeners, they would have received far less. Being what we are, we will inevitably try to enhance and increase this already great listening pleasure. Staged, as it is, in a particularly improvident decadence, the music must be carefully managed. Jazz is beyond the generative social-dance area, and we must find another means if we are to preserve it and extend it.

With the powers against composition are the satisfied New Orleans record listeners who feel that anything "done to" jazz would spoil it. They are as formidable surely as the others, the progressives who endorse the product of their currently active musicians, well satisfied with the trend which they feel improves each year—perhaps each month. The two groups constitute a sensitive front, semi-informed, an extremely effective potential against composition rising from the New Orleans base.

Before we further support and define composition in jazz, let us assume a listening hierarchy among New Orleans adherents united in compatible tastes. Moving in subjection to a great style (where there is a composition-potential), my musical exaltation takes on a much deeper significance. In many lesser New Orleans records extraordinary items[4] are scattered throughout. If I must choose between such a record and a record of great compositional value without these precious points of arrest for the listener, I am compelled to choose the latter because of the traction generated by its structure. The process is not easy; there is always a playing style of high order and the fabric is always closely bound into the distinguished New Orleans complex.

I have stated my respect for the revivalist band. Where does it stand in our listeners' hierarchy? The experience is fixed: a simulated playing style can do no more than approximate the original. Anything else would cease to be the virile and characteristic playing style that we know as New Orleans. But an extended revivalism consistently created within the New Orleans style would be another matter: an accomplishment well within the capabilities of a composer. But is it playing style only that is paramount? Given a *pleasurable* playing style, will not

4. Either remarkable material impossible to notate or fragments of melodic excellence.

highly charged melodic lines, finely integrated with each other, carry us further in the realm of serious music than playing-style-dependent music? There is no question of where musical worth lies when we move from the vitally rhythmic style of the average contemporary dance music to the frequently dull performing of classic music. My involvement despite my perennial amazement at the brilliant scatterings in all New Orleans records remains with compositional extension. However, I have yet to hear a revivalist record accomplish such an extension. It must be remembered that the revivalist bands still work composerless within the practice of improvisation. But whatever the import of a composer-revivalist band, this music would be the only live music played in the old tradition and the only means of winning a future public. If nothing else, it would prepare it to appreciate the original New Orleans records.

Another reason for a composed jazz is the necessity for a new public. A performing art is radically undermined if its public leaves it. We must remember that a performer's art needs a fairly large listener base to achieve a satisfactory existence. In the past this was furnished for jazz by those classes which always react keenly to the popular arts. The attraction held by New Orleans jazz for its audience did not emanate from the significance of its melodic content. It was kinetically suggestive like all popular music, together with the extraordinary rhythmic quality brought to this dance music by the Negroes. The classic music of today seldom produces these effects in a wide audience section. It can readily be seen that when this type of music is truly significant its attraction will encompass many categories of listeners. On the other hand, today, and no doubt tomorrow, the popular audience will find solace in some popularized progressive type, or even a dilute Dixieland containing progressive traits. A new audience must be built, an audience seeking more permanent values, within reach of the classic tradition. Although in the best of jazz the popular and the permanent meet to their mutual enhancement (and there are listeners keenly aware of this), the supporting listeners as a whole fall into either the popular or classic group.

A classic audience will lean to melodic content, for it is this element that retains the permanent factor of the music. If it were not so, then classic music would not have its massive authority for listening. Where the playing style remains and the melodic content deteriorates, the interest can only be one of superficial admiration. Although a classic audience may not fully appreciate a playing style, this is the only audience that can conceive the musical demands that will save jazz.

The best entrance to a style that does not exist around us is through a music which is primarily well-composed; one, surely, in which the general pattern is rather obvious. A composition depending upon melodic lines of great invention in which there is no obvious outline pattern is also difficult to assimilate.

Examples can be seen in the fugue of classic music and in many New Orleans records. Generally speaking, we tend to follow a markedly bold composition whether we appreciate it or not. On the other hand, we enter a playing style, as we enter a bath, and we are static until the beauties of the style emerge. They appear when we are sufficiently occupied with the music to facilitate our appreciation of both the composition *and* the playing style.

A composed jazz has therefore the best chance of winning a classic audience. The anti-composition, anti-revival New Orleansians will of course receive it more slowly if at all. However, acceptance, reluctant but substantial, may come from this sector if the composer is as successful as the improvising players.

Period opinion has often been wrong. J. S. Bach's musical thinking was so disoriented that his works were shelved for a hundred or more years. The art concept growing in the minds of his contemporaries and persisting throughout the 19th century excluded the manner of musical composition found in Bach. An age builds up a criterion which gradually assembles in the consciousness of composers and listeners alike, conditioning their whole art outlook. Composers strive for a mode of composition and listeners entertain an expectation that may be ruinous to art and sound art-practices not yet firmly established.

The change in music may very well have been timely. On the other hand, we question on the grounds of appreciation the complete temper of the period that ushered in the Romantic movement. However right it may have been in instituting this creative change, it surely was unaware of the import and value of the music it was displacing. And few of the creators of the displaced material were capable of evaluating that material properly.

Does this appreciative apathy exist today in the same climate that encourages the phenomenally fast jazz development? Was the period of change in jazz so brief that a maturing development was displaced by an acceleration bearing traits that are detrimental to present-day improvisers? Does the jazzman today find himself burdened with advances in harmony, an entirely different style of melody, and a totally new aspect which he is led to believe must be taken into account as the *new* tradition?

The progressives have not as yet established a new tradition to supplant the old. So long as the old is active (as it is) in both New Orleansians and Dixielanders, we can say that the progressives are no more than a trend. True enough, New Orleans retains a very small segment of practicing musicians. Dixieland and the revivalists are about holding their own with the progressives and pre-progressives (like the Hawk and others). What we have is a group of musicians playing furiously, with the greatest possible freedom, in the old attitude. It is in no way a music at the end of a sophisticated era that has been unable to maintain the material pulse of art.

Too, we must remember that there are periods of transition as well as development. In the periods of transition we have witnessed a negation of both complexity and extreme harmonic diversions. Polyphonic complexity gave way to monody with accompaniment and simple harmonic progressions appeared. It is the attitude that we *must* benefit by proliferating modes, held by both the modern academic and progressive jazz schools, which is fallacious. There may be artistic determinants requiring certain turns in art, but the artist should always feel free to curtail any practice no matter how arrestingly resourceful, newly arrived, or venerable it may be.

In any event, such a limited activity as New Orleans jazz would, for the modern academic composer, probably be a negation of all his personalized aspirations. While this is his individual and group problem, still it seems important to restate and demand recognition of the fact that certain conditions or climates have always attended the founding of music historically satisfying *to all levels of musical taste.* Compelling in its early stages, it has remained so in its complicated ones. All well-founded academies can be traced back to improvisational activity, and jazz cannot be classed as a different species of music any more than the academy up to the time of Bach can be marked as different compared with its own improvisational and semi-improvisational periods. Bach is very far from music consisting solely of improvisation, of course. Post-Bach development, however, I suppose, could be said to introduce a different species, built, we must remember, on that of the old.

I believe that we Americans were, and may still be, placed to create the greatest music in the best classic mode since Bach and his predecessors. However, if we allow much more time to elapse before we commence the labor of developing our heritage of jazz, it will no longer be a feasible project. Early New Orleans will be too far removed. Certainly jazz is the most significant music of recent centuries.

The Importance of Dance Style in the Presentation of Early Western Instrumental Music

The Music Review, November 1955 (England).

I

There seem to be two divergent schools of thought as to the proper way to play the music of Bach. The most strongly presented appears to advocate this proposition: superior taste is invested in a playing called "pure." This is in contradistinction to the "Landowska" approach which is called theatrical, pounding and shot through with affectation. In the interest of a clearer critical climate I would like to point out that theatricalism or plushy effects are not always indications of a total lack of present art; and that criticism based on such an exclusive approach is fallacious as criticism and deceptive in that it is only a half-truth.

Let us say that any specific way of playing, whether in the way of a contemporary culture or in an approximate conjecture of a past culture, may or may not include theatricalism, extreme sobriety, the personal idiosyncrasy or the flat surface, but in each case the approach must be set on a *typical* basic style. Basic style must be thought of as a positive entity, typical of a specific place and period. Without first acknowledging a basic typical style, the criticism of the presence, or lack of presence, of any one of many interpretive characteristics is secondary; that is to say, a significant musical presentation is not necessarily achieved by an outright dismissal of theatrical or Romantic attitudes. Let us also say that the recourse to academic high polish as a substitute for *typicalness,* is a negative approach and as presently practised reveals the paucity of our academy when turned in upon itself. An academy capable of extending art phenomena is, of course, something else again.

As a first proposition it might be stated that perhaps no license is too great if it helps to bring out the intensity of a music. If greater artistic emotion is aroused more by one device than another, it seems to me that even proof of a different and past interpretation should not deny present listening pleasure merely for the sake of establishing a quasi-authentic reconstruction. The so-called historically correct, or authentic, attitude is artistically meaningless if a contemporary listener is robbed of artistic enjoyment for the sake of available historical accuracy.

In performing material of the past, interpreters seem to lean either towards an emotional spiking of the music or cautious attempts to reconstruct the authentic. However, so rarely is a performing artist absolutely within the one school or the other, that it seems almost useless specifically to identify him thus. What we are forced to examine, therefore, are the extreme aspects presented by both approaches. Here neither one nor the other, of what might be called *permissives* as opposed to *authenticationists,* seems to hold a good answer for the problems presented. Possibly there is only one final criterion in this matter, namely taste; this is suggested with a full sense of how untrustworthy taste may be—even the most sensitive. In any case, the authenticationists seem to overlook the fact that any period is historically incapable of reacting emotionally to past artifacts as did those of the past—the difference in art climate being the determinant factor. On the other hand the permissives seem to overlook the fact that the past does not easily reveal itself and that too much contemporary drama can easily hide it.

The authenticationists forget that if all art performance is necessarily suitable to its day, reconstruction of that day cannot be said to be factually complete unless we have considered, among other things, not only the size and acoustics of the places of performance, the traditional ways of playing, but also the art conditioning of the listener to the art values of the day. Thus, it is unreasonable to present a restricted reconstruction in the name of authenticity, and set it up as an art value when we know we must have altered, or are deliberately altering, one or more past characteristics. It is even more unreasonable to presume to do so without recourse to some compensating alternative.

If, in the performing arts we have observed all known conditions of past performance of a music and then find everything has been taken into consideration save our own contemporary emotional engrossment, we have failed to present the one important factor, active enjoyment, which should be held in common by both past and present listeners. Presuming we have this opportunity to choose that art with which we wish to live, we can choose both our manner of living with it and, in the case of the performing arts, the way it best comes to life for us. It is only after going through a process of listening to all available renderings that we discover the one which we enjoy the most—and that should be the one with which we select to live.

For the listener who enjoys a Bach organ toccata in a Stokowski orchestral arrangement, it is hypocritical to forego this pleasure. To those who admit to this preference for the latest instrumental colouring and who concede that such taste is the result of contemporary musical exposure, it seems to me it must be demonstrated that a less debased taste is more enjoyable. We cannot enjoy an interpretation simply because it is more authentic. Genuine listening pleasure can only be significantly aroused by introducing satisfactory art substitutes for lost factors; it cannot be achieved by erecting an art ethic based on authenticity for authenticity's sake.

We cannot bludgeon unpleasurable listening into satisfactory listening by the mere imposition of an art ethic which depends solely upon authentic reconstruction of instruments and intellectual attention to written material. Such activity, by itself, provides no platform for significant listening. If the fact that Bach transcribed Vivaldi's violin concerti, not to mention his continuous occupation with the alteration of other men's scores, passes without arousing adverse comment from present day authenticationists, it would seem that the re-ordered presentation of an enduring music should not be critically attacked for the fact of unusual presentation *per se,* but rather criticized in the degree that its presentation seems to bring out, or not to bring out, significant quality for us.

On the other hand, the permissives must remember that to reach into the past, listeners must put in time doing what might possibly be considered dull listening, in order to receive a more significant response in the future. Since saturation in a time art cannot proceed by fleeting glances, hours of attention must be devoted to the process. Real exposure to any music is a matter of deliberation on our part—of deliberately setting aside time for artistic incubation. In its more remote aspects this presumes some patient boredom.

If our enjoyment of a past music lies in the something created which we no longer create, then, where it is possible and *pleasurable,* an approximation to a past playing style may well augment this satisfaction. That empathy which invites us to play a past music in the first place, also encourages and fosters the desire to play in a past style. In fact this very going back to the past seems to indicate that we are in search of a satisfactory something existing then, which we do not have now.

II

The harpsichord, as developed in the seventeenth and eighteenth centuries, is unique as an instrument. Upon a mere striking, any pre-nineteenth-century music seems pulled into the style implied in the written notes. It might be said that this limited mechanism of plucked strings immediately provides a precise unadulterated version of the original tone-quality which prevailed at the time. I believe the harpsichord to be one of the most significant instruments of Western culture. Its pre-eminent significance derives from the very fact that unlike the violin family or lute, what is crystallized in its own construction cannot be tampered with.

The fact that modern harpsichord reproduction provides for more pedals than those obtaining in the eighteenth century or that these new acquisitions are over-exploited by certain harpsichordists, adds up to, at least, a minor premise for rejection of either the instrument or the interpretation. The mechanical variation and differences which fluctuated in the harpsichord field during the long stretch between 1600 and 1750 cannot be seriously influential in our choice of mechanisms when playing the music of these epochs. Carried out logically it would become ridiculous, leading us to shift from the use of Bach's own instrument in order to play authentically the Frescobaldi of a hundred years earlier—an exchange which most certainly Bach himself did not essay.

Parenthetically, since a large proportion of early music was set for the voice, I should like to suggest that the singer possesses the instrument capable of the most varied performance of all and should show evidence of it. The present so-called "round-tone", so determinedly a result of our vocal training, has lost all *typicalness;* it has become merely a voice with no place in history. This is part of our Platonic trend in music making, which tries to build an instrument into an archetype serviceable for the music of all time. This concept of the possibility of an ideal voice is part of the same conception which holds it possible to promote an ideal music, namely, a music which is detached from any period. We find even Bach himself avoided the ideal in this sense by keeping distinct certain styles of his own time—for example, he specifically indicates the identity of a French *courante* as distinct from that of an Italian *corrente.*

A voice intonation, when strong, engulfs the thing to which it is attached; its message to us is so immediate that it can relegate the music itself to a background position. Thus it might seem in the case of a very positive intonation, eminently suited to one *genre* and conversely ill-suited to another, that it would be better for the music if the singer could fall back on this academic, round, negative approach. But, is round-voiced intonation our only and best possible solution? Need we, when training a voice, wash it clean of all its natural qualities? Perhaps when extending this voice, we should, on the contrary retain the natural inflections inherent in it and allow the characteristics of speech to find themselves in the singing. We find for instance that an intonation somewhat similar to Yvette Guilbert's, without, of course, her diseuse paraphernalia, has more affinity with early music than any accepted intonation of to-day.

III

The contemporary performer is definitely a product of a limited musical situation. Upon graduation from any academy, he can only hope to gain a status of creative individuality by seeking to express unique emotions suggested to him by the music he is playing. He does this by bleeding the climax tones within a melodic passage, or more disturbing still, by interfering with the basic beat. Of course Romantic music, composed for Romantic performance, if not so treated appears vapid and anaemic. However, when the academic performer tries to mitigate his Romantic excesses in the playing of Bach, he always appears to be merely holding in. Generally speaking, the Romantic attitude is so ingrained that the slightest moderation of this tendency convinces most academicians that they are playing within a restrained classicism whereas, on the contrary, they have merely achieved an overall effect of colourless Romanticism.

However, freedom-of-movement within style means we must first give up the notion that the handy directives found in critical and scholarly comment of this or that past era can be followed as style directives *per se.* They are no more than a *road* to style. Second, we must also give up the notion that by limiting ourselves to precisely reconstructed ancient instruments we will automatically engage in a fine playing style. The pursuit of purity for purity's sake—authenticity for authenticity's sake—does not lead to a vital art centre.

When a musician enjoys reading-over a composition, including halting and repeating certain chords or intervals to retaste the flavour of their strong appeal, he may be said to be engaged within a climate of *home-rendering.* For a musician, or music appreciator, set in this climate of home-rendering, for whom listening pleasure is derived directly from actual interval juxtaposition, it is quite true that a vibrant playing style may well be a disruptive interference. But in the climate of the concert hall, it seems to me, we are right in anticipating something more than coldly conveyed intervals. We expect at least the minimum of platform experience, a performance element added to the notes and extra to the delights of leisurely interval perception. Although it is true that these intervals may be partially smothered under performance impact, nevertheless, such a situation is not rectified by merely limiting the projection to a home-rendering—no matter how circumspectly agile.

IV

Is there, then, any criterion of interpretive approach which harpsichordists, and others, can securely follow? If, in fact, there are worse alternatives, is there no better alternative than the approximation to the authentic? It all depends on what the critic expects when he walks through the concert hall door. If, in a concert hall, he is satisfied with but little more than the home-rendering presented with professional continuity plus the concomitant rigid requirement that every note, turn and trill be value-intact, then only their lack of presence will lead to adverse criticism. On the other hand we have the critic to whom the best of such playing is all very well at home, but for whom the concert performance must project an excitement beyond a well-practised home-rendering.

There are very few directives that have come down to us about *playing-style.* The correct reading of symbols for ornaments certainly is not a playing-style directive; it is rather a problem of the correct placing of extra notes. *Playing-style* is largely determined by the accent and phrasing we give, not only to these ornaments, but to the whole music. It involves the use of *legato* or *staccato,* the adherence or non-adherence to strict time, and above all the place where we decide any of these things should happen. Obviously, no directives *describing* a style can actually show us how to reproduce it—words hold too great a latitude of meaning.

If within our own experience every decade brings about marked changes in playing-style, it is reasonable to suppose that distinct deviations cropped up within the seventeenth and eighteenth centuries. If to attempt an approximation to every turn of playing-style during a century or so, is, on the face of it, a hopeless task, nevertheless, it seems we might gather the *whole* period into one playing style and by steeping ourselves in it, slowly acquire a manner of playing that is both significant for modern ears and suitable to the *whole* period. In other words, in order to build a positive way of playing, we should try to embrace at a distance a whole two hundred years, rather than try to interpret the ancient, sporadic directives appearing from time to time on and about written music. Certainly Bach, when playing a Frescobaldi fugue, must have played it as he, Bach, would have played one of his own compositions, not as Frescobaldi might have played it one hundred years earlier.

Perhaps if we wish to establish a criterion for a significant playing-style, a basic point of departure can be found in the fact that instrumental music of the seventeenth and eighteenth centuries was a dance-derived music. Many critics concede that not only the suites with their sarabandes, courantes, bourrées, *etc.,* but the whole fabric and formal structure of Bach's music takes its inception from the dance. According to Parry,

> The connection between popular songs and dancing led to a state of definiteness in the rhythm and periods of secular music . . ." and ". . . dance rhythm may be securely asserted to have been the immediate origin of all instrumental music.[1]

In this connection we perceive that if this early dance music had merely been seized upon and exploited by the academy of the time, rather than, as in actuality, having been the academic mainstay, the dance influence most certainly would never be so unmistakably present throughout the whole bulk of the work of so late a composer as Bach.

It is inevitable that the musicians' playing-style conjoined to the written music of one generation will not be carried along in its original rhythmic import by the next generation of performing musicians. Actually, the dance-playing-style of musicians comes out of playing *for* dancers. When and if the dance activity is discontinued, the musician must go far out of his way to continue or re-implement the old playing-style. The composer, on the other hand, is in a position to do this very thing. If, for example, a type of music, growing out of improvisation connected with playing for dancers, is written down by one of the contemporary improvisors, the composers of the next generation, although not having had the benefit of working with dancers can, nevertheless, as readily compose within the same old dance style as did the improvisor-composer of the generation before.

Away from dancer-activity, the performer's contact with the traditional stylistic "feeling" for dance time is soon broken; his feeling for time has degenerated into a mere keeping of time and the keeping of time may be no more than the punctilio of the metronome. Although such metronome time-keeping can be accomplished more or less easily, unfortunately the doing need not release feeling for time, or what can be called *beat.* It is significant *beat* that is the vitalizer and life-giving force not only of dance music but of all dance-derived music, and according to Parry, even "all instrumental music."

As soon as a playing style leaves the imperative of improvisational or quasi-

1. Sir Hubert Parry's "Dance Rhythm", *Grove's Dictionary of Music,* 1904. Vol. I, p. 657.
Curt Sachs, *Rhythm and Tempo,* p. 281, W. W. Norton & Co., N.Y., 1953. "Eager to replace the roving shapeless monody by steadfast forms, they [16th and 17th century musicians] had no models in the flowing polyphony . . . of the early century. . . . The dances of the time, clear-cut . . . were the only point of departure."

New Encyclopedia of Music and Musicians, p. 36. "The entire development of modern musical style has been affected in various ways by the influences of dance-patterns and the *dance-spirit*" (italics mine).

Thompson's *International Cyclopedia of Music and Musicians,* "Dance Music". "Independent dance-music for instruments must have existed throughout the Middle Ages, but practically none of it has been preserved, since it was handed down from generation to generation as an improvised art."

improvisational playing for dancers, it suffers change. So the effort at recovery of a period playing style, which itself is the result of lost contact with dancers, presents a real problem. Thus it may be we will always flounder in the interpretation of seventeenth- and eighteenth-century written music if we fail to imagine the dance impetus which first instigated the style. Few keyboard artists are concerned with this attaining of a significant dance beat. The practice seems to be either one of avoiding strict time or pedantically following it; implied beat is rarely present. We have seen how musicians when playing from written music, soon acquire a slackness in the performance of a past dance playing-style whereas a new decade of composers, who continue writing in such a style, lose none of their force. It is my contention that the significance of all dance-derived music not only is seriously authenticated by strict dance pulsation but is further emotionally heightened by adherence to it.

What possible way have we of getting back to the playing style of a past dance music? Well, for instance we know that all present dance music is played in strict time and that each contemporary dance has its own beat or stress. Perhaps, to reconstruct a past style for playing the minuet, we should play many minuets in strict time. If we get a pronounced feeling from one rather than another this feeling can be incorporated in the playing treatment of the other minuets which in their written state are perhaps not so appealing or pronounced in style. This keeping of strict time and the search for points where a beat may be stressed might establish dance-conscious contact with the whole period. Moreover, such application of a dance-playing-style to written music will have value in itself regardless of the intrinsic value of the music. Although the resemblance between our reconstruction of the original dance-playing-style and that style itself would be hard to establish, at least there would be a dance-like time element common to both.

V

The Bach period adds to the difficulty of style-finding by being a period which itself was undergoing change. At that time although there was no such demarcation line between a Romantic way of playing and a non-Romantic way as obtains now, nevertheless players and composers themselves throughout this period must have alternately stressed for one reason or another either the style which was on the way out or the style coming into vogue. It is interesting to speculate whether this revolutionary situation existing around Bach, permitted Bach at once to adhere to a style of composing and writing which was on the way out, and on the other hand to be involved in a playing-style peculiar to his own moment in time!

Taking this historical state of flux into consideration, it seems to me it is now conceivable that the playing-style which prevailed *previous* to the advent of Bach, was, and is still, a better vehicle for his music than the style fashionable around and about his own time. An approximation of the driving dance-style of the period *previous* to Bach's might invest the playing of his music with style—especially if the current playing-style of the later era was well on the down grade or decline. If the concert playing-style had deteriorated in intensity—as compared to the composed music itself—the earlier playing-style will actually correspond more in art stature to the written Bach composition. In this way we might find a more significant playing-style for Bach's music, that is by feeling out the *previous* era, and at one stroke by-pass controversial dogmatic reconstructions of the so-called authentic Bach style. In any event the ephemeral character of all playing-styles up to the advent of recording devices has always denied the removed player direct access to the playing-style of any era save his own.

It may be suggested that if, as I believe, the dance rhythms in Bach's compositions derive from an earlier *improvised* dance music, it is inappropriate to annex the earlier playing-style; that it is not suitable for a music which in its compositional texture has changed somewhat from that of authentic dance music. This may be true, for no doubt there are elements in the music of Bach, coming as it did at the end of the era, which would be hidden by a too positive early dance-style—elements which might be lost in a welter of dance beat and intonation. But it is only when we have found some intimations of this original dance-style that we are equipped to mitigate their over-forceful expression. As it was emotionally charged strict time which was a strong element in the generation of the original improvised dance-playing-style (out of which grew the written compositions) so might a re-emotionally charged adherence to strict time give us real intimations of this lost playing-style.

Thus, if we grant vibrant art does not spring from the ranks of our present conservatories, and surrounding critical comment is equally constricted, perhaps by always thinking in terms of dance beat we would be making an enormous contribution towards the bringing to life of Bach's music—bearing in mind that Bach himself may not have performed in this way. As a concert interpreter of that era, he may have kept strict time or he may not.

In fact, it is conceivable that Bach himself would have been as incapable of putting his finger on the characteristics of his playing style, if strict time had already been abandoned, as we would be in explaining the stylistic involutions of our own Romantic era. The eighteenth-century performer using notated material must have undoubtedly engaged in a playing-style filtered through the direct communication of the era. However, whatever literary comment has come to us about the playing style of the time comes out of a climate forever lost to us. So often the pervading thing, so taken for granted at the time that no need is felt to mention it, is the one thing we need to know.

In conclusion, let us say, then, that for all instrumentalists and more particularly harpsichordists, it is possible to express as much performance excitement as is felt necessary without tampering with the metronomic keeping of time one *iota;* and whatever mannerisms are used further to express excitement, the practice is legitimate—*so far as the practice itself is concerned.* Let us remember that every performer who strives to express the spirit of the music always alters, perhaps in an imperceptible way, the indicated note values—as well as strict time. It is only by doing one or the other, or both, that he feels able to give life to the music at all. But however he may alter it, it is a fallacy to believe that an expressive effect can *only* be achieved by altering the so-called mechanical effect of strict time. Strict time may not offhand appear to create a foundation for freedom, but inevitably it makes more telling those minute quality variables in interpreting written notes and those dynamic stresses which tend to vivify not only the feeling for beat, but the melodic line itself. I cannot emphasize strongly enough my conviction that as soon as strict time is even minutely tampered with, the performance loses an important retaining virtue and the performer is tempted even further to dissociate himself from the beat in order to express himself.

It seems to me it is the attitude that a flamenco guitar player holds towards his music that contemporary harpsichordists should seek. In their commendable effort to put over performance excitement they should employ other means than contemporary academic indulgence in over-expressed free-time feeling. Especially is this so in a *toccata* or *fantasy* where many derived *genres* of music shoot in and out. Here, like the flamenco musician, the harpsichordist should project excitement through the feeling he imparts of keeping time at all times—no matter how many instances of changed beat are indicated. Especially must his *rallentandi* be definite. Stylistically speaking, the *toccata* and *fantasy* are always difficult to handle and should only be attempted by those seriously conscious of the classified *genres* separately.

VI

A concrete evidence of a concern with style closely linked to dance-rhythm is Wanda Landowska's contribution to the playing of seventeenth- and eighteenth-century harpsichord music. Long before the appearance of the present generation of harpsichordists, Landowska gave to the music of Bach a performance whereby the notes seemed to flow improvisationally out of her style, rather than the usual situation of a *romantic* or a *dry* style applied to the accurate reading of notes. This was a liberation of Bach from the tyranny of what had come to be expected in the recital hall. It gave to his music a grandeur that only the instrumentalist conscious of *creating before* an audience, not *playing for* an audience, can give. It gave us intimations of a grandeur of playing-style—playing style which, in its own way, can be as self-reliant as written music. To what degree Landowska has succeeded in achieving this fusion of art is debatable, but there is no doubt it was she who showed us that such a thing was possible. To say that Madame Landowska never errs in taste would be far from the truth. But her faults are quite separate from the fact that she instigated a style which is an inspiration in itself.

All in all, I think it can be said the difference between the Landowska school and all other schools of harpsichord playing is the difference between the presence and non-presence of a highly charged dance concept of music. The drive and potency of her style, when and if it is controlled by strict time, is very striking as opposed to the uncharged maintaining of *tempo* so generally exhibited by other harpsichordists. The advent of the present *purist* criterion has climaxed this approach by gauging all interpretive values on the basis of *authenticism.* It is to Madame Landowska that we owe our thanks for filling this negative hiatus with her vibrant style of playing which is so definitely positive without in any way being Romantic.

In summing up, let us say we have no real directives for playing a past music other than the preserved music itself. If casual critical comment of that time suggests that the playing was somewhat different from what the notated material would lead us to suppose, this still gives us no real insight into the actual playing style; for a corrective adjustment here and there, though naturally of some help in recapturing a past truth, is not a directive sufficient for establishing an easy pulsating style. This being the case, I have suggested, therefore, that a deep respect for dance-activated rhythms may be the most perfect approach in bringing to life those written compositions we are trying so hard to recapture two hundred years after their inception.

Jazz and the Dance

Let me add a few words to Stanley Dance's excellent defense of the dance element in jazz. It is not so much the fact that a band sounds better at the Savoy Ballroom than somewhere else or better than non-Savoy bands with an inferior beat. Our concern is with the difference between a music which is either actually played for dancers or retains a strong dance spirit, and a formless music arising from material without a naturally strong ground-beat or one which has returned to the emotions fostering this beatless music.

I believe that Stanley Dance could strengthen his argument by not pinning himself down to particular observations. For even in European music of roughly the last three hundred years the far-reaching influence of a past dance persists. Not only is the same form and style evident in the ectypes of a former dance music but the very spirit of this dance persists throughout a music's later development as in the case of the symphony.

Whether cool jazz has the strong beat found in other jazz or not, it is dance music to the core and those who say it isn't seem unaware of how completely different it would sound were the dance not very much alive in it.

Jazz Monthly, September 1956.

The Problem of Form

Jazz Monthly, July 1957.

Max Harrison in speaking of form in jazz says: "Any movement which tries to create jazz without improvisation and the blues is doomed to failure . . . ," and then goes on to suggest that a future jazz will come from a composer like Ellington. True enough, Ellington in his early pieces works through improvisation towards a loose composition, as did Jelly Roll Morton, but the markedly Ellingtonian works consisted of far less improvisation. In fact his ambitious works consisted of none. I cannot see why Harrison feels that improvisation, according to his own acknowledgment, is such an important means in jazz today.

Harrison, among many others, seems to feel that a "fixed and ineluctable chord sequence" is a limiting factor to jazz. We must remember that the three changes under a blues melody do not in themselves give any real feeling of beginning and final termination. It is the superimposed melody which does this and which can be altered to fit any desired result. Harmonic changes, in one sense, can be looked upon as no more than a convenience, making it possible for extraordinary group improvisation. It makes musical sense out of heterogeneous thought.

The use of other harmonic changes or harmonic coloring is quite another thing, certainly extending the musical horizon. Such changes and coloring also furnish movement in those sequences where the melody remains on an organ point. While an expansion in music is an enrichment, restrictions are not always evidence of a too severe confinement. Eliminate the melodic lines of long sections of Beethoven's symphonies (certainly not a music in its infancy) and we are left with merely the tonic-dominant-subdominant underpinning. It would be naive from such a foundation to presuppose a melodic paucity.

In examining a long hot solo on the blues we find that the beginning of the first chorus differs from the beginning of the subsequent choruses. Also we find that the ending of the last chorus differs from the endings of all preceding choruses. The beginning of the second chorus of Miley's solo on the *Black and Tan Fantasy* does not suggest an initiation itself but rather a continuation of something. By making certain radical melodic changes at the terminations and beginnings of each chorus within a jazz piece together with the right choice of register and style, a feeling of repetition can be obviated. By this means we get a strong impression of an integrated composition rather than a series of variations on a given theme.

As Harrison acknowledges, a solo on the three changes can pass through a maze of harmonies. But if these harmonies are also suggested in the accompaniment, and in this way follow the course of the melodic line, the linear freedom of the two lines (melodic and harmonic) is lost. The biting and desirable consequences of the clashes bound to occur between these lines are reduced in favor of a smooth and less exciting quality.

Harrison holds that the adherence to the twelve- or thirty-two-bar chorus is limiting. As to the thirty-two-bar chorus, it is merely a da capo aria which has served classical music. Too, it is made up of eight-bar sections, a numerical length to which most Western music, whether simple or highly developed, seems to adhere. But let us examine the twelve-bar blues. The feeling of the blues as sung would be violated if we tampered extensively with the two-bar break. But an instrumentalist, on the other hand, so completely integrates the melody and break that he can easily drop a bar at the end of one chorus and for that matter a bar or two at the beginning of the next chorus without anyone being aware of it. The *addition* of bars is very easy. We must remember that the end of each blues chorus plus the beginning of the next chorus contains a total of six bars on the tonic. If the accompaniment does not indicate a termination, the soloist can easily bridge these six bars by a contraction to three. In ensemble improvisation it is not feasible, for freedom in improvisation can only work within a set-up that can serve both a song and an instrumental rendering. Music itself, however, does not depend on such regularity although it is only in composition that an ingrained regularity can be deliberately broken. As it is within the province of the melodic line either to point up our feeling for these groups of twelves or completely obliterate the feeling for their periodic character of twelve, I see no urgent advantage in breaking such a convention at this time.

RESPONSE

Max Harrison

There was no actual contradiction in my article when I named improvisation as one of the essentials of jazz and then suggested important future developments would come from jazz composers. Improvisation has been present in all of Ellington's best work but it has been set to the greatest advantage by his individual and accomplished scores. The Ellington bandsmen would have made a remarkable group in any circumstances but it was Duke himself, with his composing gifts, who moulded them into such a unique force in jazz. Similarly Milt Jackson and John Lewis could contribute fine improvisations to any group, but Lewis has essentially a composer's mind and it is his arranging and composing that forges the extemporisations and ensembles into such unusually satisfying wholes.

As to the limitations of repeated chord sequences, I agree with Mr. Dodge it is possible to construct a long, non-repetitious line over a repeated sequence, but however free-flowing the melody and however many passing chords it may imply it is still ineluctably based on the twelve- or thirty-two-bar unit. Thus however much melodic freedom may be achieved no advance or enlargement in formal structure is possible.

This brings us to Mr. Dodge's astonishing point about the Beethoven symphonies. In support of his view of the relative unimportance of the harmonic basis of a work he suggests that the removal of the melodic line from long sections of these works leaves nothing but a comparatively simple harmonic support. It is not a matter for discussion in *Jazz Monthly*, but Beethoven's great achievement in the craft of composition is his development of form and this is largely a matter of the structural use of harmony—itself an enormous subject. I am not suggesting harmony is a thing apart—in a good work melody, harmony, form are fused into one—but to see, or rather hear, nothing more in Beethoven's great architectonic achievements than "merely the tonic-dominant-subdominant underpinning" suggests the ownership of a somewhat unusual pair of ears.

In his final paragraph he would have been right if he had spoken of the prevalence of eight-bar sections in most older Western music but the older classical ways of thinking has been disintegrating since the time of Wagner. Mr. Dodge will find it more than difficult to hear neat sequences of eight-bar sections, each divided into four-bar phrases, in the work of composers like Berg

 Jazz Monthly, September 1957.

and Webern. And although jazz is technically a long way behind straight music, they are not so common in the improvisations of Parker or Navarro.

FINAL ANSWER

Roger Pryor Dodge

In answering Mr. Harrison's reply to my letter, I want to take up that point which I feel to be the most important one in our discussion: my statement in defense of the use of the three changes in jazz.

Mr. Harrison correctly paraphrased my thought that long sections of Beethoven's symphonies are on no more than the three changes; however, further on he suggests that I must be deaf to the actual architectonic progression of melody and harmony if I only see a tonic-dominant-subdominant underpinning. There is a difference between long sections and complete works. And besides, I never suggested the relative unimportance of the harmonic basis. I merely suggested that the presence of an inspired classical melodic line can exist on but a few changes of harmony.

In clarification of this point, let me say that my contention is that the use of three changes in the blues is far from exhausted. I can understand that for the unsupervised jazz band monotony may result, but with creative supervision I foresee the possibility of major works developing from the use of no more than the three changes. I say this because I am not satisfied with the creations of those jazz composers and jazz players who consider the three changes outmoded. Neither do I see any reason for adhering to a development vitiated by the inept use of classical music, to say nothing of a further debasement of jazz by concessions to popular music.

I feel it important to point out that even Beethoven, who possessed such great harmonic invention, did not strain the use of such resources, but composed long sections of his masterful works on no more than these three changes. And let me say here that I am not taken back to merely a simple three-change music when listening to passages in his symphonies where they are employed.

Any disparagement of the three changes implies that all such music gives little more than the impression of a shuttling back and forth between the three chords. Because they underpin a section or, for that matter, a whole composition, does not mean that our impression of this music is one of meagerness. While I am certainly not proposing a three-change limitation on jazz, neither,

Jazz Monthly, December 1957.

on the other hand, do I feel that the ills of non-Ellington and pre-modern jazz should be laid at the door of the three-change proponents.

Parker and Navarro solos are not the only ones to deviate from the neat division of eight bars into four, etc. While the music used usually falls into simple arithmetical divisions, most good solos of the 1920s and 1930s cannot be divided this way.

Maybe I did lift Mr. Harrison's statement about improvisation out of context. However, I gather from what he says that he believes in improvisation plus guidance, that is, composition.

The question of twelve- and thirty-two-bar tunes needs more elucidation than is possible in the space of a letter.

In re-reading Mr. Harrison's original article I find a few items I must strongly disagree with. 1) That jazz is a minor art. 2) His derogatory use of the term "*string* of twelve- or thirty-two-bar sections" (italics mine). When I think of the *string* of dance tunes in Bach's numerous suites and *string* of eight-bar phrases prevalent in classic music I begin to wonder. 3) "Solos like that of *Bird Blues* are as much in the pure tradition [of blues] as anything by Dodds or Bechet." The term "pure tradition" seems too inclusive if we are to think of the blues as Bessie Smith and Ma Rainey sang them accompanied by either a Dodds or Bechet. Mr. Harrison's "pure tradition" could include anything in which there could be shown a continuum with the past. 4) "Folk tunes always have a freer structure than jazz pieces" and "that men like Armstrong and Bechet, with their enormous improvisational ability, have been responsible for typing down their people's folk music and leading it towards disciplined structure." I know that the self-accompanied blues singers played and sang blues of 11, 11½, 12, 12½, and 13 bars but it was not Armstrong who introduced disciplined structure into this music. It is a natural requisite, it would seem to me, of ensemble playing. Armstrong et al. break this discipline up as much as possible with their long pick-ups and their melodic thought running through the conventional two-, four-, and eight-bar conventions.

Objective Criteria

I fail to see that John Reddinhough (June *Jazz Monthly*) has made any point at all for an objective criterion of art. He makes no distinction between what contributes to the creation of an art as reflected in the work—"sincerity" or "background" or what you will—and some yardstick of our own we are to hold up to his work.

It's all very well to "fall back" on some supplementary theory to fortify our judgments, but I cannot see how this knowledge is going to help us make our decisions in the first place. There are thousands of sincere artists saying nothing and others with every apparent advantage of background who are also saying nothing. So where do we start?

Mr. Reddinhough seems to dismiss each of his enumerated criteria if they proved embarrassing alone while seeming to feel that taken collectively they will do for a start. He says: "The trouble, so far as I can see it, is that it is almost impossible to present one's likes and dislikes objectively, and yet unless one tries to do so criticism becomes nonsensical." I think he is wrong, as so many critics are, in trying to fill a vacuum with just any argument at hand. It never occurs to them that perhaps this critical vacuum can never be filled. I find that his reasons for dismissing any one of the enumerated criteria are unimpressive. He seems to believe that sincerity—outside of the diseased mind—is one criterion. Following this argument, I see no reason for him to call Stan Kenton "phony." Kenton as a person may be phony, or horribly sincere, but in either case his work would appear to be the same. Reddinhough also says: "Thinking only of the music . . . rather

Jazz Monthly, October 1957.

than the attitude of the musician, leads a critic such as Hodeir to dismiss a sincere and good jazzman such as Mezz Mezzrow." Certainly Hodeir would not dismiss Mezzrow if he thought he was a "good jazzman." Hodeir doesn't see Mezzrow, sincere or not sincere. It seems that Reddinhough assumes that Hodeir accepts *as a fact* that Mezzrow is a "sincere and good jazzman" but that in spite of this he, Hodeir, perversely does not like his music! This does not seem just a slip on Reddinhough's part. I see too many critics treating an opinion as a fact—particularly their own opinion.

I firmly believe that there is no criterion for objective judgments at this time. When Reddinhough says: "If one expresses an opinion, one does so expecting either agreement or disagreement. If agreement is forthcoming, one assumes that this confirms one's opinion, and if this is so then some objective basis exists between at least two people." If it is a question of agreement that an artist should be sincere, *yes;* but if the agreement is to concern the intrinsic merits of some specific work of art, then *no.*

Reddinhough further says: ". . . a work of art should mirror a certain fullness of life, it should show inventiveness within its particular sphere, it should be sincere." This is all very well but who is agreed upon what constitutes "fullness of life," "inventiveness," and above all, that abstract "sincerity"?

I believe that instead of involving ourselves with abstractions we should try to isolate those specific elements that contribute to our highest pleasure in any field of art. Should the consensus of critical opinion accept our findings as workable values, then, and only then, can we draw any conclusions as to the nature of those ingredients which create significance in that art. This method, while it opens up possibilities for evaluation, does not provide us with any convenient yardstick to measure absolute worth: we still have to reckon with the *quality* of the elements in question. That a work under consideration reveals the presence of any single one of these elements proves nothing. Certainly there appears to be no infallible formula for creating art, nor any objective criterion which exposes mediocrity nor, for that matter, which definitively explains excellence.

Reddinhough ends with: "Those who insist that all aesthetic judgments are subjective have only to look into the roots of their own judgments to see that they are based on some sort of implicit standards." I myself see no other way to arrive at any valuation other than to look for the qualities that have already moved me. Clarinet playing that affects me like Dodds's I will like; something new in clarinet playing must also move me before I can formulate any judgment as to its worth.

Bubber Miley

Jazz Monthly, May 1958.

It is twenty-five years now since James (Bubber) Miley died. A quarter of a century is a long time. One of the few younger jazzmen of the era not brought up in New Orleans, Bubber's whole career was confined to the 1920s, that is, if we stretch the decade a little over a year to include the Jelly Roll, Carmichael, and Oliver dates, the period of his Mileage Makers, and his final date with Reisman in June 1931. These few newcomers, who either showed extraordinary excellence or contributed something new to jazz, were Bix, Lang, Teschemacher, and Miley. They were brought up on the "hearsay" of New Orleans; they learned from records and from local practitioners who themselves were once or twice removed from their own source, and only later did they come in direct contact with passing New Orleans men.

A player is conditioned early in his career by his first contact with jazz and only later absorbs whatever virtues and disciplines he can appreciate along the way. No "revivalist" attitude of dedication to the past had arisen in those days. Merely the passing influence of the giants—Oliver, Dodds, and Armstrong—on a young player is not enough to mature him. Naturally such men will be admired and even worshipped, and they certainly exert great influence, but it is the actual milieu of a growing musician that truly conditions and forms him. If his surroundings are shot through with bad influences, his own creative apparatus will inevitably suffer from them, even though he himself feels no special admiration for either the prevailing style or its exponents.

For the Negro, a good beginning in music is a mother who sings at her chores; but later the desire to sell himself in show business leads to all kinds of professional expediencies. These may prove very harmful. The quick jump from blues singer to nightingale, like Florence Mills and the hundreds of lesser Florences, the off-color songs burlesquing Negro traits (interesting in their way but not conducive to great jazz), the admiration for the blossoming arrangements, the playing up to tourist trade with simulated jungle interpretations set to popular tunes, are all ruinous to art. These impurities only bring about a confusion of criteria to the detriment of both their own music and their adaptations of borrowed material. This pattern of opportunism becomes so strongly established that the examples of the giants can only be followed at the risk of tenuous results, depending, of course, on the sensitivities of the listening musician. We must remember that what is admired and what is accepted as a guide can be two quite different things. Among the non-New Orleans men there was no attitude of prostrate emulation such as existed in the Chicago school, where in the beginning the white boys were so cut off from the sources of jazz that they were forced to sit down and absorb the best from records before exposing themselves to a milieu that might prove stronger than their will to musicianship. For Armstrong, only three years older than Bubber, to sit in with Oliver's band was a priceless advantage, something denied any player gigging around in Harlem.

The atmosphere of the late 1920s in Harlem was probably one of the worst possible atmospheres for a young musician. What with semi-arranged jump bands, jungle bands, and the emergence of the arranged big bands, a player had no decent background in which to grow. These vitiating influences were so powerful that the players received little initiative for individual choice of direction. The frantic playing that we have now at such places as the Metropole, the late Stuyvesant and Central Casinos, or wherever the older players congregate, was not in evidence then. Collective improvisation had given way to the arranger, and did not come back into vogue until much later—a little hectic then perhaps, but still and all collective. Bubber had none of this collective improvisation, at least on records or in his professional work. Although the Duke certainly featured him and together they gave a stamp to a pre-1930s Ellingtonia, the Duke was not good for Bubber. I do not say that gigging around in Harlem or that playing in any other Harlem band of the time would have been better for him, but I am thinking of how Bubber's great talents would have blossomed in an atmosphere of less jungle pastiche. He and Tricky Sam helped the Duke play a kind of jazz which Orrin Keepnews berates as an "emphasis of the Harlem clubs on a pseudo-savage motif in their floor shows and music . . . effective in its aim of drawing white audiences. . . ."[1] But the

1. Notes on the sleeve of Charlie Johnson Riverside LP.

spark of genius can always start a blaze in the murkiest surroundings. Bubber had that spark. Even so, the murk was never completely banished.

Bubber's own style was far removed from that of the popular music he was playing. It was packed with changes of timbre that included some of the most searing tones. But only when his melodic structure was in a category of music removed from the popular song did this searing intonation have real musical meaning. Otherwise he resorted to a series of clichés kept on tap for ordinary occasions. These clichés were a result of Bubber's mannerisms; he used them more and more to distort the melodic line of popular tunes. They consisted of sudden interjected barks or strident dissonances which come at us violently. He makes these interjections in order, it would seem, to break down the bland easiness of a popular tune. Built into his earlier compositions such as the *Black and Tan* they took their place with great musical meaningfulness. But I must say that when arbitrarily strewn about a catchy tune they become extremely boring. The less strident style of other great jazzmen seemed more appropriate to popular tunes than did Bubber's. They either were frank in their off-straight or luxuriant renditions, or they melodically varied the original line. If they did not achieve true creativeness, at least they did not sink to horseplay.

Bubber's playing was rarely relaxed. His was a heated performance, constantly under great tension. His intensity hardly allowed for any natural flow of musical emotion. He would be so occupied with a white-heat performance that if he could not come through with an inspired improvisation he would resort to his overworked clichés. But Bubber usually had plenty of inspiration throughout his career—and not only in the manipulation of his rubber mute. For as inspired as I find Bubber's wa-wa and general delivery, I find his musical line equally great—a musical line that survives the wa-wa, as I believe any notation will show. To categorize Bubber merely as a growl man who brought the wa-wa to its greatest height does not do justice to Bubber Miley the musician. There is some excellent wa-wa playing by Cootie Williams and Sy Oliver, but when their work is reduced to paper, without the benefit of the wa-wa, it turns out to have little to say.

Though the growl and wa-wa method derived from Bubber's plunger mute may have been instrumental in creating his melodic line, still taken alone, they were at best crude contrivances towards the creation of real music.

In the minds of some of those who know Bubber's music, and for most of his critics, both favorable and otherwise, his performance is all growl and wa-wa. Although this style of playing must eventually be judged by those who did not grow up with it, what can *we,* who did, say these few decades after its popularity? As intonation in general goes through a great many changes of style, some of which disappear only to come back into favor at a later period, we may ask if there was anything in the extreme use of mutes by Miley et al. that would warrant their being taken up again?

The wa-wa mute was certainly a strange phenomenon of early jazz. In some cases it was an effective screen for inventive paucity, while in others it was of no musical avail at all. But in the hands of real musicians like Bubber Miley, Joseph "Tricky Sam" Nanton, Jabbo Smith, to name a few, it fathered a most incredible melodic line. This line, because of the nature of the technique, is not always clear-cut, but when it can be put down on paper it sustains itself in a remarkable way, in spite of being deprived of the spontaneous quality inherent in a performance.

The wa-wa is least effective with popular songs. It can best be used on them if the improviser transforms the tune so that it enters another category of music. That is, he must first alter its catchy easiness. Otherwise the intensity and the sharp explosions of dissonance coming from a Miley horn seem incongruous.

The trouble of using instrumental devices—in this case an array of jazz mutes—to produce emotional extremes is that they carry a charge which is actually unmusical. Their tones are not the natural tones of an instrument. They are highly charged with emotion and their impact is powerful. To many people they spell jazz; and although subconsciously any listener is affected by the design of melodic content, yet the admirers of instrumental devices would, if put to the test, choose muted sounds in a poor melodic design before the most excellent design less dramatically embellished. But however great the response to the poignancy of these devices, in the end they make for a mere orgy of emotion, a state soon effected by the psychological law of diminishing returns. In contrast to this inevitable let-down, the tranquil satisfaction evoked by a work whose formal design is supplemented by a rising emotional curve gives us complete and enduring pleasure.

It may seem paradoxical that the one type of playing regarded as not much above "musical noises"—namely, wa-wa growling—has contrary to general impression inspired certain musicians to create a melodic line superior to all but the greatest open-horn trumpeters. In line with this, let me say that I have always staunchly believed that it is the melodic line which is important in jazz, and not the delivery. "Progressive" critics speak of both contemporary jazz music and that of the past in melodic terms, whereas the older critics have incessantly stressed tone and playing ability above everything else. Of course the most apparent difference between jazz and classic music is in delivery, although the ultimate worth of jazz resides solely in its melodic line. We must admit though, that when a really lusty band is playing, it is of little importance whether or not any specific instrumentalist's work can be isolated for attention. However, the critics, whether they praise or disparage Bubber for his wa-wa, still seem completely oblivious of the melodic line that threads itself through his apparent noises. A plunger mute constantly manipulated will qualify every tone that comes to our ears; but we need only listen to the average growl music to find that while it can be intensely piercing it may still say nothing because of lack of structure or inspiration in its melodic line. Whether one favors a growl style or not, the fact remains that in Bubber's work we find, throughout his growling and wa-wa, creative improvisations which are highly melodic.

Bubber was a very great musician. Although certain critics have praised him unreservedly, I'm sorry to say that the books on jazz have given him a raw deal. True enough, the most perceptive writing may not always be between the hard covers of a book, but unless the many scattered magazine articles are collected and bound, the casual reader will have no other source of information than the more easily accessible books. In some thirty-five books in English, only two or three writers are highly favorable to Bubber. In only a third of them is he mentioned at all. Ulanov clearly shows Miley to have been an important man in Duke Ellington's band.[2] It was the Duke himself who said: "Our band changed its character when Miley came in. That was when we decided to forget all about the sweet music."[3] It is only Panassié, who in saying: "I do not single him [Bubber] out [for wa-wa] . . . but because of the intelligent way in which he used these same effects"[4] seems aware that we are contending with something more than wa-wa effects.

I must say that Miley's dynamic growl, intensity, and highly dramatic way of playing always struck a sympathetic chord in me. I must admit that for quite a while my admiration for his playing kept me from appreciating any style less vigorous and less dependent upon such effects. But now that that time is past, I *would* like to make a re-valuation of some of Miley's records.

An early record of Bubber's gives him plenty of chance to spread himself. Made in 1924 with organ accompaniment and billed as the Texas Blues Destroyers, he has both sides, *Down in the Mouth Blues* and *Lenox Avenue Shuffle.* There is no clear-cut playing here as in Armstrong's early efforts, but rather an indecisive manner with plenty of feeling for the blues. The growl and mute give it a mixed intonation characteristic of such playing, and is indicative of the direction he was to take with so much masterfulness later on. These two pieces do not have the simple approach of an early New Orleans trumpet nor do they exhibit any positive sign of a coming innovator.

According to Bubber the theme of the *East St. Louis Toodle-Oo* came from the sign of the Lewandos cleaning establishment in Boston. The sign translated itself in Bubber's mind into the music of Fig. 1. Ulanov[5] corroborates the fact

2. Barry Ulanov, *Duke Ellington* (New York, 1946).

3. Shapiro and Hentoff, *Hear Me Talkin' to Ya* (London, 1956).

4. Hugues Panassié, *Hot Jazz* (New York, 1936).

5. *Duke Ellington.*

East St. Louis

Fig. 1. Oh Le-wan—dos oh Le-wan—dos

Fig 2. Gouge of Armour Avenue

Fig. 3. East St. Louis

Fig. 4. Gouge

Fig. 5.

Fig. 6 East St. Louis

Fig. 7 Mama Goes Where Papa Goes

Fig. 8 Omaha Blues

that Bubber liked to look at signs, and to play around with the names on them. I can imagine a different result had Bubber walked the streets of New Orleans with *that* city's music in his subconscious ear rather than the music of Harlem. And surely there would have been even a greater difference in his music had he brought this material to Oliver rather than to the more sophisticated Duke.

The germ of this Lewandos theme, however, was not Bubber's. Charlie Green first recorded it in 1924 for his solo in Fletcher Henderson's *Gouge of Armour Avenue* (Fig. 2). However, Bubber's development is different: he extends the theme while Green repeats it. But it is not only a matter of both players starting with a minor triad with a long-drawn-out fifth and then going their slightly different ways. After eight bars we find in Figs. 3 and 4 the identical thought in both players, a sure sign of the persistence of Green's or someone else's solo in Bubber's mind. This kind of solo became especially popular with trombonists, and was probably common musical knowledge ever since Green had first played it, or perhaps even earlier. Besides Green we have both Armstrong and Ory using it before Bubber. The Duke provided a base somewhat similar to the ground theme of Bach's Passacaglia in C. Although this theme (Fig. 5) sets off the composition, it is not jazz.

After Bubber finishes the first sixteen bars of drawn-out tones, together with some fast triplets, he introduces a middle section (Fig. 6), which is taken right out of the blues and Negro popular song. Maybe he heard the exact phrase, I don't know, but this *kind* of phrase is common to Negro music. A few dotted eighths and sixteenths, culminating in the syncopation of the next bar (Figs. 7 and 8) is common in many blues and Negro popular music as published by Clarence Williams. Bubber had a tendency to straighten out such syncopation with direct "on the beat" phrases. This syncopation came from ragtime, while Bubber's tendency came from jazz. The rest of the record is a series of solos and arranged choruses ending with Bubber's restatement of the theme.

But the common knowledge of the theme or themes does not detract from Bubber's creation. His composition only goes to show to what good use he put the material. This is real musical development. I found this record a remarkable stride towards composition in jazz. In its early version, with Bubber playing, it is an indication of where composition *might* have gone at that time but the Duke had other ideas.

Bubber spoke of a spiritual, *Hosanna,* that his sister sang. It was none other than what might be called the chorus of Stephen Adams's sacred song *The Holy City.* The tune is eight bars long, but two bars are taken out just before the last bar and the remaining six drawn out to a twelve-bar blues by dividing each bar in two to make the *Black and Tan Fantasy.* I always felt that the *Black and Tan* was a remarkable record. In my *Hound and Horn* article (1934)[6] I was mistaken in attributing the four printed trumpet solos on the *Black and Tan* as being those of Bubber Miley. I did acknowledge the substitution of Tricky Sam in the first chorus of one of the solos but I did not know that the first two trumpet versions were by Jabbo Smith. Bubber's two choruses on Victor are among the

6. Reprinted in Ralph de Toledano's *Frontiers of Jazz* (New York, 1947).

most remarkable plunger solos in jazz. In fact the versions of Tricky Sam and Jabbo Smith are extraordinary too. The only fault I find with this recording is that, when stripped of the Miley effects, some of the phrases tend to be popular in feeling. What is remarkable is his first attack after the four-bar organ point (Victor). While to some ears it is a series of wa-wa, on paper it is a most inspired and rhythmically detailed continuation of his first burst out of the organ point. The second chorus is a tense creation more or less to the first turgid rush of tones. It is a complete creation in spite of its stemming from the Adams original. Bubber made a two-chorus solo for me in which the C organ point is held for seven bars, ending with a truly remarkable last five bars. The second chorus which starts off something like the Victor solo veers away from it, and although interestingly different tends to become a little popular. (Let me say here that in estimating a player's or a composition's real worth, I cannot accept the presence of popular material. Though true enough, the greatest players are constantly occupied with popular material, I still find by far the completest satisfaction in their non-popular improvisations.)

The *Black and Tan* has a dramatic ending by Bubber, a type of solo in which it is the duty of the player to halt the even flow of variation after variation by a trumpeting that gives an exciting life as well as an intimation that the end of the piece is at hand. The tag of Chopin's *Funeral March* used here at the end was unfortunate. Other later versions of both the *East St. Louis Toodle-Oo* and the *Black and Tan Fantasy* degenerated in character. The *Black and Tan* especially was given a "going over" that was indicative of the Duke's bad taste and lack of appreciation for the jewel he really had in the early *Black and Tan.*

One of Bubber's greatest solos is on the *Yellow Dog Blues.* Usually a soloist chooses the more melodious strain for his major effort, but in this case Bubber was given the verse. For the singer, the patter effect of a verse is an excellent lead into the chorus, but it is a section of a popular tune that the jazzman does not usually tackle. To insure variety the popular bands of the polka-waltz-tango-foxtrot days were glad to get a tune with many different "strains" to it, while the jazzman, on the other hand, gets all the variety he needs from his own inventions. Using this patter-type strain Bubber created a solo which appears quite different from the original verse without actually departing from it to any extent. I am inclined to believe that the very lack of melody so appropriate in a verse gave this solo that particular quality needed for his treatment.

Both *Got Everything But You* and *Tiger Rag* are excellent solos. The former is very melodic and well constructed, while in *Tiger Rag* the melodic structure that he has built up sets off his peculiar style most advantageously. It is an exciting segment of this record.

On *What Is This Thing Called Love,* with the Reisman band, Bubber plays one of his rare subdued solos, using his effects sparingly. It shows how well he could play off-straight, with a minimum of personal effects. His breaks behind the vocalist are restrained but succinct.

On another Reisman record, *Puttin' on the Ritz,* we find one of Bubber's most successful solos on a popular tune. This time he at least succeeds in obliterating the catchy aspect of the tune. When we take into account the fact that he was only sitting in for a recording date, the result was more than satisfactory.

The solo on *Rockin' Chair,* with the band Carmichael assembled for this specific date, is typical of Bubber, but again, it is a matter of the over-treated popular tune. What is remarkable on this record, so far as Bubber is concerned, is his dialect singing with Carmichael. It is rhythmic and dramatized with questions and suggestions in the manner of Negro children.

Another solo very different from the usual Bubber creation is that on *Ponchatrain* with Jelly Roll Morton's Orchestra. We get glimpses of what he could do with a blues when his thought ran to a lyric treatment somewhat removed from jazz of blues tradition.

Both the *East St. Louis Toodle-Oo* and the *Black and Tan Fantasy* have captured the imaginations of countless listeners. These two pieces have attracted the attention of people ranging from those who recognized them as good jazz to those who, without knowing much about jazz, were still fascinated. What is this large audience appeal, and has it any relevance to any extra values these pieces may have as examples of great jazz? I find that, although some critics mark them for special mention, on the whole they are judged solely by the value of their jazz content and not as compositions. I grant that a large audience picks out a composition for special attention for reasons that are frequently hard to explain. Sometimes it is evident that these reasons are extramusical, or if musical, then for the appeal of some decadent or cheap effects they contain.

On the other hand I do believe that there is a state of development sometimes achieved by an early music (as I term jazz), and nearly always striven for by a developed or classical music, which gives it an appeal beyond that of the music out of which it grew. I grant too that there is much music—too much in fact—that though reaching this stage of development, still remains extremely poor music indeed, and in our case very poor jazz. Much of Ellington and many arrangements of the 1930s and 1940s had a vogue because they were inspired compositions, but unfortunately the inspiration of their arrangers did not extend to making these pieces first of all good jazz.

The *East St. Louis,* as I said, starts with a non-jazz theme arousing our anticipation of Bubber's solo, and is then repeated a few times under the solo. This introduction, together with the particular construction Bubber gives his solo, makes the record the extraordinary thing it is. After the opening a certain amount of music has to fill in before Bubber closes the piece, and this stretch is taken up with a poor joggling tune reminiscent of *Sister Kate* and performed in uninspired solos and trite arranged sections. (Tricky Sam's solo is lusty only in tone.) Now in order to achieve the effect established here we need not have had the introduction Ellington used. Any number of blues could have been played in a manner to serve the same purpose, which, with the substitution of better jazz material after Bubber's solo, would have considerably improved the piece as jazz as well as strengthened its compositional quality.

The *Black and Tan* is another matter. The whole effect of the record derives from the introduction by Miley and Tricky Sam, playing a truncated version of Adams's *Holy City* in the minor, while any solo that was muted and departed from the original theme would have done as well as Bubber's dynamic version. The rest of the solos (including Bigard's which is only heard on the early 1932, 33 rpm version and also played in conjunction with Tricky Sam's on *The New Black and Tan Fantasy*) are very good Ellington-men solos; again it is only their tonal color and not their specific contours that puts over the record. Bubber's final solo, of course, gives a lift at the end, and was always played pretty close to his original by succeeding players. I suspect that even Chopin's tag also helped to make this record unique. Leonard Feather referred to the *Black and Tan* as the piece where Bubber growls Chopin's *Funeral March*. I would say that the special appeal of the *Black and Tan* comes mostly from the atmosphere created as well as by the well-placed solos. The beginning, which is the essence of the popular conception of the *Black and Tan,* does not wear well, but Bubber's solo and the structure built up by the following well-placed solos give the real support to the composition as a whole.

I should add that the *Yellow Dog Blues* comes close to equalling the special compositional value of these two other records. Strangely it was only recorded once. I feel that such compositional achievement raises the whole level of the material used, but only when the material itself is great does music benefit. Then a whole composition attains the positive and cohesive qualities of a great song.

When we consider the Mileage Makers and Miley's band for the *Harlem Scandals,* which I heard, there is no evidence that any band of Miley's would ever approach the distinction of Duke Ellington's. Although I credit Miley for some of Duke's best records, I know that by himself Miley could never have brought his inspiration to the point of equalling those records. Whatever fault I find in Ellington, and as much as I wish that his collaboration with Miley had borne better fruit, I must admit that he gave Miley's inspirations a needed setting, which although poor as jazz, at least pointed the way to the development of the compositional elements I have been stressing. Otherwise Bubber's contributions would only have been limited to excellent choruses set in commonplace arrangements. They certainly belong more in the context of Ellington's musical creations than in any other arrangers' practicing in Harlem.

As far as Bubber's biography is concerned let me set down these few items. James "Bubber" Wesley Miley was born in Aiken, South Carolina, on April 3, 1903. His father, a carpenter by trade, was Valentine Miley, and his mother's maiden name was Eva Arthur Dangerfield. Her father was Wesley Dangerfield, a descendant of the Cherokee Indians. Bubber's first name comes from his paternal grandfather, James Miley. Bubber's family brought him to New York when he was six to live at 414 West 52nd Street.

The Mileys were a musical family. His father played the guitar, filling Bubber's home life with the blues. His mother and three older sisters all sang. After Bubber's death his sisters, Connie, Rose, and Murdis, formed the South Carolina Trio and were known as the Six Hundred Pounds of Rhythm. They sang on the radio and at the Roxy Theatre.

When a young boy Bubber sang in backyards for pennies and with a friend, Alfred Sanford, went down to Broadway to sing and dance for small silver. His first job was errand boy in a grocery store. He was a boy with little interest in school until the ever-busy truant officer put him in P.S. 141 on West 58th Street. Here he was taught music by a German professor, first trying the trombone before settling on the cornet. The professor was delighted with his musical talent; and Bubber played in the school band.

At fifteen he joined the navy to serve a term of eighteen months. He left with an honorable discharge to join a small band known as the Carolina Five. Its personnel was: trumpet, Bubber Miley; clarinet, Cecil Benjamin; soprano sax, Johnny Welch; violin, English; piano, ______. They played in Purdy's Cabaret and Dupre's Cabaret on 53rd Street. Bubber played at Connor's Cafe, 135th Street and Lenox Avenue, during the winter of 1922 and through 1923, then went on a tour of the south with a show called the *Sunny South.* After that he played with Mamie Smith's Jazz Hounds at the Garden of Joy, 140th Street and Seventh Avenue, and later at the Waltz Dream Dance Hall, on 53rd Street. In 1926 he joined Duke Ellington's band at the Kentucky Club and played with the Duke until January 1929. In the latter part of May he left for Paris with Noble Sissle for a month's engagement at the Ambassadeurs. On his return he played in Zutty Singleton's band at the Lafayette Theatre with Charlie Green. During the summer of 1930 he had three recording dates with his Mileage Makers. Later he played in Allie Ross's band at Connie's Inn where the famous Snake Hips Tucker was dancing. In January 1931 he joined my act in Billy Rose's first show, *Sweet and Low.* Here he played the *East St. Louis Toodle-Oo* for about four months. He doubled at the Paramount Theatre with Leo Reisman's Orchestra for a week in February, playing the *St. Louis Blues* dressed as an usher. On May 7, 1931, he played for me at my dance recital with Mura Dehn.

He played *East St. Louis Toodle-Oo, Black and Tan Fantasy* and *Yellow Dog Blues.*

Bubber was backed by Irving Mills to build a band of his own. They played in Irving Mills's show the *Harlem Scandals,* featuring the comedian Tim Moore. The show opened at the Lincoln in Philadelphia, where a doctor told him he was tuberculous. However, he stayed with the show and came to New York to run the week of January 23, 1932, at the Lafayette Theatre in Harlem. The next week Bubber played in the band at the Harlem Opera House, and at the end of the week he had to leave because of the inroads of the tuberculosis. His condition was now acute. After a short time at home with his mother (209 West 62nd Street) he entered the Metropolitan Hospital on Welfare Island on April 18th and died a month later, May 20th, 1932.

I was fortunate enough to meet Bubber Miley at the time when he was playing around from one job to another and I was devoting myself to jazz dancing—using any jazz music that fit my purpose. I had an act in the Billy Rose show with a trio that included a trumpeter from the Claude Hopkins Orchestra, by the name of Clarence Powell. Soon after the show opened my musician was eliminated and I was forced to use the cornet player in the pit, none other than Jimmy McPartland. Although we could have managed well enough for the run of the show, I was looking forward to building up my act, and for that I needed a growl man, which McPartland was not. So I hired Bubber Miley and we worked out the season with him. Thus began my friendship with Bubber.

I had also created a dance to Armstrong's *King of the Zulus* which Bubber learned. Besides another *Black and Tan,* he also made a record of the Armstrong number for me. It was done in a miserable little studio of the time and did not come out too well. It was a difficult feat to make a whole record with only piano accompaniment and no support from a band, switching, as I had him do, to his French horn for the Ory solo. The record is full of clinkers: we had no chance to warm up or make any second takes. However, it is an interesting sample of what could be done with Bubber. The whole affair was quite an experience for me.

Bubber was a man with a good sense of humor. He had an infectious and winning personality that was serious underneath. He was a heavy drinker, to which the Duke's published remarks bare witness. Bubber told me that like many jazz musicians he used the weed, for the wonderful inspiration it gives a player. Moderately used, he said, it would condition a player for a whole evening, so that the ideas were always on top. When improvising at rehearsals he always wanted the three voices of the chord clearly stated as it helped him a great deal. I had a little reed organ he used to like to play on while humming the blues. In spite of his quietness Bubber had a natty appearance and sported an Auburn convertible. One night driving home after the Billy Rose show his car stalled in Central Park. Bubber just got out, slammed the door, and walked away.

Where does Bubber really stand in jazz music? It is all very well to praise his unique contribution of the growl and for the excellence of his presentation of it. But for any final judgment we must measure his product against the best of jazz. Personally I feel his real contribution, as it fits into the total picture of jazz, is his melodic innovations, not the way he came about them. He created something different from his predecessors, shouts of dramatic statement that were unique for the time. Such material can well measure up to New Orleans and, sparingly used, could have been a worthy counterpart of it. Take, as an example, his bursting thought after the long organ point on the Victor *Black and Tan.*

What Bubber would have done had he lived is anybody's guess: my own would not be too optimistic. The atmosphere of the 1930s, though cleared of jungle noises, certainly was not conducive to any *new* trend. It was a period of development of existing material, good as well as bad. How Bubber's style would have fitted in with the advances that the Armstrongs, Hawkinses, Bechets, and other highly developed men were making I don't know. As the wa-wa by that time had become discredited, Bubber's compensating clichés may have become more pronounced.

Coming in any other period, a discredited innovation like the wa-wa would be of little interest to the future. But different as Bubber's music is from New Orleans style that preceded him, it nevertheless belongs to that extraordinary period of the 1920s, the period of the first recordings of New Orleans jazz, of the meager disseminations of strolling players, of the white Chicago adaptations, of the first of a long line of meaningless arrangements, the emergence of the Duke and the apprenticeship of the great giants to come. Everything about the 1920s is interesting, including Bubber's wa-wa. The period demonstrates what happens when an indigenous music is transplanted into artistically sterile ground and is commercially trimmed to popular taste. Bubber was one of the products of this period, a victim of it, but still a great man.

The Cuban Sexteto

It was back in the middle 1920s that I heard my first Cuban sexteto record. Then, like most novices, I was struck mainly by the astounding drumming, but as the years went by I came to appreciate the singing as something extraordinary in itself and containing the most glorious music. Now Riverside has issued the only record I know of that gives any intimation of the character of this music to present-day American listeners. *Festival in Havana* is a recording of a music that preceded and gave birth to the Cuban sextetos.

In the notes on the Riverside record Odilio Urfé speaks of the later bastardization and distortion of this music by popular taste. Here I have to disagree with him unless he excludes from this blanket dismissal those early sexteto recordings of the 1920s. Along with the present wave of ethnic recordings we are treated to much talk about the superiority of primitive folk music over its urban development. While I'm all for the availability of ethnic recordings, I cannot go along with the further-back-we-go-the-better conception of folk music. Maybe I feel this way because I belong to a day when we played our jazz records until they were truly "beat." Our enthusiasm was certainly not for their folk integrity alone.

Moreover, I still find in the urban growth of jazz an advancement carrying benefits not found in its primitive prototype. An attitude of contempt for its urban development would dismiss all the instrumental jazz we know: Oliver, Dodds, and Louis, all of boogie-woogie piano, together with Bunk's glorious band. And I will remark right here that

The Jazz Review, December 1958.

Sexteto Habanero, Cuba 1930. *From left*: Andrés Sotolongo, Miguel García Morales, Felipe de Nery Cabrere Urrutia, Gerardo Martínez Rivero, Guillermo Castillo García, Carlos Godínez Facenda, José Interian Portillo. The top pennant is their distinctive emblem flag, awarded to them in 1925 on their winning the first national competition on the Cuban "son," playing "Tres Lindas Cubanas" by Guillermo Castillo.

my listening pleasures have gravitated to certain jazz records made during a short period in the 1920s, and extending, of course, to those later discoveries in the same vein, rather than to any singing, solo or ensemble, that preceded this music. In Cuba during the 1920s there was a short comparable period when the sextetos were of the most listenable character before they sank into banality. By listenable I mean that the music is not only charged with good melody, brilliant intonation, and playing style but that the development of its material is made intriguing and exciting.

Here let me insist again on a point I have often made before: it is my theory that much of our great classical music has its roots in certain rare fertile periods of highly developed folk music. In line with this, the Cuban sextetos, I hold, have reached this stage of development and lie ready for the notation-composer's use. My interest in both jazz and the sextetos goes beyond the deep pleasure I take in listening to them; I am concerned with discovering and bringing to notice their many inherent qualities characteristic of an art ripe to be taken over by composers in the classical tradition. Further, I feel that should their potential ties be neglected we would be much the poorer for their loss. This theory of mine has slanted all my criticism, causing many of my readers, whose interest in jazz stems from widely different reasons than my own, to misunderstand my approach. But if they will consider that I see jazz as a part of the whole body of Western music and find my deepest pleasure in the early works of the classical periods, I think that they will come closer to my meaning.

While modern jazz still retains the method of improvisation of early jazz, the Cuban mambo in its synthetic development grows from no such indigenous roots. The only musicians with any improvisational freedom in Cuba are the smaller groups—sextetos and trios—whose musical practices were never emulated by performers on more conventional instruments. One exception, however, is the trumpet; its best example can be heard on the Riverside release. Performances on conventional instruments taken over from the military bands and *danzon* orchestras always show the hand of the arranger, whether in assigning to the various pieces their harmonic and rhythmic roles or in creating more ambitious compositions like those by our own big band arrangers. As the sextetos, unlike the New Orleans improvisers, never employed conventional instruments they could not develop a school of instrumental improvisation ready to be taken over by the popular bands.

The most infectious part of Cuban music is the *montuno.* By concentrating on this infectious section certain musical leaders of popular bands in fashionable musical spots of Havana established the mambo. Their prototype was the *montuno* of the *danzon.* The peculiar character possessed by the style and form of a *danzon* gave way to a more popular and modern approach, utilizing, amongst other older types, the music of the sextetos. However, it would have taken the integrity of a revivalist's fervor to establish anything of solid merit.

The mambo as played in New York consists of nothing more than a Cuban rhythm section together with large choirs of trumpets, trombones, and saxes playing an arrangement derived from modern jazz. It is completely written out and as characterless as is big band scoring in jazz.

There have been some indications that Cuban rhythms are creeping into American jazz. To do what Gillespie did with Chano Pozo is mere novelty. Any great drumming—East Indian or African—combined with the melodic virtuosity of jazz would always prove exciting. When two types of music are near neighbors in the same city marriage is inevitable. But the glory of Cuban rhythm cannot be fused with American jazz merely by seating a Cuban drummer in a jazz band any more than jazz could be injected into Cuban music by the reverse process. There would be no real union. On the lower level, if Cuban and jazz players were to sit down together year after year in small Harlem joints the outcome will certainly be momentous for the Cuban-jazz amalgamation. But imagine a snare drum–bass drum type of drumming in Cuban music! In the United States only Baby Dodds has given us a style that seems to belong by right in a jazz band. The fact that our snare drum–bass drum outfit has never been really integrated into jazz playing leaves room for a style of drumming that fits as naturally into its context as do the drums of a Cuban band.

But to come back to the two musics of our subject. My attachment to them as well as to classic music derives from the perennial interest I find in them, an interest I do not find in other folk musics. As an example, the dance music of Chambonnieres, which is certainly classical, not folk music, is a genre made possible by the status of the earlier improvised music of his time. It seems to me, moreover, that had this improvised music been notated it would appear very little different, if at all, from Chambonnieres!

The sextetos are definitely a music of professionals. They were concerned with creating a form of music to be listened and danced to, and with the necessity of staying within the confines of a more compact ensemble, than did their predecessors. Of all the pre-sexteto records that I have heard I find that none of them gives the completely relaxed and pleasurable effect of the sextetos, nor does this relaxation ever lead to any let-down of interest and virility. In consistency of folk tune integrity, the Riverside release is pre-eminent. But this unspoilt state is characteristic of most primitive art and does not necessarily make for satisfactory daily fare. The great importance of the sextetos lies in what they have done with this folk material, an achievement that far outweighs any compromise they have made with popular taste. The ensemble balance within a sexteto and a New Orleans jazz band did not exist in their prototypes. It is the specific nature of the instruments and their particular duties when working within a traditional form and accepted style that transforms basic material into a higher work of art.

For example, the Cuban *danzon* is a form with exciting possibilities but its evolution out of a sedate past never gave birth to any truly great music. A great deal of material out of the sextetos can be seen in it but the whole character of melody and treatment is that of a written composition in a popular vein. If we can forget the paucity of its melodic content, we will find that the styles peculiar to its different sections and their specific sequence are highly intriguing. One or two ballad-type insertions followed by a *montuno,* each of which is initiated by an introduction—abacadac—is the structure of the *danzon.* But it was not destined to flower into the great music found in the *son* of the sextetos.

The Cuban rumbas and congos on Riverside have a luxuriant sound due to the many voices which, played at fast tempos with fervid drumming accompaniment, grip our attention. This is music for a festive gathering—musicians and spectators—either in a packed room or in the open, and its whole spirit grows out of their back-country cult rites. But when we are not on the scene and are not brought out of ourselves by the spirit of the occasion, a subtler and somewhat more involved approach wears better from day to day: something like what Hamlet told the players, ". . . for in the very torrent, tempest, and, as

I may say, the whirlwind of passion, you must acquire and beget a temperance that may give it smoothness." It is the inevitableness and relaxed momentum of the solid rhythm section of marracas, guitar, tres (three-stringed guitar), and string bass behind the singing and bongo playing that give these sextetos a stability not found in the previous music. Though the sextetos too were part of the life of the cantinas and dance halls they achieved a style that is self-sufficient apart from this context. Most ethnic recordings strike me as more reports of happenings I was not fortunate enough to witness rather than a presentation of a piece of music in its own right.

I cannot claim to be an authority on the sextetos and do not have the knowledge to point out the differences between the tune genres. The sextetos of the earliest recordings called their music "sones." Whether the difference is in melody or treatment I do not know. The word "rumba" I never heard used for this music. The groups consisted of six men—guitar, tres, claves, bongo, and bass. Three or four of the players sang. The singing on the sexteto records is a mixture of solo and ensemble. It is loosely knit, somewhat analogous to the freedom of the melody instruments (trumpet, clarinet, and trombone) of a jazz band. The tunes lend themselves to full-throated and rich singing either in complete solos or injected breaks, giving, thereby, a rhythmic outline to the melody. The bongo acts as a free voice, pointing up the music where the player feels it is needed rhythmically.

The different groups usually have their own particular introduction after which the theme may be stated by the tres. The voices enter easily with the rhythm section carrying incessantly on. After finishing the first phrase the voices enter again at what appears some seemingly arbitrary point. The incessant march of rhythm makes these intervals between voice entrances interesting and absorbing with an effect of artistry only attained in a most advanced classical music. Although the folk tunes themselves are as short as most European folk tunes, the specific setting given by a Cuban rhythm section avoids the usual terminations of the European treatment which does no more than exhibit the little tune. Many of the tunes have phrase endings which inspire a long held tone of no apparent definiteness. As sung we get a smooth impressive composition without attention being drawn to the briefness of the theme. The tune in a few instances appears to be cut up and laid on a rhythmic webbing. The interdependence of the two forestalls any feeling of brevity and the effect exceeds in grandeur anything possible to the little tune itself. The protruding rhythm between the melodic sections makes for a most ideal form of early composition. *La Mujer Podran Desir* by the Sexteto Occidente is an exquisite example of this laid-on choral singing with protruding rhythm between sections and is but one of the many ways of presenting material of compositional status. It goes beyond the mere repetition of the tune.

After the first tune has been sung a few times the original choirs go into a repetitive coda that has become in many instances the glory of the Cuban sextetos, the *montuno,* and a contribution of great importance. It gives the *son* a most balanced proportion. We appreciate the contribution of the *montuno,* if we compare the sextetos with the small trios that abound in Havana, playing at tables singing popular and folk tunes similar to those used by the sextetos. The solidity given the whole piece by the sexteto style is tellingly different from the mere singing of folk or popular tunes. The transition from opening *son* to the *montuno* is often a most inspired indication of great delight to come. There is a controlled lift of tempo, especially at the introduction of the *montuno* that imperceptibly accelerates to the end without giving the least impression of being hurried.

The sextetos had a great vogue throughout the 1920s. Judging by their records the two greatest groups were the Sexteto Habanero and the Sexteto Nacional. There were some excellent records by such groups as the Septeto Matamoros, Estudiantina Oriental de Ricardo Martínez, and the Sonora Matancera, but I could not judge their continued worth by the few discs available. Just as the country dance music of the 16th century became fashionable and was subsequently used by composer-improvisers to be glorified and preserved in the dance suites we know, so the primitive music of the pre-sexteto days became popular. We must remember that the popular "taking up" of a music is not the same as a popular influence being exerted on a music. Folk musicians who find themselves "taken up" and then become professionals, deriving their living from their art, have no desire to participate in the current popular music they may be displacing. The conditions of their new set-up may prove so beneficial that they are enabled to create an art which even out of context of its original locale is still self-contained.

Whatever the popularity of a folk music, its wide appeal is not because of its depth of meaningfulness. Imitators without the ability of its innovators will naturally appropriate more and more of the popular element until a new genre has been created. This new genre will strike a chord in a larger public who will respond to it as a music after its own heart. Then gradually the original folk artists find their popularity waning until they themselves start incorporating some of the more salable elements into their music. At first they so transform these elements that they lose none of their delight; eventually they degenerate into something no different from the cheapness of their competitors. Or, if the performers maintain their integrity, find themselves so outmoded that they rarely work. Except in times of revival these players become the sad cases we read of—King Oliver, Johnny Dodds, Jimmy Yancey, and the Sexteto Habanero.

The music of Cuba differs entirely from jazz. What happened to instrumentation in this country is unique and resembles no other music deriving from Africa. In African music melody has been the province of the singer, and rhythm the province of strictly rhythmic instruments. The rhythms in the sung melodies never showed the freedom and inventiveness so characteristic of jazz. The few melody instruments either played rhythms hardly more extended than the voice or played rhapsodically around the voice. This African element remained basically unchanged in Cuba. On the Riverside release we hear both solo and ensemble voices singing simple tunes with a rhythmic lilt in the tune itself. The soloists participate in a semi-declamatory manner that has its own rhythmic and rhapsodic style. We also find this in Flamenco and Cuban Punto Guajiro and in the Spanish Saetas. It is either strident or tends to monotony. In its proper setting I find it most enjoyable but a little goes a long way on records and is best as an ingredient of a composition rather than a thing in itself. The melodies the Negroes have evolved out of Western and African music are beautiful and simple with a richness that can only be described as pure gold. It has not become overly fixed as has the folk music of Europe, but allows the singer to sing on and on without ever becoming monotonous.

Although there may be found both folk and non-folk tunes in the sexteto records, we can be assured that on the Riverside release we are listening to truly great folk music. From the scanty specimens of Cuban cult music that I have heard, only in the Abakwá Song by male chorus and drums (recorded by Courtlander) did I find any indication of what may have been the prototype of the music on Riverside. Much of the cult music appears African in character, but the Abakwá Song is certainly an early example of what eventually flavored the sexteto sones with their Afro-Cuban character. The form of the music does not keep to a definite number of bars—twelve, thirty-two—and through variation gives us what appears to be a continuous creation. Blues and gospel singers follow the tune, their inflections are more subtle than bold, and it is only from the words alone that we get any real compositional progress. The same can be said of the inflection of Cuban singing, except that the boldness of jazz instrumentation is here realized by a freedom in laying out the composition. The integration of Cuban rhythmic instruments does more than merely provide a solid web of chordal rhythm (guitar, tres, and bass); it is integrated throughout. This contribution, which I feel to be significant, no doubt was made by the sextetos. The early sextetos were not much more than copies of the choirs we hear on Riverside, but with fewer men. In the earliest record I know, the guitar family was present while the old marilbola had already been displaced by the string bass. I find the *montuno* on the sexteto records more relaxed and a more integrated part of the whole rather than the frequently used hurried ending consisting of a repetitious phrase.

The personnel of the Sexteto Habanero changed at times but I believe this line-up represents it at its best. The director Gerardo Martínez Rivero (3rd voice) played the bass, José Jimenez Echavarria (1st voice) played the claves, Felipe de Nery Cabrere Urrutia (2nd voice) played the maracas, Carlos Godínez Facenda played the tres, Guillermo Castillo García played the guitar, and Augustine Guiterrez Brito the bongos. When we hear a trumpet it is no doubt José Interian Portillo. Except for Martínez, Guiterrez, and Interian they are all dead. Martínez and Castillo divided most of the solos between them. Martínez is still carrying on with his Conjunto Tipica Habanero de Gerardo Martínez. Castillo's voice is one of the most wonderful of its kind. It is deep but when he sings solos he raises it a little, giving the impression of reaching for the tones. José Jimenez, although he sang no solos, had the most celestial voice. In the ensembles his voice always trailed and lingered on at the end of a phrase. That wonderful ability to ease into an ensemble and fade out at the end they had to perfection.

Their singing has the beautiful relaxed drag found in Bunk's band. Combined with the rhythmic sections it gives a feeling of solidity that makes all other types seem thin. Their delivery of the popular tune *Mama Inez* is a revelation. Comparison of their version of *Mama Inez* with that of another inspired group, Ignacio Pineiro's Sexteto Nacional, only emphasizes their supremacy and makes us realize the heights singing can rise to; while further comparisons between the Sexteto Habanero and others treating the same tune reveal that besides their supremacy in quality they are superior in their extraordinary improvisations of solo incidentals.

Because the beauties of singing are not easily handed on, the sextetos of secondary importance never had the interest of the few great ones, something not true of the early jazz groups. Singing depends more on rare vocal quality than band playing does on rare instrumental ability. Thus, the followers of Ma Rainey and Bessie Smith are extremely limited, and unless we go to the more stark school of singing stemming from a Leadbelly or a Blind Lemon Jefferson, the contemporaries and followers of the Rainey-Smith high point are nobody to adulate truly. The full-throated singing quality of the Cuban Negroes both in ensemble and in many of the solos usually does not have the peculiar folk quality of a blues singer, though many of their voices do possess a nasal folk intonation. Generally speaking, however, their songs sound more like lusty Western singing but without the bad effects of academically trained voices.

Cuban trumpet playing is a most beautiful off-straight style that either plays the tune in simple fashion in answer to the singers or indulges in what may best be called cascading cadenzas. When these cadenzas are simple and not overdone, as with so many players, they are a most useful adjunct to the music. In the choirs the trumpets are most virile. The sextetos, when they add a trumpet player, actually have seven men, but as a rule they keep the name "sexteto."

Although the number of sexteto records made during the 1920s is small, these few preserve for us a rich store of music. Whereas the Riverside record stands up amongst the world's ethnic records in quality and esteem, I am inclined to believe that because these sextetos superficially appear similar to the cheap music of the 1930s, they are not appreciated by the casual listener like the older music. The same situation holds with jazz. One who likes the church revival music of the 1900s might consider the music of Oliver's band inferior, especially in the pop tunes. Although some Cubans consider these early records to be collectors' items, on the whole they have no great standing. Let us hope the Riverside release starts a revival of Cuban sexteto in the United States, now that we have such a large audience for Latin-American music. Enthusiasm for them could lead to an lp re-issue from Victor's vast store.

Popular Singers

Jazz Monthly, July 1959.

Popular songs are enjoyed for the most part by people whose only standard of criticism is the fact that the music gives them pleasure: they "know what they like." The appreciation of any music above the popular song level suggests an awareness in the listener of other standards.

Because the devotees of popular singing exercise no real critical judgment, this genre has had to make its appeal through the most elementary means: it has always expressed the basic principles of dance music, namely, rhythmic patterns and melody. While obviously the exercise of taste excludes what is cheap and banal it may also be responsible for the introduction of standards that can eventually undermine any art to which they are applied. In avoiding the tawdry, criticism may overreach itself and place a false value on elements that lead away from the spirit of the music. It is just because no critical canons are brought to bear on popular songs that they retain as a body the elements of highest charge, however flat and uninspired any one specific song may be. But popular song is not static. As the genre varies in style from one period to another, so does its artistic significance. Whereas in the 19th century the classical composers had little truck with the popular song, today they pride themselves on their acceptance and practice of it. This is partly understandable on the grounds that classical music, in response to criticism, eliminated its accretion of over-ripe and decaying elements, with the unfortunate result that it has shorn itself of the very elements that make music—those elements that perennially hold and excite us. A frank return to a music which, because it must hold the masses, dare not eliminate those vital musical elements, is evident in this regression to another musical level.

The popular singer is a creature who is not satisfied with merely giving musical utterance to a song. For the voice is the greatest instrument in the world, with a flexibility that knows no bounds, and it is the infinite possibilities of the voice which make of a song something more than words and notation. The popular singer uses these possibilities to the fullest.

The trained lieder, madrigal, or operatic singer is a product of a technique which concentrates on the well-rounded delivery of all tones from chest to throat. Flexible chest tones are more difficult than palatial tones and so need a curriculum of exercise to loosen up and give fluidity to these deeper tones. In the course of this training the expressive powers inherent in the voice can be lost. The singer becomes so conditioned by her training that she finds she cannot free herself from this "vocal studio" criterion, and once she accepts it she never dreams that her voice is capable of any other function. There are few trained singers who are able to take advantage of their schooling without allowing it to ruin their art.

Fluidity in chest tones always struck me as incongruous; it is a contradiction of nature, for it is from chest tones that we get bursting volume, not scale passages. There is pompousness in the chest tones that lends itself readily to interpretation, but their use does not constitute a method or a basic manner of singing in itself.

The traditional popular singer is untrained. Starting with a given voice and great intention, and the desire to express some specific quality, coloring or emphasis that is peculiar to her alone, she develops a personal style. Most singers begin by imitating the manner around them; few go any further. But a singer of talent soon develops her own unique style. They are always concerned with charging the basic tones with an immediacy they would not otherwise possess. Singers even alter certain tones to give a song a new and distinctive quality.

These personal stylings by popular singers endear them to millions, and also call down on them the abuse of the followers of classical and semi-classical music, who apply to popular singers the same criterion as to schooled singers. Now any art expression, including popular singing, is influenced by some quality that pervades the times. Elements of this quality creep into instrumental music as well as into singing, and they influence each other. There is nothing negative about the product, though it may result in an unhealthy musical expression. Just as classical music goes through a cycle of development, maturation, and decadence, so changes can come about in singing, changes which introduce elements actually unfavorable to the art.

The influences on popular songsters at the present time are many and complicated, coming, as I see it, from three sources: the mainstream of popular singing, the style of blues and jazz singers such as Louis Armstrong, and the close connection between the music of singers and all jazz music since bop. Of course all along its course the mainstream of popular singing has been influenced by frankly outright jazz singers of one persuasion or another. Now I must admit that whether or not I go along with all the developments in singing since the bop era, I am certainly against the influences of voice culture on popular singing. An academy which furthers the composition of music is one thing, but an academy which undermines expressiveness in voice is something I am against, and heartily so. Bessie Smith did not sing scales and practice tone development, nor did Louis Armstrong, nor did Georgia Peach the gospel singer. Nor for that matter did any number of popular singers back in the days before voice culture invaded their field.

The popular singer starts with a given voice and through *singing* develops it, not by merely humming or singing to herself, but by actually singing at performance pitch. As the melodic line is very often more or less improvised, the singer only does what comes naturally. She is never faced with some ridiculously high note, nor with scale passages such as we find in Bach or Verdi. Yvette Guilbert, the greatest of European popular singers (who also excursioned into the music of the past), says in her book, *How To Sing a Song,* that she taught herself breath control by singing more and more of a song on one breath until finally she found she could sing a whole verse. Certainly we need an academy to train performers to sing classical music as it was intended to be sung, which usually is somewhat beyond the natural range of modern singers. Two centuries ago singers added their own embellishments to a song, a practice that was a training in itself.

But when an academy presumes to dictate the whole gamut of tone coloring, then the result is that every vestige of natural expressiveness is washed out of the voice, as is happening now and which seems to have become the standard style. A school-teacher aesthetic has been imposed on the taste of singers and public alike. A singer with a large voice like Chaliapin is accepted by the academy, even though he uses the full expressiveness of that voice; but should a lesser singer draw on these same vocal resources he will be looked upon with chill disfavor. Just lately I went into a shop in search of a cut-out record of Adalgisa Giordano, a contralto who sings with all the expressiveness I believe a voice should reveal. The salesman said: "Oh, you mean the contralto with the funny voice." This sums up the attitude today towards the injection of meaningfulness in singing.

But to come back to popular singers: in the new musical comedies we find them employing a style that echoes more and more those songs that fall between the classical and the popular—the genre of light classics and operettas. With the present trend in musical comedy towards integration of plot, music, and *mise en scène,* the songs used have a burden of seriousness put upon them that they are by nature too slight to carry. The result is the sheer pretentiousness of a lesser art trying to be a greater. When story line and music are made of paramount importance to a production, then, to lighten the budget, the mimes (comedians and musical personalities) are dispensed with. But without the leavening of these performers we are left with dullness and banality; so to make up for the deficiency we have resource to the trained voice, the trained dancer, and dramatic coaching. Nothing significant is contributed by the performers themselves: they merely obey orders. With trained voices they deliver vacuous ballads and quasi-jazz numbers; their own spontaneity is lost and forgotten.

The bane of all musicals has always been the puerile love ballads, aimed it would seem at the most moronic of teenagers. But there was a time when they were made endurable by our anticipation of something good to follow—the ribaldry of the comedians or the jolt of an exciting dance number. Now we sit through them with nothing to look forward to but a return to the interrupted action, much as we wait during a TV commercial.

There is a class of singers who were never able to make the grade in opera but who made a place for themselves and even gained some renown with such semi-popular songs as *The Road to Mandalay* and *Asleep in the Deep.* Songs like these were the mainstay of a popular tenor's or baritone's repertoire. But after *Oklahoma!* the voice culturist came into his own. The great operatic bass Ezio Pinza, and later Lawrence Tibbett, applied their talent to the banal songs of *South Pacific,* hybrids of classical and popular styles. To me there is nothing more incongruous than an operatically trained voice singing Hammerstein's songs—they cry out for a male equivalent of Eva Tanguay.

As for the orchestral backing to the singers in a musical show, there is no more trivial music than that issuing from a pit orchestra. A popular song has its rightful place in the sun, be it by a man who hardly knows one note from another on the piano or by one who can read an orchestral score. But the arrangers of popular songs are hard put to it when they try to make them into music coming from the pit that speaks for *itself.* Even the arrangers for jazz bands failed at *this.* The *composers* who have entered the field of pit orchestration, for all their knowledge (maybe too much), have hardly improved on the arrangements made back in the 1930s. My advice is to leave arranging alone; let a hack write the parts underlying the melodic line, and be satisfied with what a Palace Vaudeville pit orchestra could do—no portentiousness, no tricked out key changing, and no over-snappy rhythmic cadences.

Today every singer, whether in a band, night spot, or musical show, is taking lessons. With the polishing of the popular voice we find, and will continue to find, composers writing a music which takes advantage of the voice's newly acquired but negative flexibility. The singer, on the other hand, is being reduced to the ignoble position of interpreter, whereas before she was truly a creator in her own right.

Of the music dealing with Negro life, the "coon" songs at the turn of the century were typical. Of the recorded music dealing with this subject I find May Irwin's singing the greatest. Her *Bully Song, Moses Andrew Jackson Goodbye,* and *Frog Song* are delivered in the authentic dialect of non-professional Negro singers. The injection of spoken phrases, so common in popular singing, combined with the pathos she gives to certain sung phrases, makes her one of the greatest examples of the first decade of this century.

Another great singer was the versatile comedian Bert Williams, who in such a song as *Nobody* tells a most revealing story. It is really a recitation in song, something I believe this kind of music should be. Another popular singer was Al Jolson with his nasal, brassy, showmanship-voice. Jolson's style of putting a song over was carried on by such men as the entertainer Eddie Cantor. The detached showmanship of Jolson's delivery in *mammy* songs was a little hard to take, but it won him millions of listeners. There was a vulgarity in his exaggeration of his own style which, together with an early Dixieland attitude he had towards whooping it up with a jazz song, had nothing winning about it.

Sophie Tucker was a perennial favorite. She was the last of the big-voiced singers who exemplified the Ted Lewis idea of jazz, before the advent of a new era of singing and before the emergence of the Negro popular singers.

Until the band singer Rudy Vallee appeared on the scene nobody had ever tried to sell a song by achieving outright rapport with the audience, especially with its female contingent. He was the first to exploit the voice of intimacy, a style made possible in large auditoriums by the innovation of the mike. Next came the tragic torch song, with Libby Holman doing a Joycean "stream of consciousness" act. Singing reached a stage of almost embarrassing intimacy, a pouring out of amorous confidences of desire and despair that pointed to only one consummation. At the same time these various innovations of style were being presented there were singers following a middle course in shows, night clubs, and band fronts.

There were also such singers as Florence Mills who were discovering a *voice* quite the opposite of a blues singer's; and for whose delivery a trained or natural-born voice was necessary.

Then we had the singer in the Harlem joints and Negro vaudeville who sang in a lusty manner songs that were neither blues nor accepted popular songs. Both Ma Rainey and Bessie Smith used this style, though with a difference which sets them completely apart. We should also mention the great Ella Fitzgerald, who sang with Chick Webb's band and who took it over after his death. Ella was the queen of this type of singing. She had plenty of rhythm and voice and her talents in this medium were boundless. There was nothing of the torch song in her early records.

From here on I must admit that my partiality works in reverse. The changes in blues intonation, through the unhealthy influence of Billy Eckstine and Billie Holiday, resulted in a whining intimacy, a merging of blues with the torch song. This whole period is tied up with modern jazz. Modern jazz, whether it be the bop of Gillespie or the cool of Miles Davis, says one thing in a moderate tempo, another in fast tempo, and something else in the slow ballad or modern interpretation of the blues. In the slow tempo of the instrumentalists' work I see an undesirable affinity with the singer, which I am sorry to have to call decadent.

On the platform or on TV Billie Holiday projects a complete, complex, and absorbing personality, of which her voice is but one facet. Disembodied, as on a record, her voice sounds almost drivelling. It has that strong personal inflection, before mentioned, which is by no means negative but because of its very positiveness carries in the wrong direction—it negates the purposes of art. A negative (academically trained) singer cannot do this as she only gives her voice the pitch and time value notated. What is said in the tones of the song itself a singer of *voice* repeats, nothing more, nothing less. Billie Holliday *is* something, her singing definitely *says* something, as it should, but I do not go along with it.

Sarah Vaughan is more in the present-day post-bop mainstream of singing. A far less distinctive singer than Billie Holiday, she offers us rare delivery, voice, and sincerity rather than any true artistic creativeness. She has nothing of the *diseuse* about her nor any of the other qualities I look for in a popular singer. She is the Rudy Vallee of modern jazz, acceptable to those who like the songs.

There is a new style execrated by jazz followers, namely, rock 'n' roll. Whatever its faults, in its up-tempo numbers it is a reaction against the ballad. It gets its effects through rhythm and shouting, and in the voice of Elvis Presley I find it a more healthy trend than the one that leads off into the dim smoke-filled intimacy where the torch-singer moans.

The art of the French *diseuse* has perfected a style of delivery which I firmly believe should pervade all singing. Depending on the music, the application of the art of *diseuse* is either played down or used to the fullest extent. Today popular music in general certainly needs singers possessed of this gift. It is an acquisition which, in the hands of a great talent, gives singing some of the quality that free-wheeling comedians used to bring us. If our singers can no longer supply it we may as well hum the tunes to ourselves as listen to the "just plain singing" of any entertainer who can hit a note and looks well in a low-cut gown.

Throwback

The critical brickbats hurled at Elvis Presley's uninhibited performance come out of the same arsenal of opprobrium as those used against the jazz players back in the 1920s. The spirit of brashness that so scandalized the decorum of the classical music reviewers was equally beyond the grasp of the more popular writers. The prevalent conception of popular music as something light, catchy, or sentimental, could not be stretched to include the ill-mannered intruder from the bayous and the barrel-houses. And now that jazz has lost its obstreperousness and settled down to a respectable middle-age of bland crooning, old time Dixieland, and cool avant-garde extensions, the critical pundits are busy sharpening their long-unused invective and whooping it up against this new disturber of the peace, young Mr. Presley.

Personally, though I find Presley artistically disturbing, at least he has injected into entertainment something long lacking—the roistering qualities of the rogue with provocative hints that he himself may not be exactly a cozy family man. Without going into the private lives of Mae West, W. C. Fields, Joe Frisco, or Eva Tanguay, we caught a leer or a gleam in their eyes that seemed more than histrionic, and were cheered to feel that the "dark gods" were not dead after all. Opposed to these inspired performers, we have our crooners who in attaining a style of respectable nonchalance succeeded in undermining virility.

Young people are by nature exuberant, and although they have in turn idolized the Rudy Vallees, Bing Crosbys, and Perry Comos, they have never lost their re-

The Jazz Review, May 1959.

sponse to the dance as pure rhythm. First there was their "scandalous" exhibitionism in the Charleston, then their jitterbugging antics to their version of the Lindy Hop, and now their enthusiasm for rock 'n' roll with Presley as Pied Piper. Whatever may be said of his influence on present-day teenagers, I personally see a far more healthy release even in overdoing a dance craze or in idolizing the young man who is its symbol than in passive submission to long heart-rending séances in the darkened enclosures of movie theatres.

All this talk about bumps and grinds would lead one to imagine that Presley had invented them; jazz dancers have always used bumps. Even the well-respected Agnes de Mille TV show presented them for family viewing. Though Presley may have made them livelier, it is a little late, it seems to me, to be objecting to bumps per se.

The question of what is good for teenagers is another matter and not my concern. What I am concerned with is the attitude of the critics whose business it is to evaluate our popular entertainment. I would like them to show a little more appreciation of sheer theatrical vitality and less acquiescence to what is maudlin and insipid. In reading their reviews I always got the impression that they were lauding well-known performers whether or not they had really enjoyed the fare offered; it was too much to believe that they actually liked them. Certainly, I argued, it would be rash to condemn publicly a "personality" whose reputation represented a large investment in expensive ballyhoo. But their denunciation of the most popular (for teenagers at least) performer today proved me wrong. They really did like the performers they so lavishly praised.

In the past we have had some lively theatre. There were periods when the great mimes attained such renown that their characterizations have come down to us as representative types and are a part of our folklore. The creations of the *commedia dell' arte* persist even today in all the respected ballet companies. Though the prototypes of our modern pierrots, harlequins, and the like were far from being the innocuous creatures we know now, their very strength as characters and their very persistence in our imagination derives from their origin as irreverent, bawdy, wayward beings akin to something deep within ourselves. Harlequin was far from the trim little creature we see today in the Biedermier version Nijinsky made so famous; his original was no moral example for 17th-century teenagers to follow. No more was Harlequin's friend, Brighella, a cutthroat if ever there was one. Allardyce Nicoll says of him, ". . . his lips themselves are sensual, his eyes are sated, evil, avaricious." Such a character would need special dispensation to go about the city after his act on the stage.

But these are the characters that made art on the open-air platform and later on the stage, and to which mankind has always responded. Are we to believe that they were created merely through histrionic insight? Or rather might it not be possible that the actor and his role were one and the same being? These characterizations are immortal because they were drawn from the deep level of Dionysiac lawlessness and from the gamut of life wherever life had a sting. In the accepted singing and performing types today they have no counterpart. Of course the *commedia dell' arte* had its innocuous characters too. In its *innamorati* we get a foretaste of our modern love-duets, but nothing has come down to us from them in either character or name.

Elvis Presley brings to us one of those fascinating characters of the old tradition, which by its very impropriety might go far to shock us out of our complacency in accepting the puerile fare that is being handed out to us from the stage and screen and in our living rooms. In spite of his blues, gospel, and hillbilly background, Presley has succumbed to the practice of the balladeers—to make sentimental trash of all popular music. But when he sings the songs that make him do a lot of "wiggling and quivering" as he calls it, and such songs as *Heart-Break Hotel,* I would say that he shows a singing talent far above the needs of the total character he projects. What he has, however, is a delivery that tinges everything it touches with enough popular spirit to engage him to millions. Although Presley's is certainly a singing role, he must be taken for his act, not his singing per se. This act includes some dance movements of great strength, as those in his movie *Loving You* showed. His short dance sequence in this picture shows his unusual ability in this direction and that, without his having all the necessary elements that combine to make a great dancing talent, he does have the stance of a very great performer. Behind most art expressions there lurks one or more influences. Bo Diddley states that he was Presley's model; the Presley act before and after Bo Diddley no doubt *was* different.

If his talent is small in comparison with a Charlie Chaplin, a Fanny Brice, or a Bert Lahr, he is still in their tradition. He has no kinship with the tuxedo-ed M.C. or the business-suited emcee-crooner-gagman who have superseded them. His approach is refreshingly alien to any tailor-made compliance to society. He is the present-day gadfly of the entertainment world.

Jazz Dance, Mambo Dance

Any attempt at exhaustive description of an unfamiliar art is more likely to confuse with a welter of vivid and high-sounding terms, than to convey the essence of the thing itself: such is the danger in writing on the Jazz and Mambo Dance. Although I certainly believe that we would profit by a detailed recording of both, it is not to my purpose to present one here. Rather, I will discuss what is already familiar and point out some of the problems that arise when a dance of "doing" is transformed into a dance of "presentation." I hope to show that the idea of a musical parallel to them is fallacious and breaks down on analysis.

The arts of performance, specifically dance and music—we dance, we sing—are certainly authentic arts, but by their very nature they transport their "doers" beyond any intelligent consideration of what they are doing while they are doing it. Self-criticism by an artist is objective examination of his finished work; it does not take place simultaneously with creation. Thus any real self-criticism of these arts must come through a later observance of the performance as reproduced by screen or phonograph. Even when the folk-popular dance reaches its highest pitch of excellence self-criticism at the time of performance does not become a part of the art.

In its early state the dance of "doing" became a major art; but one which loses all sustaining power and satisfaction when uprooted from its native soil. Music, because it was more easily transplanted, did not deteriorate under these conditions, but was able to develop and sustain itself without, like the dance, hav-

The Jazz Review, November 1959/January 1960.

ing to rely on expedients introduced by the individual performer. Music that serves the dance prepares itself during its improvisational phase for the coming of the composer, who will first work in the spirit of improvisation and finally go beyond it into extended forms. Until music's inevitable decline it was this individual performer-composer who charged its material and forms with greater and greater significance, but in the beginning he did not institute any new ways, new forms, or formats.

The dance as a "doing" lacks composition, except in so far as a few over-all formations are resorted to for the sake of convenience; nor is composition the dancer's concern. But there comes a time when, under the eyes of spectators, the dancer is made conscious of the effect he is producing on others, and his "doing" is no longer for himself alone. If he leaves the dance floor for a platform and exhibits his steps to an audience he is faced with the problem of how to present his scant material. As his previous concern with the act of "doing" has switched to a new concern of presentation he must have recourse to a purely optional framework to hold the interest of his audience. Within this framework he sets his dance, he composes; but throughout, his practices must be optional rather than, as in jazz music, proceeding through the means of a conventional framework. Unlike the musicians, he finds that he had no material in a state ready for presentation. This explains much of what I shall speak of in connection with the presentation problems of the Lindy Hop.

I take it as a premise that art springs from the folk subconscious. Without trying to explain my reasons for holding this theory of the folk origin of art, I should like to add that I find folk expression alone far more significant than any individual expression that fails to take advantage of folk material. How much of this basic material any artist may choose to retain in his own work is a matter of personal taste; but observation has taught me that only in the very late stages of an art is it advisable for him to rely on purely optional means of his own. Folk expression seems to be the result of a built-in mechanism, like a bird's nest-building instinct, which provides a powerful and unself-conscious driving force for the creation of strong and healthy art forms. Once the drive has spent itself these forms become static—they no longer develop. Anything further becomes what I call the use of personal options which needs a rare aesthetic acumen to guide and preserve it—at least in today's world.

Let me say that when a sophisticated art deriving from the folk aspires to a major position it will only be successful if the folk art from which it sprang had itself already developed far beyond a simple back-country expression. Otherwise we would merely have a new folk flavor shot into the arm of academic art, not a genuine new art springing from the folk.

Let us look at some differences between music and dance. The folk subconsciously developed the song form, which with the greatest ease encompassed any time length from eight to sixteen bars, lasting, in the blues, for instance, as long as forty seconds. With the possibility of variety through different instrumental solo passages and further variety added by the imagination of the players in the ensemble, together with a nod from the leader signaling the last chorus, we have a piece of music easily four minutes in length that conveys to the listener a feeling all of a piece. This entire procedure can come about with little thought. In the dance the longest stretch is a combination of single steps, which put together constitute what, in dance terminology, is called a "step." None of these steps actually takes more than two bars, while most of them take only a half or whole bar. The Charleston, for example, takes one bar; Boogie-Woogie, two bars; Jig Walk, a bar and a half. While music is thought of in tunes, dance is thought of in steps; and between the two there is a great disparity in time length. In performance we expect from both dance and music a duration of two to three minutes. A piece of music easily meets this requirement in its natural and subconscious development within a conventional frame, but in the dance, the dancer himself must substitute an optional frame in order to fill out the time.

A dancer normally uses only the steps at his disposal, and in any order. His dance never gives the onlooker the same feeling of the progress he has when listening to a tune. Our only cue to the approaching end of the dance comes from the music. This is evident when we watch any floor full of dancers, whether they are dancing in couple or completely apart. Any dance composition, however brief, is the optional invention of either the dancer or choreographer. But whether optional or not, the choreographer *must* eventually intervene. Just as one sequence of solos and ensembles in a band contributes more to the feeling of a whole than some other, so one sequence of steps will give a greater over-all impact than some other. While in one instance we are working with the arrangement of large blocks of integrated music, in the other we are working with the arrangement of short steps.

In music, I believe, there is a carry-through of mood inherent in the melodic process itself, while in the dance any prevailing mood derives not so much from the sequence of steps; but up to now such movements' expression superimposed on the material by the dancer. Of course bodily movement and stance can vary regardless of what steps are used, and these enforce the pleasure we derive from rightly qualified steps; but up to now such movements have not crystalized or been codified in any grammar of the dance.

And so I say that a group of "pick up" musicians with *no* rehearsal can give us sessions of music that have over-all unity, while unrehearsed dancers can only start dancing with the music and keep going until the last note. Very often in the Savoy, groups of dancers used to circle a pair who would dance furiously for only about fifteen seconds, and then be followed by another pair. It was terrific dancing—the greatest—but it gave no feeling of inherent unity; nor, because of the dancers' very brevity, did one demand this.

This unorganized state is of the nature of the dance, and this is not said in disparagement. From the time a dancer walks onto the dance floor he is in a state of exaltation in which he threads his way through the throng or remains in one place, repeating over and over a few steps in no set sequence until the music stops. This for him is a complete experience, both aesthetically and physically—as exhilarating as any physical exercise: taking a long walk, skiing, or fighting waves on a beach. A few steps done with little variation will satisfy for a whole evening—and evening after evening. The remarkable thing about this activity is that its artistic significance carries to onlookers, and they, through the power of empathy and their aesthetic sensitivity participate in the dancer's experience. They are also conscious of a silhouette in space, something hardly intuited by the dancer.

The spectator at a dance hall can receive the greatest kick by just watching the dancers in context of the whole whirling assemblage, picking out here and there a couple for his particular attention. From his vantage point he may see wonderful things, or he may get only tantalizing glimpses of something terrific; but altogether the impact of the music and the beat of hundreds of feet provide him with a completely satisfying experience.

The difficulties start when a single dancer or a single couple is taken from the crowded floor and exhibited on a stage where the poor individual or the couple must start dancing and continue long enough to satisfy the spectators. As soon as the dancer is removed from his dance floor to a stage the spectator feels he can demand a kind of performance he never expected to witness when he was a bystander in the dance hall. Though the dance may be precisely the same as it was on the dance floor, something is missing. But with a band, granted it played in a concert hall as well as it did in the dance hall, we would find that it still had lost none of its original impact. In fact even on the far-removed medium of records it still reveals all of its old dance-hall glory of invention and playing style: *nothing* is missing. The dancer, however, rightly finds that he *must* arrange; also he usually feels the necessity of injecting an overdose of showmanship (in fact he is likely to seem inept if he doesn't). Whereas the band, without changing anything in its performance, withstands the change to the other side of the footlights (or to a record), the dancer has to cater (and I believe he should) to what is expected of a "presentation." Here the optional dictates of a choreographer must come in, and with the steps and talent at hand, create a presentation which, while showing off the dancers' abilities, projects some feeling of over-all unity. This is the constant preoccupation of dance production and of the choreographer who uses strictly dance material. (There is also a choreography of miscellaneous material not necessarily dance in nature, which is another matter.)

After presenting these theories, let me mention the long line of steps derived from Negro dancing. None is over two bars long. More than two bars are called combinations. Out of what may be hundreds of steps a few, for one reason or another, have been taken up, though they may not necessarily merit the special esteem they enjoy. The greatest step was the Charleston; it is truly generic in character. When done to a Charleston rhythm in the music it could be infinitely varied without losing any of the quality that we sense to be *Charleston.* Its step was but one bar long. Next came the Black Bottom, which consisted of a small combination lasting two bars in which the dancer played up to the name by spanking his bottom a couple of times, although the dance is said to derive from wading in deep mud at low tide. Then there is Truckin' which is nothing more than a walking dance in which the dancer bends and straightens his legs as he walks. Combined with the Suzy-Q it makes, when done by comedian Pigmeat Markham, an extremely fascinating dance. All of these steps had music written in their name and were featured in musicals.

Although the Charleston was done as a couple dance in ballrooms, it wasn't until the Lindy Hop came along that we arrive at what has become a permanent national dance. With its many highly integrated steps—a necessity for a "led" couple dancing with break-away—we have a popular dance on which many variants have been grafted, such as the style deviation of the jitterbugs, the detached attitude conjured up by bop music in the Apple Jack, or the rock 'n' roll dance that has been sweeping the country.

The Lindy Hop is the only dance which has both cross-rhythms and more than two time values. Besides the steps which are synchronized with the musical phrases in the Lindy, there are steps which cross the rhythm of the music in the same fashion as polyrhythms in music. The extra time value, besides that of the commonly used slow and quick steps, is found when the dancers do a double-quick two-step in place of one slow step. We get cross-rhythms in the Foxtrot and the Peabody as a result of the breaking of the tight hold music had on dance during the 19th century. Then all steps were tied down by the tyranny of the musical first beat. When ragtime came along with eighth-note and sixteenth-note beats between the two main quarter-note beats (2/4 time), dancing in common time was liberated. Interjected notes tend to mitigate the power of the strong first down beat thereby acquiring a near equal stress between these down beats, a procedure that led to the character of popular 4/4 time. With syncopation and cross-rhythms in the music of ragtime, the main beat became definitely 4/4 time. Dancers started to sway from side to side in a jog trot (two-to-a-bar) in the rag dances that preceded the Foxtrot. All feeling for bar lines had vanished.

This led eventually into the One Step and later into the Peabody. For a short while the Foxtrot was a matter of four steps to the bar in the quick stepping fashion of a fox, but this gave way to dancing single steps consisting of two-to-a-bar together with two-steps, as in its present state. But the previous step-and-music identity was completely broken down, opening the way to cross-rhythms, of steps taking one and a half bars. This freedom was never attained by Cuban dancing, but was characteristic of the Tango through its own course of development.

This complete breakdown of adherence to the first musical beat made possible the introduction of cross-rhythms in the Lindy. Whereas many of the cross-rhythms possible in the Foxtrot are due to the aimless meandering of the dancers, in the Lindy the presence of cross-rhythms was a matter of precise steps, either definitely cross-rhythmic or definitely monorhythmic. A polyrhythmic character between music and dance is clearly felt when the 3-count (1½ bars) step is being used while with the use of the 4-count (2 bars) step the dance is coordinated with the musical phrases.

And so with this rhythmic freedom and the aforementioned double-quick two-step which, by the way, also came about in the Mambo and was called the Cha, Cha, Cha, we arrive at a highly developed folk-popular dance. With a variety of steps in couple position and a short space during the break-away for individual invention, the Lindy becomes a most complete dance, incorporating in itself any number of other steps. These steps were adaptable to the widest range of tempos. Of course the Lindy Hop, like any other great dance, possessed its own highly significant style and stance. Style and stance went beyond the mere deportment of any great exhibition dancers, whether Spanish (gypsy or classical) or our own ballroom performers. Although all great Lindy dancers possess a style and distinction of their own, within this style there is complete freedom of movement, opening the way to great bodily invention.

Out of the Lindy Hop or in conjunction with it came what the dancers called "the routine." These routines were a matter of changing from the couple position as led by the male to a team position (such as is done on the stage). Steps done in this position naturally had to be set and learned. Each couple made up its own routine or used that of others. The dance allowed for the use of the entire step gamut of their repertory. With the introduction of the new material the dance took on an entirely different character from the couple dance of the Lindy Hop: the "swinging" feeling of the continual break-away and return was broken up. The routine was like the development of a symphony or one of the alternate strains used by a jazz band. Interpolation give a special lift to the original theme when return is made to it. The same kind of lift is given the dance when the couples return from the routine to the right-turn break-away. Using the routine was the first sign in the Lindy Hop of creating a counter-section of a different character, something to set off the original Lindy Hop to greater advantage.

The Big Apple was another set dance, done in circular formation, in which the dancers either followed a given procedure or responded to the commands of a caller. It was a sort of square-dance in the round.

Another couple dance which had a short life was the Shag, a 3-count dance (1½ bars). A description makes it sound like a lot of hopping, but when it was danced well it became actually a smooth dance which at the same time retained the strong beat of solid hopping, the smoothness being achieved by the gentle swaying bodies above the quick footwork.

Sadly enough, out of this whole dance mania, this sweating it out night after night all over the country, plus the presence on the scene of innumerable great dancers, all contributing their own style and personal invention, none of it developed into the professional stage dance. Stage dancing with its pocketful of musical comedy steps remained unenriched by the Lindy Hop. While teams explored the possibilities of the Waltz, Polka, Tango, Rumba, Mambo, and what not, nobody explored this really great dance for what it could contribute to the stage. Under the aegis of Herbert White the best dancers of the Savoy Ballroom were formed into Whitey's Lindy Hoppers. They danced in movie houses, in motion pictures, and in Broadway musicals; they were featured at the New York World's Fair. But nothing really came of all this and the Lindy Hop was never integrated into show business.

While the great Lindy Hoppers stood on the sidelines, a new breed of dancer, fortified with ballet and modern dance training, took over show business and danced to some form of jazz music. The new dance has none of the style, refined or not, of the Negro dance. With its few movements derived from jazz it became a choreographer's idea of what dancers with ballet or modern training should do to jazz music. All of TV, movie, and musicals are loaded with this type of dance. The days of the good old dance man with his pocketful of steps now seems a golden era of the stage dance.

*The jazz dance, other than tap, has had very few eminent representatives on the stage. The best and among the earliest were the Berry Brothers—Ananias, James, and Warren. The whole act was built around the extraordinary ability of Ananias Berry. Aside from his technical proficiency his strut itself was marvelous—and it was quite a feat to build a composition around the few possible ways of strutting. But the Berry Brothers did, and their dance, *Papa De Da Da,* was the greatest of its kind. Another famous dancer was Earl "Snake Hips" Tucker, the originator of the *snake hips* dance, a title inadequate to describe the wonderful thing Earl Tucker made of it. His was a loose-jointed body which together with his great ability as a dancer made this dance with its short pantomime sequence impossible to imitate. Very often a dancer will capitalize on a special gift, which, incorporated into his artistic creations, makes his dancing unique. It is always fatal for even the greatest dancers to attempt what they are not cut out to do—that is, physically fit to perform.

There were many other teams who depended on either excellent comedy or excellent dancing (tap or otherwise) to show themselves off, but except for the great comedy acts, no dancers ever possessed the greatness of the Berry Brothers or Snake Hips Tucker.

Tap dancing, a category within jazz, has a complete academy and many talented dancers, who, with nowhere to turn and lacking the imagination to foster Negro non-tap dancing, found themselves sucked into the tap convention curriculum. As an accessory to singing, clowning, or even other types of dancing, tap gives a definite punctuation to the rhythm, but although its academy is thorough and complete it tends to undermine a dancer's inquisitiveness because of the fund of material it presents him with. It is precise and fun to do, but its charge is always that of astounding virtuosity, which, once seen, continually diminishes in interest and bears little repetition. Without the personality of a Bill Robinson, accepted tap procedure has become pretty dull.

JAZZ DANCE AND MAMBO DANCE

There is a complete confusion in our understanding of both the dance and the music of the Mambo. We often hear the chronology given as Charleston, Black Bottom, Lindy Hop, Big Apple, Apple Jack, and finally Mambo. We hear loose talk of the influence of Cuban rhythms on modern jazz either through the introduction of a lone Conga-drum artist into a band or by jazz musicians switching to Cuban rhythmic instruments. Whatever the future holds in the way of Mambo jazz dance or Cuban jazz music, at present the two musics and the two dances are quite different.

When Americans expert in the Lindy Hop, white or Negro, do the Mambo at the Palladium there is no trace of the Lindy Hop or any other dance in their performances. The differences between the Mambo and the Lindy Hop are as complete as the differences between a Viennese Waltz and a Tango, though the Waltz is no longer truly Viennese nor the Tango Argentine. As exports both have undergone a process of attrition during the past fifty years, and have lost much of their original character. But dancers get their kicks from the differences between Tango, Foxtrot, and Waltz, not in trying to amalgamate them. And so it is with the Mambo and the Lindy Hop. Their music is completely different—at least it is now. And the delightful feeling of relaxed "out and in" momentum, like the swing of a pendulum, in the right-hand break-away of the Lindy Hop danced at moderate tempo is not to be found in the break-away in the Mambo.

The Mambo is primarily a development of the Rumba, with its box-step, its pivots in both directions, its break-away, its distinctive placing of feet and distribution of body weight. The basic Mambo of couple-dancing is a natural evolution of the box-step which does not change the two-step rhythm or anapest of the Rumba, but flattened its box-like character into a forward and backward movement. The pivots remained the same, the breaks for all practical purposes are the same; the change was merely in the flattening of the box. With the flattening of the box, many proficient dancers have shifted the whole step from the musical count of one to that of two (4/4 time). The Rumba like all 19th-century dances, was tied to the musical bar, in this case 2/4 time; the Mambo, with the many beats of the rhythm section, was better expressed in 4/4, and dancers no longer felt tied down to the musical first beat. Many of them start on beat two or even beat four, which are the off-beats as in jazz. This shift in rhythm is an accomplishment that Lindy Hoppers never attained. They merely used either the first or second strong beat (first or third musical beat); they never shifted their strong beat to the weak, the two or four, of the music.

Another innovation in the Mambo is the complete break-away where the partners dance facing each other most of the time. Although the break-away is complete physically the partners follow the two-bar sequence (a step on the left foot—one bar—presumes its alternate on the other) in which the girl can either pick up her partner's step quickly or use a step of her own choice. We do not see this close connection without physical contact in the Lindy Hop. Lindy Hoppers sometimes parted and did a boogie-woogie step toward each other, but constant separation through a great variety of steps was never a part of the dance.

In the separate position, arm movements become a very significant part of the Mambo. The reaching out and bringing back of the arms, somewhat analogous to arm movements in flamenco dancing, becomes a very conscious occupation. The arms move either in unison or alternately to and fro, and the movements achieve great style.

As in the Lindy Hop, the dancers in the separated position often perform antics of the most extreme kind, doing either picked-up or remembered steps or movements of the wildest abandon dictated from within. During these excursions the girl maintains the two-bar pattern to guide her partner's return.

When a dancer in the Lindy Hop or the Mambo abandons the style of either and performs personal expressionistic movements he is doing neither dance. And until this abandonment becomes organized and a procedure common to both dances established, they remain different. I would say that the greatest significance of both dances is attained when these abandoned movements, more frequent in the Mambo as danced at the Palladium than in the Lindy Hop, are absent or kept to a minimum. The great *style* of the Mambo is rarely retained in these exhibitions, and the dance seems to fall apart. For the Mambo is a great dance, certainly in a class with the Lindy Hop. But I find that the pattern of movement in the Lindy Hop, especially during the periods when complete break-away is not attempted, helps to discipline the dancers in a way that the Mambo, with its tendency towards abandon, does not.

THE PRESENTATION OF THE JAZZ DANCE

Great difficulties are encountered in taking an art, a "time art," out of one milieu and placing it in another. In the major arts we have an endless repertory of works suitable for exhibition. Even improvised jazz can stand up under the scrutiny of a passive and critical audience, but dance, aside from the concert dance, has neither material nor improvisational procedures which give the semblance of completed works. Attempts to make a "show" of jazz dancing have always been abortive, in spite of the terrific impact of dancers graduated from the Lindy Hop. Short turns climaxed with their best work have been programmed, and aroused great enthusiasm from audiences, but such shows are geared to give the greatest punch possible to an audience with little understanding of the dance miracle before their eyes. There is nothing substantive which can be watched with relaxation and with keen interest.

When Herbert White presented his Lindy Hoppers at the New York World's Fair, he chose a framework which seems has since to have become standard in the presentation of Negro dance. Whitey had to provide a fairly long program, though not a complete evening. He had no repertory of works, and could not present the dancing of the Savoy Ballroom in its original form as a stage presentation, however enjoyable it was to watch in its own milieu. Whitey took their best dance, the fast Lindy Hop, and built a group dance around it, using all their aerial acrobatics as a climax to his program. The dancing and the acrobatics together carried a terrific effect. This was one instance where a white-heat performance carried impact of electrifying force, felt even by spectators who had no acquaintance with the Lindy. But unfortunately, the format of jazz dance history provided very tame entertainment indeed. Still for that time, and for the conglomerate audience that passed through the World's Fair, such a program probably amused and held the attention as well as, or better than, a more serious and varied use of the many possibilities in the Lindy Hop.

Mura Dehn used this same format many times without, of course, achieving the impact possible with a whole Whitey contingent. Except for those who made excursions to the Savoy people never saw the Lindy Hop by great dancers. They were usually horrified by the less professional "antics" of their jitterbug children (in spite, I would say, of the excellence of the antics), and when faced with a great Lindy Hopper they respond with instant enthusiasm. Mura Dehn's presentation usually consisted of many dancers and a small band. "History telling" was a small part of the proceedings. Among the dancers were two from the original Whitey group, Leon James and Al Minns.

With Marshall Stearns as interlocutor and commentator these two dancers have expanded the format by—among other items—imitations of Arthur Murray's teaching some of these steps, followed by an exhibition of how they should be done. The demonstration of step chronology became the mainstay of the performance; actual dancing was cut to the minimum. With only two performers, a long program of jazz dancing is grueling, even though they present only bits and snatches of their material. However, at the old Savoy Ballroom we used to see dancers keep it up all evening, because they danced without bearing down too hard on their physical stamina.

It is sad that of the multitudes of fine dancers, which far exceed the number of jazz musicians, should survive no more than these two. It is easy to understand why people are so enthusiastic over them; they are authentic survivors of the magnificance that came to life in places like the Savoy. Leon James and Al Minns are the products of those days and carry in their bodies sparks of the fire that inflamed all the Lindy Hoppers. James is a great dancer who has it in him to do terrific things. Al Minns is of the same school and has much to give in the Lindy Hop and its related dances. Although both have a fine stage presence and a talent for impersonation, I find that when they burlesque an awkward Murray pupil or even Snake Hips himself, they are not nearly as great as when doing the Lindy and jazz.

There is some talk both within Stearns's presentation and elsewhere that the jazz dance is neglected. Maybe this is so, but I am inclined to believe that it is up to the dancers themselves to present their work to the best advantage by attacking the problem of stage presentation. The dance in its natural habitat does not provide for the slightest beginnings towards presentation, as music so successfully does, although the material created on the polished dance floor has great import and certainly *can* be used as basis for further developments. It is up to the dancers, the only ones who really know their art, to make this effort, instead of relying on their own innate ability to "just dance." All the ballroom dancers back to Maurice and Walton, the Castles and Joan Sawyer, in the glorious days before lifts, developed their art, routined it, in a word, choreographed. They were very great dancers but they did much more than "just dance." On whatever level we place exhibition ballroom dancing, at least it was well presented. Let us see a far greater dance make use of the same advantages. The jazz dance and the Mambo dance are expressions of certain segments of our society which have now infiltrated the whole. Let us see its progenitors emerge as creative exponents of this dancing and carry it to new heights of excellence.

DANCE STYLE AND MUSICAL STYLE

To the question "which came first, dance or music?" there is no hard and fast answer. Some dances show an obvious rhythmic individuality; some are made up of combinations of two rhythms, such as the two-bar steps; while some steps, although they give a positive impression of identity, have nothing about their rhythmic basis which suggests a musical phrase. The Boogie-Woogie two-bar step (boogie-woogie music has no actual relation to the dance step) is an example. The famous single step, the Charleston, suggests the rhythm 4/4 but does not explicitly state it. For example the Charleston step has four definite beats within its length of one bar. If the music is not fast a dancer can hurry his second beat to coincide with the second strong syncopated beat in the music.

The Charleston rhythm, on the other hand, pervades all music, and for our purposes is predominant in ragtime. Nevertheless the Charleston rhythm had not, up to the time of James P. Johnson's composition *Charleston* (introduced in the musical *Runnin' Wild*), become unmistakably identified. While ragtime sheet music and piano rolls are records of the past, the elusive elements of the dance are lost. Whether the dancers actually used a Charleston rhythm before James P's piece I do not know. Since that time and with the help of a music strongly accenting this rhythm, the dancers still do not actually follow it, but give the impression that they do only because we associate the step with a music having this rhythm. Maybe James P. made a point of constantly holding to this rhythm because of the impression he got from the dancers, or very likely with the music in a slower tempo some dancers did more than just give an impression of the rhythm and did accent it. Certainly the dance was not evolved to fit James P's composition; rather James P's derived his music from his impression of the dancers, with the possibility that some of them actually may have followed the musical rhythm. Thus our visual impression of the step is influenced by the Charleston rhythm in the music so that whatever the dancer is doing we assume he is still following the musical rhythm.

Generally speaking, music and dance are free agents going separate ways, and a great deal of what appears to be an affinity is a matter of association on our part. Any music today is a highly complicated and self-sufficient entity. As an adjunct of the dance its rhythmic potentialities result not so much in rhythmic patterns corresponding to certain specific steps, but rather in electrifying the beat itself. This beat liberates the dancer, fortifies him in his performance as the beat of a drummer liberates the melody players and gives them such assurity that they venture into extended improvisations without losing themselves. The copying of drum work by the melody instruments, or of dance steps by the music or vice versa is not the crux of the situation. The crux is the liberty and inspiration one can give the other by vitalizing and solidifying the beat.

With the swing bands, the Lindy Hop reached its climax. They emphasized a beat highly inciting to dancing, whatever values the music had otherwise. It is the feeling for swing which differentiates the Lindy from the incessant head-over-heels character of Mambo music. The Lindy Hop did not develop further after its free association with swing music. The advent of bop was an affair of music alone, though of course it did evolve out of a strictly *dance* music. Thereafter, though the music continued to develop, it did so entirely on its own, and not as a stimulous to the dance—a common phenomenon throughout musical history.

As bop became the vogue, with its new and different beat and mood, the dancers merely interpreted it. But the interpretation of specific musical works in the concert dance is not the same as a group interpretation of music. The latter, of course, is far more self-sufficient, although dependent on the music. In the history of dancing we do not find any suggestion of a long process of growth and development, though in Western dance some growth is evident in ballet. Music has a greater potential of growth and given the right conditions can go on and on for a long time putting out heavily new shoots from its trunk. Dance usually evolves into an interpretative art, after which it only changes its style to interpret better the spirit of the music. In this process it loses any self-sufficiency it may once have had. Although music as an art eventually takes precedence over the dance, this must not bind us to the fact that at the time the dance was completely independent and at its height they were equals.

The subsequent Dixieland wave in jazz spread from certain young players who, back in the 1940s, revived, to the best of their abilities, the early New Orleans music. Its whole value and potency is due to their having returned to the very source of jazz rather than having exercised their talents on later musical styles. I believe that only when the new "jazz dancers," those of musical comedy and TV, feel a deep enough urge to return to the real jazz dance will they be able to establish something vital as *dance* and strong enough to be the basis of a future art. Certainly taking steps from ballet, modern dance, or even the tap academy, infusing them with a few jazz movements, has resulted in an insipid stylistic hodge-podge that has done no justice to the dance but much damage.

Of course jazz dancing was at a disadvantage from the beginning because it never really became a commodity, which as a "presentation," it must. Presentation dancers now face two tasks: they must revitalize the dance by going back to the source, and they must take up the problem of presentation.

The audience is ready. Are the dancers?

Write That Thing

The Jazz Review, November 1960.

I am writing this in the hope of enlisting the sympathies of revivalists, Dixielanders, or whatever the proper name, in helping me with a project I feel to be important to jazz.

I have never been one of those critics who believed in holding to the status quo of some particular period—usually, with such critics, the jazz of the 1920s. I have always believed that jazz should have taken the historical course, that is, a course like the one that gave us the great music of the Western world. This course consisted mainly in the development of improvised 16th-century dance music, first into simple pieces, later into suites, and finally into more extended compositions.

Since composition is constantly being made use of in modern jazz it might be argued that jazz is following the historical procedure. But while Western music developed slowly at first without too marked a change in the melodic content, jazz ran the gamut of scalar development (with the model of classical music always before it) that had taken Western music four hundred years, *without developing in construction.* As for my views on modern jazz, I have stated them before and although I naturally keep finding new arguments against it, this is not the place to air them.

However I think it is important to say that I feel the jazz of the 1920s, however great its value in the musical scene, failed to develop its potentialities. Instead it continually changed for the worse and I believe that what will finally emerge subsequent to these changes will not have the great and unique character that distinguished jazz in its early stages. Clearly the historical procedure as applied to jazz must start at a much earlier stage than the one at which the modern jazz musicians began applying compositional techniques.

As much as I enjoy listening to the present-day Dixielanders I feel that they are at a dead end and have long since proved their point: that it is possible for contemporary players to come close to the jazz of the past. Of course besides proving this point they do fulfill a need in the same way a classical orchestra does when it plays a concert of music of the past. But my belief is that the best of these bands could be still better and their music, as a whole, could be raised to a higher level of significance.

I have already made a start in an attempt to apply the historical procedure to jazz with four pieces, but unless one has sympathy for the project to begin with, he is not going to see that anything is proved by what I have done. And rightly so, for the growth from improvisation to notation will be, in its beginning, hesitant as it must be if it is going to be of any value at all. We must remember that the compositional efforts of the early lutenists and harpsichordists were not world-shaking, and as a matter of fact did not compare with the brilliance of the actual improvisational performances. These early compositions were hesitant; there was nothing bold in the result, only in the venture. In this connection I quoted in the *Hound and Horn* (Summer 1934) the impressions of the 17th-century writer André Maugars on hearing Frescobaldi (1583–1644) play: ". . . to judge of his profound learning, you must hear him improvise." But if it were not for the men who, for whatever reason, notated their compositional works we would never have had the music of Bach, or for that matter of any classical musician since then.

Admittedly there have been some new possibilities with the invention of phonograph recording, but again this is not the place for me to go into arguments as to why I feel that the possibilities of recorded improvisation have not displaced development by means of notated composition.

When I go to a live jazz concert today I can never be treated to anything but the present, unsatisfactory stage of its development. And if there is a Dixieland band present, even if it could play some of the classic jazz pieces well, the chances are that I will get renditions "in the style of something else," somewhat similar to a pianist playing a little tune in the styles of Beethoven, Mozart, or Bach. As apt as such performance may be, they are a far cry from the great music of these composers.

If there are any players, composers or arrangers who share my assessment of the development of jazz to date and who feel as I do about the promise for its future that the historical process offers, please get in touch with me through *The Jazz Review.*

Uses of the Past: A Reply to Martin Williams and Douglas Pomeroy

Roger Pryor Dodge and
Jerome S. Shipman

Note. Messrs. Dodge and Shipman's original statement appeared in the Feb. '63 Jazz Forum; *individual replies by Martin Williams and Douglas Pomeroy in the April-May '63 issue of* JAZZ. *Further comment from musicians, critics and laymen is invited.—Ed.*

Rather than discuss which New Orleans material compositional methods could be applied to, we would prefer first to find out whether there is agreement with our belief that jazz should move into notation and then composition (using the term as it is understood in classical music). We feel that improvisation has its useful periods, but that after a time in each of these periods it develops serious limitations. What has happened in jazz is that when one (improvisational) style declines, a new one arises to take its place. So far jazz "composition" during the flowering of any style has amounted only to a more or less skillful combination of improvised solos with written material; when the ability to improvise tellingly in that style becomes lost, "composition" in the style stops too. We believe that even if improvisation in a style is no longer as vital as it was during its height, it is still possible and desirable to develop, through notation and compositional procedures, the forms and material that were produced. The alternative is to be left only with what exists of the period on records; records which must become increasingly strange to generations of listeners further and further removed in time from the original music.

Jazz, October 1963.

We tried to suggest in *Jazz Forum* that classical music, like jazz, had its era when improvised music was better than what was captured in notation (Frescobaldi), but classical music patiently developed the forms that grew out of improvisation; the dances improvised on lute and harpsichord became little suites, and the process culminated, after two centuries, in the great instrumental music of Bach. In any period of jazz bustling with creativity (and there have been several, although we might prefer one, Martin Williams another, and Douglas Pomeroy a third), there is plenty of material that stands up under notation. This material can be played by merely adequate players and still retain its significance. Once it has been notated it can be pondered over, pruned here, added to there and eventually combined with similar material to make a larger structure which might be as satisfying in performance, in spite of any lack of spontaneity, as records of the original style. Now standing in the way of an attempt at this process are the widely held ideas that "the past cannot be recaptured," that the original music cannot be separated from its creators, that there can be no jazz without improvisation, that jazz cannot be notated. We called them critical fallacies for the obvious reasons that they are promulgated by critics, and that there are cogent arguments to show that each of them is fallacious.

Martin Williams and Douglas Pomeroy both seem to feel that what jazz critics write has little effect on the development of the music. They are right if they are thinking of a *direct* effect, for it is certainly not the case that a critic writes an article and after it is published jazz moves in the direction he proposed. The process is more subtle but none the less real, since for any musician except the folk artist the continual arguments he has with his musician friends and critics are a force which influence his thinking and the direction his art might take. A young musician interested in New Orleans jazz might very well be discouraged from trying his hand at composition (in our sense) in the style if much of what he reads leads him to believe that it can't be done. Fallacies ought to be exposed for what they are, not simply dismissed because they happen to spring from the mind of a critic rather than a musician.

We think that it is both possible and desirable for jazz to turn to composition, that several periods of jazz could in this way be developed further than they were, and that jazz as a whole might benefit from a body of written music accessible to any competent group of musicians. If Martin Williams and Douglas Pomeroy agree in general with these basic ideas, we could go on to discuss further points raised by their comments provided, of course, that enough readers are interested in the question.

Tradition in Ballet: *Les Sylphides*

Every art, in its development, goes through certain stages in which it attains its purest form of expression. An example from Western music would be the great dance suites of the 17th and 18th centuries, culminating in those of Bach. Where should we look for their counterpart in our Western theatrical dance art, the classical ballet?

The essence of 19th-century ballet still emanates, in all its fragile charm, from the hundreds of surviving prints of the period; its spirit lives in the *danse d'école* as handed down from Legat and Cecchetti. Today there are two expressions of this heritage: and re-creation of 19th-century ballet, using the hindsight of today, and the reconstruction of long two- and three-act *ballets d'action* with little hindsight. An exquisite example of the former is in Fokine's ballet *Les Sylphides.*

I believe that if every practicing choreographer were to have a try at making a *Les Sylphides* of his own, a ballet in which he exerted his talents to the full in an effort to top Fokine's production, then out of many such efforts there would emerge a few great enough to take their place in ballet's classical heritage. If there was any validity in the early re-creation of *Les Sylphides* from old material, the same approach to the material is still valid. Today there is as great a need as ever to fulfill those potentialities inherent in the art of the ballet. Much art of the past, cherished in its own day, has little appeal to the taste of a later generation; and the old *ballet d'action* with its second-rate or inappropriate symphonic music is a case in point. The absolute dance *is* the dance; it is the only part of a ballet that time does not wither.

The Dancing Times (London), January 1964.

There is such a thing as absolute music—the great symphonies are that—and there is the kind of music we find in opera and other compositions using a narrative framework. But whatever its category, it is always music. However, it cannot be said of the *ballet d'action* and the forms stemming from it that they are exclusively dance.

One of the chief problems in ballet construction has always been how to stretch a composition out to give the absolute dance sufficient program length. That was why the 18th- and 19th-century choreographers resorted to the use of plot. Though today's choreographers compose with equal ease in both narrative and pure dance forms we have tended to drift away from Noverre's belief in working within a story frame. There is, of course, a present vogue for choreographic compositions exhibiting a wide range of possibilities—pantomime, dance, movements, business, and stunts. But the dance I speak of is pure dance.

Certain types of *ballet blanc* would be aesthetically more satisfying if mere originality ceased to be the choreographer's preoccupation. Rather than strain for uniqueness of *mise en scène,* he does better to develop such creations from the standpoint of pure dance. With success, something enduring comes into the world. We must remember that the initial reception given to *Les Sylphides* was mild in comparison with the enthusiasm that greeted the torrid productions of *Scheherazade* and *Cleopatra. Les Sylphides,* with its more subtle appeal, took its own time to win wide acclaim—but it is still with us. It is not the fault of the work itself if sometimes it seems too much with us and our interest wanes with each repetition. *Les Sylphides* needs the company of a body of comparable works in the repertory to save it from staleness. Alone, it is in danger of being done to death.

To sum up, we may say that *Les Sylphides* is not a ballet expressing a unifying idea. Rather, it is a series of episodes danced in traditional ballet costume. The parts are so finely balanced in their relationships, and each is so inventive in itself, that we derive from them a feeling of artistic unity. The great musical dance-suites do nothing more; and in both instances there is no limit to the possible variations on this pattern. *Les Sylphides* outlives every other of Diaghilev's creations: there is no end to what can be done with it. The idea of a *ballet blanc* did not originate with Fokine; *Les Sylphides* was, rather, an inspired adaptation within a classical archetype to suit the demands of this century's taste. It in no way closed the door to further efforts along the same lines. In this respect alone it differs from all other ballets; it is truly a germinal work.

My hope is that for a long time to come other choreographers will follow through the door Fokine has left open. The repertories of ballet companies throughout the world could well absorb a bevy of beautiful, new works destined to hold the stage as long as Fokine's masterpiece.

Reviews

GEMS OF JAZZ Vol. 4
Decca Album No. A-249

Stardust	18251
Well, All Right Then	
Lost in a Fog	18252
I Ain't Got Nobody	
It's the Talk of the Town	18253
Nagasaki	
I've Got To Sing a Torch Song	18254
Night Life	
Blue Interlude	18255
Once Upon a Time	
Somebody Loves Me	18256
Pardon Me, Pretty Baby	

Since Hawkins's slow tempo is sentimental in style, he adds little to the sweet, slow popular tune. Jazz has a sweet side (it need not always be dirty-hot) but nevertheless the sweet side of Hawkins does not seem to occupy a significant place within the sphere of our best jazz. Such popular recordings successfully present Hawkins, the great instrumentalist—musically they have nothing much to say. On the other hand, in brighter tempo we find a continuous "ahead" movement, a constant brilliant attack of phrases. Although these solos are built in a somewhat involved manner and are not particularly melodic, there is no doubt that what melodic content they have, needs, and gets, the force which Hawkins alone is able to give them.

Jazz, August 1942.

Well, All Right Then starts out with a Hawkins solo beginning with a phrase from *Four or Five Times.* The second chorus, of this first solo, is especially good. Here he gives the effect of two saxes, his own and an echo. His next, after the piano, seems especially fine. There is nothing more satisfying than a great technician playing a good solo. Here polish and surety bring their own reward, a reward in listening that some of the more primitive low-down solos do not present.

It's the Talk of the Town features Hawkins in a weaving solo that will appeal to those who like to hear the song plainly stated and treated with the few nuances such a likable old tune gathers in its old age. In *Nagasaki,* fou bars of very good orchestration usher in the piano solo. "Red" Allen's trumpet solo, after his singing, is in his best style.

In *I've Got To Sing a Torch Song,* Hawkins slips in and out the tune in a most delicate manner but unfortunately brings out the most sentimental phases of the tune. For those who like this tune, attention is arrested and Hawk's beauty of tone and nice figuration are pleasantly present.

In his album notes, Leonard Feather comments on the two surprises presented by *Once Upon a Time*: the trumpet playing of Carter and the first major appearance of Teddy Wilson. Wilson has never appeared to better advantage since, except, let us say, in *Just a Mood.* Carter plays much in the vein of *later* Armstrong.

On *Pardon Me, Pretty Baby,* George Chrisholn's trombone sounds good. His melodic treatment of the tune, after clarinetist Jimmy Wilson's hot and raspy release, is excellent. There is nice phrasing in the beginning of Carter's solo. In the Carter-Hawkins dust, Hawkins's last statement, descending as it does while Carter lifts the answer high up, finishes off this duet in fine fashion. Certainly a moderate fast tempo brings out the best to be found in many improvisers.

In conclusion, the Hawk's fans and those who are partial to the run of the year popular tune will find all his slow solos most delicately woven and beautifully phrased—future "long hair" saxophonists will really enjoy reading them from notation. But the great Hawkins, both in instrumental technique *and* in musical content, is to be found in quick tempo work, work which may seem slightly dry in texture, but which is none the less very inventive and very forcefully delivered.

BIX BEIDERBECKE	(English)
Margie	Parlephone R2833
PAUL WHITEMAN (BILLIE HOLIDAY)	
Trav'lin' Light	Capitol 116
ART HODES BLUE THREE	
Tin Roof Blues	Jazz J.101
Diga Diga Doo	
EDDIE CONDON AND HIS BAND	
More Tortilla B Flat	Commodore 1510
Lonesome Tag Blues	

The posthumous work of any artist is always of great interest for those who know that artist's work well. Unfortunately, *Margie,* like a lot of tunes that Bix played, remains *Margie* even after Bix plays it. However, in the ensemble chorus is heard one of his great runs. The solo is inventive but it is easily distinguishable as Bix. This is its importance.

Bix's name is revered by many jazz enthusiasts and he deserves it. But except for a few solos of his, which are musical in a Chopinesque sort of way, the bulk are remarkable, not so much as improvisations, but as examples of Bix's tone. So much has been said about his tone that little can be added. For me, the very notes themselves have a liquid and seductive quality rarely drawn from brass other than a French horn. The sound of his cornet is probably the furthest removed from the sound of the virtuoso classical cornet. The difference approximates the sound difference between a squeaking violin and a violin with a mute. To mute, however, is to mechanically contrive, whereas Bix's tone is owed to the phenomenal sensitivity of his lip. There is no set-up to the sweetness of a mechanical contrivance, but a sweet tone from the lips alone, and in the hands of a player of good taste, can hold great variety.

Trav'lin' Light (Paul Whiteman with vocal by Lady Day (Billie Holiday) featuring Skip Layton, trombonist) presents two opposite examples of sweet interpretation, one by a soft-spoken trombonist, the other by a vocalist with fine quality to her voice. Billie Holiday is in the tradition of Ella Fitzgerald and Fitzgerald could have been a real blues singer if modern "hot and sweet" singing had not forced out the place of blues. Billie Holiday's work on this record, although sweet, is good. We can see the results of her previous blues singing. On the other hand, Layton's trombone is a good example of what Bix's sweet cornet was not. There is tremendous difference between Layton's beauti-

Jazz, September 1942.

ful but flat tone and the Bix tone that occasionally could be called sweet, but at all times was subtle and brilliant in its coloring.

The new label Jazz has a reissue of *Tin Roof Blues* and *Diga Diga Doo,* formerly under Bob Thiele's label: Signature. On the side *Tin Roof,* Rod Cless's solos seem good but so immediately strung together they harm rather than help each other. His clarinet solos should be spaced so as to gain contrast value. Art Hodes's piano refreshingly differs from that of many other pianists. When he plays his special style, the solo is always interesting. It always fits in. I find more drive to the side *Diga Diga Doo.* Hodes's second solo is good, even exceptional. This record of *Tin Roof Blues* shows that musicians can, when using a good blues, "take it easy." The blues do not necessarily have to be pointed up to become significant in recording. Cless's treatment of Brunis's original chorus is an understatement, but in spite of this, I enjoyed it. This face is a tribute to Brunis and jazz.

On Commodore's twelve-inch issue are two sides of good stomp blues. The sides ride right along, the soloists following in the same order on both. On *More Tortilla B Flat,* Pee Wee is good, both in his solos and in his four-bar breaks towards the close, and Kaminsky plays well. *Lonesome Tag Blues* has Brad Gowans giving two good choruses and Pee Wee adding another one to his credit. The second chorus of Al Morgan's bass solos seems better than his others.

GEMS OF JAZZ Vol. 5
Decca Album No. 324

ART HODES AND HIS ORCHESTRA
Georgia Cake Walk 18437
Liberty Inn Drag

ART HODES AND HIS ORCHESTRA
Get Happy 18438
Indiana

JIMMY NOONE AND HIS ORCHESTRA
The Blues Jumped a Rabbit 18439
He's the Different Type of Guy

JIMMY NOONE AND HIS ORCHESTRA
Way Down Yonder in New Orleans 18440
Sweet Georgia Brown

JIMMY McPARTLAND AND HIS ORCHESTRA
Original Dixieland One Step 18441
I'm All Bound 'Round with the Mason Dixon Line

Of Hodes's four sides, his *Liberty Inn Drag* is by far the best. Starting out with two choruses by Hodes, each soloist, in turn, plays well. Cless has some nice descending rhythmic passages. Sidney de Paris plays a fine muted trumpet. Hodes ties up the whole recording. His beginning put one in the right mood for whatever is to come afterwards, and his closing is a preparation for a logical end. In fact, Cless's entrance is helped a great deal by the fine atmosphere created by Hodes. It is a good record and one of the bright spots of the album. Of Hodes's other three sides, *Get Happy* seemed the most appealing. For a popular tune it has an attractive beginning and ending.

Bessie Smith sings this verse in her *Rocking Chair Blues*:

Blues jumped a rabbit
Running for a solid mile.
Blues jumped a rabbit
Running for a solid mile.
The rabbit turned over
And cried like a natural child.

On Jimmy Noone's record *The Blues Jumped a Rabbit,* Guy Kelly sings this same verse, except that his last lines go—

> Yes . . . The poor fellow lied down
> And cried like a natural child.

It is a fine record. Guy Kelly's trumpet has the old intonation—something between Bessie Smith's voice and Louis Armstrong's cornet on *Reckless Blues.* His second chorus is very melodic.

Note the beautiful entrance. After Preston Jackson's excellent trombone follows Noone's low-register clarinet playing just right. From here Noone climbs up to the higher register to finish off a fine record.

Jimmy McPartland's *Original Dixieland One Step* is his best side. However, he plays too much in the background on both recordings. I'd like to hear him more up front.

ALBERT AMMONS

Bass Goin' Crazy	Blue Note 21
Suitcase Blues	

MEADE "LUX" LEWIS

Rising Tide Blues	Blue Note 22
Tell Your Story No. 2	

JOSHUA WHITE TRIO

Careless Love	Blue Note 23
Milk Cow Blues	

Albert Ammons misses his chance to make a really great record in his *Suitcase Blues.* Out of the nine choruses making up the record the third chorus, a Hersal Thomas chorus, resembling a tail-gate trombone, is played five times over. Such repetition, although typical of a type of old blues piano playing, like anything else, can be run into the ground. If only Ammons had newly improvised each succeeding time, he would have built up a record of never flagging interest. I believe in twelve-inch recording for these artists, but feel it is possible to introduce a more nicely calculated juxtaposition of solos. Ammons's *Boogie Woogie Stomp* builds to a tremendous climax. His *Bass Goin' Crazy* has enough material, so that it would be as good as *Boogie Woogie Stomp* if the order of the variation was rearranged and played in varying registers of the piano. As it is, the last chorus is a letdown without being intentionally so—that is, it is not the pianissimo so often employed with intention. Nevertheless *Bass Goin' Crazy* is a great record. Note that the rhythmic figure of his Solo Art recording of this piece has been extended through the whole twelve bars on this last issue.

Any record made by Meade Lux Lewis can only be compared with his others. *Rising Tide Blues* seems, comparatively, a little uninventive. However, his playing, clean as it always was, has so improved that listening to that alone is an intense joy. His shakes roll along with an ease at once soft and beautiful yet rigidly unsentimental. They pour out of his polyrhythms and out of his melodic figures; they are abruptly cut off by short sudden runs or they have a long rhythm all their own. In the sixth variation I notice, for the first time, a use of dynamic coloring involving both hands simultaneously. This is something the folk pianist rarely does. He may use a varied coloring in the treble but the bass is usually kept at its established dynamic intensity. The last variation has tremendous power created by its continuous chordal pulsation.

Tell Your Story No. 2 holds the same interest as did his first *Tell Your Story.* All through the record there is a high level of invention. Choruses one, two, and four are the only ones that somewhat resemble those of the previous recording. On his first recording he had a variation resembling the theme of a Hodes record, *Randolph Street Rag.* On the present record the fourth chorus is another version of this same *Randolph Street Rag* variation. A double pianissimo ending, built on a very melodic figure, finishes the record, holding the interest right to the end. This deliberate dynamic change helps to leave the listener in a complete state of satisfaction. A too earnest application of this musical device is liable to sound precious, but Meade Lux possesses the rare and remarkable attribute of absolute detachment in his playing. This detachment allows the swing and spirit of the style to govern his playing, rather than artistic intention. His intentions are so alive that the average interpretive pianist probably is bound to romanticize them. This Meade Lux never does. In his version he is always crisp and clean.

There are very few blues singers who do not invoke a sense of weary repetition, although blues singing is the background of jazz and of the best that is jazz. All that is great in jazz has sprung from the blues. Among singers, however, only too often variation means change in the lyrics, not in the music. Twenty years or so ago, every blues singer had something to contribute to jazz consciousness. Coming as they did from widely dispersed sections of the country, these singers brought to jazz a variety of musical "dialect." Not all of the "dialects" matured, but even those that did not contributed greatly to the one that did, giving it an added strength. The old blues records are valuable in that each singer contributed some little something towards jazz, not withstanding the endless repetition which makes listening monotonous. Although every jazz enthusiast praises these records highly, singing records are not played as often as purely instrumental ones. They do not hold the same interest. Bessie Smith, Ma Rainey, and a few other blues singers are the exceptions that prove the rule. There are few famous singers as compared with the scores of well-known instrumentalists.

From the first quivering long note of Joshua White's guitar to the end of Bechet's chorus, *Careless Love* is a perfect record. I cannot say enough for these first three choruses. The intimacy in White's voice when he first says "Love, oh love, oh careless love" is unique among blues singers. In quality it recalls the late John Mills of the Mills Brothers. White's accompaniment to Bechet is inventive and independent in character, especially effective in the first and sixth bars. Bechet's solo is the great Bechet—the clarinet Bechet. The emotional ending to his solo is truly inspired.

Milk Cow Blues, at a brighter tempo, is more typical in its singing. Bechet's solo is good and so also is Wilson Myer's bowed bass solo. Myer is somewhat hidden on *Careless Love.*

These last six sides put out by Blue Note, under the supervision of Alfred Lyons, are in the great tradition of all the records he has sponsored. Because of an established consistency we can always look forward to high standards in player, high standards in selected material, and distinctive supervision; we never have to listen to *Margie,* however great the player. Although he was not the discoverer of either Mead Lux, Ammons, or Peter Johnson, and although their work has been issued on other labels, and although similar combinations resembling somewhat the Hall Johnson Quartet have been recorded, Lyons's persistence in issuing the boogie-woogie pianists, the small combinations such as the Hall Johnson Quartet and Pete Johnson Trio, the use of harpsichord for boogie-woogie, all go to make his Blue Note label stand for something that we could nearly identify as Blue Note Jazz. I have often urged the recording of a "steered" or supervised jazz—steered as opposed to arranged. These records, from the standpoint of supervision, are surpassed nowhere. This does not mean that they have the greatest solos on them or the greatest contributions to jazz. It simply means that all we can ask has been done to promote good jazz. As most recorded solos are impromptu improvisations, a lot depends upon the presence of inspiration at the recording date. But there are certain things in jazz not quite so dependent upon invention which are always interesting, such as boogie-woogie executed by its greatest exponents, and the performance of such small combinations as Lyons has gotten together. When we find his music choices receiving praise from the concert critic, the best jazz critics and the run-of-the-file reviewers, then I think there is something in this choice of jazz which the praise of any one critic alone might not have guaranteed. I am not choosing one or two choice Blue Note records. This can be done under any label. I like the whole Blue Note catalogue.

ELLINGTONIA Vol. 1
Brunswick Album No. B-1000

East St. Louis Toodle-Oo	80000
Birmingham Breakdown	
Rockin' in Rhythm	80001
Twelfth Street Rag	
Black and Tan Fantasy	80002
The Mooche	
Mood Indigo	80003
Wall Street Rag	

The *East St. Louis Toodle-Oo* is surely a jazz composition. What, in future, may be said of the early recordings of this piece is uncertain but, for today, as compositional material *not* depending upon the uniqueness of a performer, it is one of the rare pieces that can be taken as composition. As it stands it is not much more than a very long solo with a following string of variations. However, it is this beginning of the Duke and Miley's that points a way to what jazz can be in composition. Miley's long hot solo, with the Duke's eight-bar introduction repeated over and over, is spontaneous composition. It is regrettable that any subsequent "improvements" made by the Duke only cheapened this fine piece. For surely, jazz, to grow into the compositional stage, must start with whatever material it has that resembles composition like the *Toodle-Oo* and from there go on—go on in an authentic line of improvement. Jazz cannot be regarded, in the mass, as offering material for the First Jazz Composition. We have a number of unfortunate Firsts. If we are not ready to improve the semi-improvised *East St. Louis,* then we are not ready to compose as do the composers in the classic field.

The *Black and Tan Fantasy* has still greater propensities toward composition than has the *East St. Louis.* With no change in tempo it gives the impression of a variated beat. It, also, is nothing more than a theme with variations, but the variations are integral to the composition so that the "you take the next" is never felt. Since the solos are complicated in character, the *Black and Tan* at this time does not stand up well when played by other than exceptional soloists. Miley's solo in the *East St. Louis* is more a theme for a trumpet, easier to read and easier to remember than is his solo on *Black and Tan.* For this reason alone, the *East St. Louis* is more likely material for composition, at present, than the *Black and Tan.*

Jazz, December 1943.

Miley's solo on the Victor *Black and Tan* somewhat eclipses this Brunswick recording, but, nevertheless, on the Brunswick, his solo has all the material that is on the Victor and is not only very beautiful in its own musical right but also has a sensitivity of feeling that no successor of his ever had.

Birmingham Breakdown and *Rockin' in Rhythm* are both extremely good Duke numbers. The former has a wonderful busyness in the orchestrated parts and represents the very best orchestration of that period. *Rockin' in Rhythm* has a fine piano introduction by the Duke. The Cootie Williams solo is simple but good. The one-rhythm-one-harmony background for Barney Bigard is typical of the 1920s jazz. Sometimes solos, apparently influenced by such accompaniment, take on an Oriental slant, but this does not happen in such solos as Charlie Green's in *The Gouge of Armour Avenue* by Henderson's orchestra and in this Bigard solo. They give good contrast while remaining good jazz. Tricky Sam's solo not only arouses nostalgia but shows a fine inventiveness.

The Mooche and *Mood Indigo* seem to point the way to *Reminiscing in Tempo* and such examples of this side of the Duke's output. The *Black and Tan* might be thought by some to be the first in such a style of music, but there is a great deal of difference between the *Black and Tan* and *The Mooche.* In the days when even those in the best Chicago style were playing *Nobody's Baby,* this kind of music by the Duke seemed much more of what was generally expected of a Negro orchestra—expected of jazz. Although everyone likes the solo, and has always appreciated it to a certain extent, the arrangement has always been thought to be the "brain" of jazz, and the Duke, as far as this was concerned, surpassed everyone else. With *The Mooche* and other things, he created something consistent which was not an arrangement and not a string of solos—in other words, a composition. But, although *The Mooche* is important to us in a certain period of jazz, it only led to something off the path that the best of jazz has taken. It must be remembered that here the soloists are not being taken into consideration. Naturally they are part of what makes the complete Duke so great, but it is on such records that the Duke slips out from under his soloists and reveals a not too great *jazz* musician.

Joseph Nanton is always praised but he is a greatly neglected musician. There is only a short solo, a chorus at most, and never an opportunity to see what his full creative powers might be. His invention is melodically radical with no intention to startle. His growl tone has great "personality." It cannot be dismissed as another "jungle style." On both *Twelfth Street Rag* and *Wall Street Wail* he has two very good solos. *Wall Street Wail* is mediocre otherwise. The polyrhythmic climb of his entrance on *Toodle-Oo* is typical of his playing.

JELLY ROLL MORTON
Victor Hot Jazz Series Vol. 5

Sidewalk Blues	Victor 40–0118
Dead Man Blues	
Deep Creek	Victor 40–0119
Red Hot Pepper	
Burnin' the Iceberg	Victor 40–0120
Pretty Lil	
Little Lawrence	Victor 40–0121
Ponchatrain	

Victor has reissued eight Jelly Roll Morton sides. I think there could have been a better choice to set off with in an album, however, there are some very worthy sides of his here. Both *Dead Man Blues* and *Sidewalk Blues* are typical Jelly Roll discs. They start off with some corny noise effects and talking which in a way is nice to hear. *Dead Man* has two fine solos by Simeon and Mitchell. There is also a nice little orchestrated part with a heavy accent on every eighth beat. The second chorus of this is backed by a trombone countermelody. Ory plays this most beautifully. I miss this sort of thing in modern records. Ory was wonderful at this sort of playing. *Sidewalk Blues* has two very good off-beat solos on trumpet and clarinet. The orchestra then goes into two thirty-two-bar choruses, playing very straight a tune that has one note to a bar. The last eight is jazzed up in both instances but the twenty-four bars preceding this contrapuntal jazz ensemble is much too long. The contrapuntal part, however, does stand out all the more because of this.

The best side in this album is *Deep Creek.* In fact it has qualities that make it one of Jelly Roll's best records. It starts off with an amazing introduction. Just in these few bars there is fine jazz writing. A statement of a theme, a Charleston beat, and then the theme's restatement polyrhythmically set forth. A few well-placed drum beats and then a crash by the orchestra. From here Swayzee has a very tuneful solo. Cato's trombone is ushered in with three chords by Jelly Roll. So simple and yet so telling. This record is full of the sort of material that prompts an orgy of image writing such as we received in the description of romantic music. When this quality in music is created subconsciously we are in no danger of receiving romantic music but instead a music that is both pure and vividly suggestive.

This trombone solo of Cato's winds around in a most mysterious way. The

The Record Changer, July 1945.

clarinet solo extends this same spirit making but another segment in this wonderful record. Jelly Roll's solo is some of the very best of his piano. Every part of it sparkles with ideas. It has a quality of holding one suspended until it is through. His entrance into this solo is great and rare. After the piano we have a most reedy solo on clarinet coming forth in a most plaintive manner. The record ends with the ensemble playing easily on to the end.

In *Red Hot Pepper, Burnin' the Iceberg,* and *Pretty Lil* there is a scattering of interesting spots. The first of these is the best of the three. Jelly Roll has some interesting work in his left hand on this side. *Burnin' the Iceberg* is a fast two-to-a-bar piece. It has a pretty good trumpet solo at the end. Except for those who get this album just for a side or two in order to fill out their Jelly Roll collection, I think for the uninitiated these three could have been replaced by others. An album presupposes some sort of reason for its selections. We are badly in need of re-issues, but unless we have a complete edition set forth in some manner (such as chronologically), an album of four records should contain the best examples of the band.

To continue, we have a fairly good record in *Little Lawrence.* It has some good solos, one on clarinet, a guitar solo by Addison backed by Jelly Roll's piano, and a dynamic solo by Bubber Miley. There is no mistaking what Bubber is about when he plays. In his later years, however, he fell into some clichés that he used too often. I speak of his sharp blast that pierces through anything around it. He uses nine of these in *Little Lawrence.* They are exciting and dramatic, but even this wears away when overdone. The musical line also suffers when it is employed so copiously. There is one long note, however, which starts before the end of one section and runs through the beginning of the next. In spite of being a single note it has a tremendous rhythmic effect.

In *Ponchatrain* I hear a tune reminiscent of the Duke. The ensemble plays the first chorus. Pinkett squeezes himself into an excellent solo. Eddie Barefield has a very sensitive solo too. Starting with a sustained note he plays a solo with a very long line consisting entirely of eighth notes. This leads into a beautiful solo by Bubber—very original, clear-cut, and well composed throughout. Needless to say it is impeccably played and is so originally presented that it upsets the general tranquility of a Jelly Roll record. Addison's guitar solo, again backed by Jelly Roll, is very fine.

YANK LAUSON'S JAZZ BAND

Squeeze Me Signature 28103
The Sheik of Araby

Signature has issued a record which gives us a fine solo by Miff Mole. The first section of his chorus, in *The Sheik of Araby,* is great trombone playing and it is too bad that such a high standard was not kept up through the rest of the chorus. Miff Mole has given me such great moments of fiery inventiveness that I find myself with the greatest anticipation when I listen to him—an anticipation that I suppose would be hard for him to fulfill. In the first half of his present solo he gives me the greatest musical satisfaction. His musical line darts ahead with piercing thrusts. Miff Mole's tone is not ingratiating but when playing in this fiery manner it is sufficient to set off his melodic line. This is probably why I demand more of him melodically, and when he does play with spirit the tone does take on more quality. He must play vigorously to create this tone.

In the following solo, Cless makes a nice little figure out of the theme and then uses it to good advantage. The ensemble in the last chorus is not too exciting. I hear Mole but not enough of Cless. The ending I could dispense with. These endings are quite meaningless. If, after the drum exhibition, there was something new and startling, then we might have something to listen to instead of this all-out tag to a piece that has ended.

In *Squeeze Me,* Jimmy Johnson gives us a fairly good but not too inventive solo. Against a vamp bass he has a few nice figurations in his right hand. The solo is all right but there is so much more in Johnson. I do not blame him too much as he has many admirers who accept his straight playing with the greatest approval. Cless's solo is excellent and played with a great deal of emotion. The break at the end is quite something in the design of the very notes.

JAM SESSION AT COMMODORE, No. 5
(DIRECTED BY EDDIE CONDON)

Basin Street Blues	Commodore 1513
Oh, Katharina!	

GEORGE WETTLING & RHYTHM KINGS

Struttin' With Some Barbecue	Commodore 561
How Come You Do Me Like You Do	

The historical jazz classic *Basin Street Blues* and the very non-jazz tune *Oh, Katharina!* oddly enough are just in reverse in so far as jazz ideas go on these two sides. *Basin Street* is played straight by the men with little inspiration, while *Oh, Katharina!* has wonderful invention in the ingenious twists given the tune. Kaminsky plays the lead in the first ensemble with continuous support by the other men. Pee Wee has a very interesting solo. He runs off into *Annabelle Lee* in the second half and is very melodic throughout. Both Kaminsky and Benny Morton are good, showing a good deal of invention. They both play a little heavily, though, and this makes it drag a little. Catlett's long drum solo is certainly heavy. Somewhere along here the tempo slows up a little so that the last ensemble drags horribly. The first ensemble and the solos mentioned are very good and are nice to hear.

In *How Come You Do Me Like You Do* the solos are set forth interestingly. After the ensemble there is a piano solo by Bowman. Wilbur De Paris excellently takes the break at double time, then finishes the chorus. The next chorus is started by Hall with very good passages in it. Butterfield has a very good break and his transition from double time to the slow blues is justly phrased. It is a good record and a few of the spots I have mentioned have good invention. De Paris sets a fine mood by starting the record with a glissando reminiscent of the one in *Jackass Blues.* I like to hear such music finding itself repeated.

SIDNEY DE PARIS' JAZZ MEN

Everybody Loves My Baby Blue Note 40
The Call of the Blues

SIDNEY BECHET'S JAZZ MEN

Blue Horizon Blue Note 43
Muskrat Ramble

These two new releases by Blue Note are exceptionally fine. If there was ever a healthy tune it is *Muskrat Ramble* by Kid Ory. Not a blues but neither is it a popular tune. The men play it with the greatest joy. The first section is repeated—it is so fine it needs repeating. The gusto of the third section is terrifically exciting. The trombone blasts out the beginning of each phrase in this section and makes some other renderings I have heard seem anemic. Dickenson has a very melodic and gracious solo. It is a little out of the *Muskrat Ramble* spirit, but I suppose we need a little change after the vigorous beginning. Bechet takes two very good choruses, growling a lot in the second. The finale builds well with De Paris playing a good lead.

Blue Horizon is an extraordinary solo record. Bechet's tone is like nothing I have ever heard. His vibrato is so pronounced that we nearly hear the very rattle of it. There is nothing forced in his tone as this vibrato seems to flow of its own account, especially on the deep low notes with which he usually ends his phrases. The choruses are simply conceived, in fact are a little too simple. On first hearing I was taken with the sound of his clarinet. Hearing it again seemed to reveal a lack of inventiveness. The choruses seemed to hinge on an unchanging pattern. Subsequent hearings, however, revealed the piece as being quite just in its composition and when one does not expect startling invention the piece takes on an overall quality of good composition. The different choruses take on just enough newness to carry the piece along without protruding in themselves. Bechet did a fine job and a job not as easy as it looks; he could have very easily become repetitive.

In *Everybody Loves My Baby* the solos are well played but there is a lack of invention in them. De Paris has the best solo although it is not outstanding. Johnson's piano is a little on the accepted side of orchestra piano playing. Maybe I am asking too much, for it is no lack of imagination on the artist's part but rather in the medium or style of such piano playing.

The Call of the Blues is another unique record. To a background of Johnson's boogie-woogie De Paris plays many choruses. They appear to be all in the same manner but the variations are so melodically conceived that the sameness of manner disappears in compositional progression. Shirley has two good guitar choruses. Some of his melodic twists are so intriguing, when playing on his bass notes, that it is a shame we cannot talk more explicitly about them without recourse to notation. Hall's choruses are not too inspired but they are, as always, intact and move along in the medium of great music. There is a background by De Paris and the orchestra to Hall's solo which is only one more of the ingenious things we hear on this side. Dickenson keeps rather close to theme introduced by De Paris. He is more mellow than Hall and coming after him makes good contrast.

The Call of the Blues is a great record. How great I cannot exactly say. It does not have, nor does it attempt to have, the quality of greatness that we may expect to come from early New Orleans or from the style that the men, surrounding Bunk, have today. The very singable music in it should make it very winning and I cannot see why it should not become popular and at the same time have nothing to be ashamed of. De Paris so played his choruses both at the beginning and at the end that the whole piece holds tightly together. The other solos in the middle vary in their style somewhat which is something we must and should expect from a later-day jazz. All the solos are played in a manner that hints of the fact that something has transpired since early jazz. De Paris and Dickenson are both extremely melodic in their own way.

LOUIS ARMSTRONG
Brunswick Album No. B-1016

Wild Man Blues	80059
Melancholy	
Georgia Bo Bo	80060
Drop That Sack	
Static Strut	80061
Stomp Off, Let's Go	
Terrible Blues	80062
Santa Claus Blues	

In this Brunswick album of Armstrong's re-issues we have a collection of not so famous but none the less good examples of early Armstrong. There are examples of his many talents from breaks in slow blues and fast numbers to complete choruses in both tempos.

This *Wild Man Blues* is quite different from the Okeh version. It is much more mellow and not as startling as is the Okeh. Neither Dodds's nor Armstrong's breaks are separated from the tune as they are in the Okeh. Dodds fuses the two-bar melody quite completely with the following two-bar break. To the musician who is aware of the separateness and difference between the tune and the breaks, Dodds's beautiful blending of the two into one pattern gives extra pleasure.

Both solos seem to have more subtlety than do they on the Okeh. Although Armstrong's solo has some fine spots, it does not have the inventiveness that is manifest on the Okeh. The accompaniment, in the main, plays through the breaks. Armstrong's passage in the stop-chorus section is reminiscent of the type of singing from which this tune derives. On the Okeh we do not get the feeling of the tune so much but we get a more precise instrumental composition. I like Louis's introduction better on this record and the phrase at the end of his solo is very melodic. Somehow through Louis's whole solo I feel Armstrong the player, his mannerisms, etc., more than on Okeh. On the Okeh, Armstrong creates something much greater than what we might expect from him. Altogether this Brunswick *Wild Man* is a truly great Armstrong and the subtlety of its line needs and deserves attention. I am inclined to like Dodds's solo better on this record than on Okeh. On Okeh his solo is a little disrupted between tune and breaks without much invention in the double time breaks whereas on this record his blending is extremely beautiful.

 The Record Changer, September 1945.

In *Melancholy,* Dodds plays the verse in chalumeau after which he plays a very silvery toned half-chorus that fairly floats in air. He plays it too straight. Armstrong darts around with a more rhythmic release and finish of the chorus. Hines's playing is sober and interesting enough. It is good because it is so close to Jelly Roll's playing.

The only singing side in the album is *Georgia Bo Bo.* On it Louis takes a straightforward solo, apparently unconcerned with the wonderful material created by him. It rolls out so easily that it hardly sells itself, and almost passes unnoticed. Ory plays the blues near straight, and it is fine, because the blues are perfect music *to* play straight. In the sixth bar he has a slide passage, the kind of Ory passage that "eats right into you." Louis's singing is simple, we do not have any of the hackneyed scat variety. Dodds's solo is in his best vein. So far as intonation goes he has the hottest solo on this side. The piece ends with the ensemble playing the verse and blues.

In *Drop That Sack,* Dodds plays a very wonderful solo. It is a twenty-bar tune. The treatment from the thirteenth to sixteenth bars is a stop chorus with the orchestra playing one to a bar. In Dodds's solo we have a four-four background up to the fifth bar and then off-beat background until the stop chorus passage. As I listen to these records I realize how interesting such devices are. In the fifth and ninth bars Dodds sounds wonderful with this off-beat encouragement. Lil sounds very good in the stop-chorus passage. Both hands have equal emphasis and we get a relief from the vamp bass. Although vamp bass on the piano does carry the rhythm, and is occasionally bearable in moderate tempos, it is generally speaking too tiring. Ory's solo is good Ory but it is the kind of trombone playing I do not like. Behind such playing there does not seem to be any intention but in this specific instance the fault that I generally find with this *kind* of playing is far outbalanced by Ory's tone. His vibrato in the stop-chorus passage is solid and dirty without any forced intention on his part of achieving this effect.

Louis's solo has great variety especially in the breaks. *Drop That Sack* is healthy New Orleans jazz even though it is not a blues. I would like to see the off-beat devices creep back into a little of present-day jazz whether it be Dixieland tradition or not.

The Tate sides are mainly orchestrated. The sections behind the soloists are extremely good and enrich these solos. On these sides I also find that although Louis has great facility on his trumpet his playing lacks real content. His solos do show forethought which is possible because of the ease with which he can use the various trumpet registers. I would say that it is the rhythmic ease with which he plays the actual musical line that invests it with interest. Thus in these solos there exists a balance between the parts and a facile execution but the total effect suffers because of the lack of melodic content.

Following the orchestrated ensembles in *Stomp Off* there are two solo choruses, one for Louis and one for Weatherford. These solos have excellent examples of simple but effective accompaniments. The orchestra starts out with one to a bar for three bars then is silent for one bar. After this is repeated there follows an unaccompanied solo for piano. Louis has about the same accompaniment which gives breadth to these two choruses and strongly impresses on the listener the pattern of the music, something which straight strumming of two or four to a bar cannot do.

Teddy Weatherford, the pianist who went to India, is a brilliant player. On both these sides his solos are brittle and crisp. His invention, however, is overpianistic and depends too much on its brilliance.

Not very much can be said of *Santa Claus Blues.* We hear verses and choruses and Buster Bailey playing fairly straight an inconsequential tune. *Terrible Blues* is a nice old record but one that could not be done today. It is effortless in manner and we get a fine feeling of that period. Bailey's solo and the ensembles move along very nicely and it is a relaxing rather than a stimulating record. Trombonist Thompson has a good break in the ensemble before Louis's, and Louis's solo has good intonation and his break in the ensemble near the end is the Louis of Bessie's records.

EDDIE CONDON AND HIS BAND

Back in Your Own Back Yard Commodore 551
All the Wrongs You've Done to Me

DE PARIS BROTHERS ORCHESTRA

Black and Blue Commodore 552
I've Found a New Baby

All the Wrongs You've Done to Me is played in a nice slow fashion. There is a very fine trombone solo by McGarity. His beginning has good swing. The rest of his solo is not up to this beginning but carries through because of his fine tone rather than because of his ideas. The drop in the second bar is excellent instrumental jazz. Such playing can distinguish a whole record. Pee Wee has a solo stressing his very dirty tone which at times becomes slightly obvious. However, he also has some very good passages.

This slow tune, *All the Wrongs,* and the two fast numbers, *Back in Your Old Back Yard* and *I've Found a New Baby,* stress the playing of the tune too much. These are nice enough old tunes but in these disks they appear to be plug records for the tunes. We all know *I've Found a New Baby* so why play it straight at all. I must still complain of the tags after the drum fanfares. If we must have them why not have something like Jelly Roll's *Smokehouse Blues* tags. Why can't we have fast tags in the fast records as absorbing as these slow ones.

Black and Blue has an extraordinary solo by Wilbur De Paris. His solo is very close to the tune but seldom touches it. De Paris plays his own invention fairly straight, and if one did not know the tune *Black and Blue* his solo would be taken as the tune itself. He has used a little embroidery of the most exquisite sort which embellishes the intervals between phrases. These embellishments hold the position of breaks but do not sound like breaks. They resemble more an extended pick-up integrally holding the sections together. Notice the octave jumps before the repeat of the first section. De Paris's melodic content is slightly sweet as is the general style today but it is exceptionally fine and musical playing of its kind.

The release is taken by the piano which is a mistake. Wilbur De Paris should have had all of this chorus to himself. The release is not an ordinary release. When I first heard it played by the trombonist Fred Robinson on Armstrong's record I thought it was his invention but subsequently I realized that he was playing it straight. This straight playing fits the trombone excellently. Wilbur De Paris carries the inspiration completely to the end of his solo which ends with a few very beautiful figures. It is very satisfying to have such interest carried through to the end of a solo. Edmond Hall has a very good chorus with Sidney De Paris taking the release. Sometimes musicians give the feeling that they play just because it is their turn and both of these releases are so played. However, this is a minor complaint when we get such a solo as that by Wilbur De Paris.

MIFF MOLE AND HIS NICKSIELAND BAND

St. Louis Blues Commodore 1518
Peg o' My Heart

MUGGSY SPANIER AND HIS RAGTIMERS

Memphis Blues Commodore 1519
Sweet Sue, Just You

There is a nice feeling in the Commodore record of *St. Louis Blues.* It is very slow, a tempo more suited to it than the sped-up versions of a later-day jazz. I find that the insistence on adhering to the tanagra rhythm a little boring in records made of the *St. Louis Blues.* Although the rhythmic section keeps up a

The Record Changer, October 1945.

good four-four beat, the melodic instruments use the tanagra rhythm but occasionally each instrument breaks away which enhances this section. Especially energetic is Miff Mole's intrusion during this second section.

St. Louis Blues is a long blues consisting of three different sections, the standard version always repeating the first. Orchestras seem compelled to play all three sections before getting down to solos or new improvisations. The Commodore men shorten this up considerably, turning over the third section to Bobby Hackett. The solos are adequate creations but are not too melodic. *Savoy Blues* is introduced for the last half of the record.

On *Sweet Sue* Pee Wee has by far the most to say in his solo. All the way through he has surprises of melodic thought. Spanier's and Mole's solos are played well but there is not much said.

The best side is the dear old *Memphis Blues.* The music proper is introduced by a cadenza of Muggsy's after which we run into two choruses by the ensemble. They give a nice twist to the accepted break in the seventh and eighth bars. Pee Wee again gives us a fine solo consisting of two choruses. The first, in his tone resembling more a bullfrog than a clarinet, has great musical insight in understatement; probably inspired by the tone he used. The musical lift of Pee Wee's first phrase in the second chorus, and then the rhythmic movement of the band with Pee Wee silent for a couple of beats, shows the greatest inspiration on his part. The rest of this chorus has great vitality and some very melodic turns to the phrases. Muggsy's second chorus is by far his best. He uses some exceptionally good jazz intervals besides playing this chorus with an extremely passionate feeling.

The whole record is good even though I especially single out these two solos. Just as such old pieces as *At the Jazz Band Ball* and *Muskrat Ramble* inspired the players with a fine rhythmic feeling, so does the *Memphis Blues* bring about this same instrumental attitude. It is the rhythmic aspect of ragtime expressed in the best of blues and jazz.

JAMES P. JOHNSON
(New York Jazz)
Asch Album 551

Euphonic Sounds	551–1
The Dream	
Four O'Clock Groove	551–2
Hesitation Blues	
Hot Harlem	551–3
The Boogie Dream	

Amongst an improvising group of varied talents the written music of Scott Joplin takes the prize. I do not want to discuss at this time what I feel about the sheet music of ragtime but I must say that in spite of its simplicity and in spite of the fact that a Scott Joplin or a player with the repute of Tom Turpin is not playing we get extreme pleasure in hearing Johnson play it. James P. was brought up on the tail-end of ragtime so that however authentic it sounds in his hands we never know what quality the original improvisers had. As I have not seen the sheet music of *Euphonic Sounds* I cannot say whether or not Johnson is filling it out a little more abundantly or not.

If Johnson is playing the music as written then I cannot see that great satisfaction would not emanate from any pianist playing such music, if added to his pianistic abilities, he had some ragtime experience, even though his knowledge of ragtime was acquired through sheet music. If, on the other hand, Johnson is not playing it as written then we should have an edition of the ragtime classics edited by James P. Needless to say Johnson is *the* man to play these sides and we can only hope that someone will eventually do an album of ragtime as good as *Euphonic Sounds.*

Euphonic Sounds is an odd rag. The three sections are very different in character, especially the second, which has the strangest melodic progressions I have ever heard in ragtime. It makes fine piano music and is a happy choice for this album.

The Dream just misses being another *Orchids in the Moonlight* or some other derivative of the real tango, and a slight turn of the "stylistic knob" would have made it a replica. *Four O'Clock Groove* is a very melodic tune. Its content is pointed up well though having little to do with jazz. Casey's chorus, however, *is* jazz invention and breaks up the excessive sweetness of the record. *Hesitation Blues* gives us the initial performance of James Johnson as a singer. His voice reminds me of the Negro vaudeville comedian in Harlem—a few jokes, a shuffle, and then a song. Although it stands up well in this album, I would rather he sang only two choruses of a simple blues rather than occupy so much space with such a long-drawn-out affair. James P. does have a sincerity in his delivery that some of the more pat singers of this genre did not have. After a piano chorus, Newton plays his best jazz chorus. Before he comes in the musical atmosphere is saturated with expectancy of him.

The sporadic entrances of Casey in *Hot Harlem* are the best on this side but the whole piece does not quite come off. Newton has a good chance to really do a very good solo in the stop chorus but does not come through so well.

Boogie Dream, the best orchestra side, starts out with an introduction definitely arranged. It is neat and compositional in character but sounds more like a dinner-music composition for classic musicians than jazz. With James P.'s first boogie chorus real interest starts. Although a straight chorus, the melody is so constructed that it takes on the character of an introduction. The boogie players also have this talent. James P. has two good choruses after which Casey takes two excellent choruses on guitar. His second chorus is exceptionally fine and terminates what is really fine on the record.

James P. starts something new at the end of *Boogie Dream,* which has little to do, so far as style is concerned, with what precedes it. It is good of its kind and, similar to the beginning, has the quality of a later-day compositional piece. *Boogie Dream,* nevertheless, is a fine record and the four boogie-woogie choruses reveal more and more inventiveness on the part of these two improvisers, James P. and Al Casey. The just movement of Pops Foster on bass and Dougherty on the drums makes fine orchestra boogie.

JOHN WITTWER TRIO

Wolverine Blues Exner No. 1
Joe's Blues

Come Back Sweet Papa Exner No. 2
Tiger Rag

KID ORY'S CREOLE BAND

Dippermouth Blues Exner No. 3
Savoy Blues

Ballin' the Jack Exner No. 4
High Society

To make good records with such a line-up as we find on the Wittwer sides is not an easy task. Although I like what these players propose to do stylistically, neither of them is up to seeing it through. Such records even by exceptional players are not, as a rule, satisfying. We can recognize in such cases the merit of the player but that merit is not sufficient to establish a truly interesting record.

Of the four sides, *Joe's Blues* is by far the best. I would say that the last three choruses are the satisfying ones. Darensbourg sings two choruses with a slight modern intonation. It is too bad that he has not got an older intonation on this side. After the singing, Wittwer takes a piano chorus which is heavy low-down piano. This chorus seems to have punch when first heard but does not wear well because it lacks melodic content. It is very good, however, and I would like to hear more of this kind of piano. Darensbourg's entrance after the piano is very beautiful. He shows in these last two choruses a good feeling for the blues. The piano supports him well in the last chorus.

In *Come Back Sweet Papa* Wittwer plays the end of his piano chorus with a good feeling for rhythm. There is a certain suspended quality that gives it interest. Darensbourg's chorus is high and his emotion in it comes through well. The drums keep going along with the same beat giving little help to the total effect. Some parts of *Tiger Rag* are not bad but they are not able to keep up with the pace demanded by *Tiger Rag.*

The band is larger on the Ory sides. On the whole there is a certain shakiness about them. I cannot say that there is great inspiration, either in the spirit with which they play or the inventiveness of the solos. *Dippermouth* is played fast but seems hurried and is not exciting. The tune sounds better at a slower tempo. In so far as it is not exciting it annoys one in attempting a hurried spirit. I find this true with all their fast sides. Ory plays one chorus of the traditional trumpet solo. It is wonderful, his tone and all. The trouble is that he repeats it instead of going on with either the original solo or one of his own. I only wish that he *had* taken the whole solo instead of Carey.

Wilson's piano is a two-handed affair. By that I mean that both hands play the same rhythm which is a relief to the vamp bass. It is nice solo piano. Some of his breaks, such as the one in *Ballin' the Jack,* have a character of Jelly Roll. It is a piano playing that has escaped the extreme quality of the Hines school. I like to hear Carey, and in the traditional three choruses he sounds all right.

Savoy Blues is the best of the Ory sides. The ensemble has a sort of clip beginning from which we run into a breathing chorus. The whole chorus is played in this deep sleepy manner and is highly satisfying. Ory plays a four-bar introduction to Darensbourg's solo in chalumeau—the smoothest of his playing on any of these eight sides. Both this solo and the piano chorus are in a good blues mood.

Ory's solo is mostly a staccato affair. His legato rough trombone is wonderful and a relief after the staccato section. When Ory plays in the staccato fashion he is throwing away his talents. Although he may be taking it seriously, it has the character of kidding which I hate to see in a great trombonist. Now that

Ory is recording again I hope to see something come forth that is really expressive of his best self. His glissando into the second section of *Savoy Blues* is fine while his use of dynamics in his second glissando does not pass unnoticed. The clarinet is up at the top and sounds good at the end.

High Society is played in the traditional manner, piccolo solo and all. Darensbourg's version is a little different but not bad. The build-up section to the solo by the ensemble is no build up at all. It is such places as these that the records show great weakness. As a general rule Darensbourg lacks authority and fire in his playing which is evidenced on this side. *Ballin' the Jack* runs along with Ory straight and somewhat staccato again. Carey's solo is good and Darensbourg's counter voice after the breaks at the end are very melodic. Altogether these eight sides wear better than one would expect on first hearing them. On the whole it is their intention that I like better than their material.

SONGS BY LEADBELLY
ACCOMPANIED BY SONNY TERRY
Asch Album A343

Good Morning Blues	343–1
How Long	
Ain't You Glad	343–2
Irene	
On a Monday	343–3
John Henry	

Leadbelly keeps adding to his rich store of American folk music. Folk anthology is growing but it is such men as Leadbelly who give it any point at all for being presented to the public at large. Only with the folk singer himself do the words and music make any sense when repeated many times, unless, of course, we personally go in for folk singing ourselves. Leadbelly's attack of the lyrics is quite foreign to accepted singing. It is this folk quality which, without detracting from their meaning, removes them from mere repetition of their literal content.

Of the two sides on the first record I find *Good Morning Blues* the most interesting. Sonny Terry on his harmonica plays one introductory chorus that makes his simple material strong and striking. Leadbelly has a good attack on *Ain't You Glad.* Terry is very good on this side and shows great variety in his accompanying material. As each verse comes up it has renewed vigor because of Leadbelly's new interest in it. This is not usually evident on most of his records. It has a spirited jump to it.

In the waltz *Irene,* Leadbelly shows more sympathy with the words than on any other of these sides. The repeated chorus of

> Irene good night
> Irene good night

is far from monotonous. In the last verse he has great feeling. I would like to see more of this in Leadbelly. *On a Monday* is a lively tune. Terry plays very rhythmically on it while Leadbelly sings better than usual. What I miss on all these sides is more of Leadbelly's guitar. More of it would have added immensely to them. In Leadbelly's singing there runs a certain monotony. Although he is prolific in genre, he does not give to any one piece great variety. Except where I have pointed out, we get the whole piece in the first chorus.

Folk singing is rightly an expression of a people, a people singing themselves. It is the exceptional singer who so raises this art that it has constant interest when *listened* to. We have only a few folk singers of this caliber while the ranks of blues singers would be very easily swelled immensely if the demand warranted it. There is a wonderful stark quality in Leadbelly's voice which might be termed "bottom" but if it is to be "bottom" I would like to hear a little moving around on this "bottom." Bessie Smith was stark sometimes, such as in the middle of *Work House Blues,* but wherever she placed her interest she continually moved her inflection hither and yon.

Although Leadbelly has been "built up" a little he nevertheless holds his own amongst the mass of blues and folk singers to be considered after the Bessie Smiths and Ma Raineys. Probably it would be more just to term Leadbelly as a folk rather than a blues singer. It is in the fast tunes that he warms up to the music most readily. Each one gathers momentum under his hand and their tunefulness runs through one's head long after. As with all singing, the knowledge of the words contributes greatly to the enjoyment of the singing. The folk delivery of lyrics is so mannered that it is hard to understand the words, especially in the fast numbers. I do not say this in reproach as I certainly would not want them articulated. On the other hand, however, I would like to see a printed text issued along with singing records. The one very beautiful verse in *John Henry* printed in the notes contributes interest to the record, especially so, as at this point of the record Leadbelly is not so clear. The album is representative of Leadbelly and does his voice and folk spirit justice.

VARIATIONS ON A THEME
Harpsichord Improvisations By Meade Lux Lewis
ART HODES' BLUE FIVE

19 Ways of Playing a Chorus	Blue Note 19
Self-Portrait	
School of Rhythm	Blue Note 20
"Feelin' Tomorrow Like I Feel Today . . ."	
Shake That Thing	Blue Note 45
Apex Blues	

Alfred Lion is again pressing the Meade Lux Lewis harpsichord sides which were never well received in the past. Meade Lux, *Honky Tonk,* and boogie-woogie are all closely related in the public's mind. The music is so highly charged in its own way that, like Bach's fugues, no individual side stands out separate from the group. Because there is nothing of a popular vein on Meade's records, one or two of his records are for the average listener representative enough.

When we add to the "non-popular" aspect of this music the fact that the instrument on which it is played is itself a "non-popular" instrument, and is relegated to the field of novelty in jazz, we have little that attracts the jazz fan. This is unfortunate because the instrument has a richness that the piano has never attained. In ensemble, especially a jazz ensemble, the harpsichord lacks sufficient strength to stand out. There is too much dynamic disparity between it and the other instruments. Over a microphone it can take its place with the other instruments and in the recording studio its variety and richness are unmistakable.

There is no reason why the harpsichord should not be accepted as an instrument for the playing of jazz. The reason that it has not been accepted is that the jazz public has been conditioned to certain other instruments. Naturally the jazz musicians choose their instruments from among those at hand and popularized these instruments. Any lesser-known instruments seem exotic and are therefore shunned as bizarre by the average listener. There is no valid reason why we cannot use other instruments, if by so doing we improve the music. The unorthodox instruments such as the kazoo, washboard, jug, and the like, found themselves supplanted by instruments of greater possibilities, so why not further experiment?

Meade Lux's *Self-Portrait* is an excellent example of a very melodic blues transcribed to a keyboard. I hardly know of another piece which is such a perfect balance of singing blues and keyboard rhythmic design. Amongst the many different treatments of music on the keyboard the richest lies somewhere between the straight playing of a song and inventive design with no apparent song continuity. Most pianists fall into playing either one way or the other; they rarely hit the rich mixture of the two.

After Meade Lux's bold introduction consisting of two groups of vigorous triplets and a termination of the blues structure, we are given two choruses of keyboard blues exposition. Following these melodically rich choruses Meade introduces a polyrhythmic chorus and from here keeps to the lower section of the keyboard. The fifth is a light airy polyrhythmic chorus that he has used more than once. There is no other boogie-woogie player that has anything like it. After this airy chorus Meade plays more of this melodically rich material further down on the bass, making this side a sober but rich piece. One of the many devices of the harpsichord is the *jeu de luth* pedal. The use of this pedal gives a fine effect to one of Meade's choruses. It makes the notes sound extremely muffled while the plucking mechanism of the instrument is quite audible. The last chorus is divided up in phrases of five or more chords each. Against the boogie-woogie left hand it makes a stirring finale. At the end of this chorus the blues is also evident.

A fast boogie, *19 Ways of Playing a Chorus,* and a little slower side, *School of Rhythm,* are both extraordinary pieces. Each chorus on these sides is packed with ideas and the silhouettes of each are clearly outlined in the composition. There is a sweep to both these pieces that comes up to the best of Meade Lux's creations.

Feeling Tomorrow Like I Feel Today has about as many surprises as *Self-Portrait.* There are two outstanding choruses on this side. One is the sixth, where he has an extraordinary passage in dotted quarters. Nowhere else will we find anything like this. The other is the ninth, where he has a series of bold utterances every four bars. The chorus ends with a rhapsodic flight of the right hand.

On these four sides Meade Lux has shown his richest imagination. Except for such set pieces as *Honky Tonk* and *Yancey Special,* the material here shows him to have the greatest imagination of any keyboard artist. Add to this the timbre and earthy quality of the harpsichord with all its accompanying devices for variety, and we have a keyboard music which can at last compare with the complete musical statement of the wind instruments.

Apex Blues is a very quiet record, in fact a little too quiet. The duet of Kaminsky and Mezzrow is straight but nice. After this Mezzrow takes two very simple but gratifying choruses. The music in the ninth bar has exceptional blues quality. Hodes takes the next chorus holding our attention more in the beginning than at the end. There is another duet section and a return to the first theme for a final. The record is nice but too repetitive. Everything is repeated "as is" which makes it drag in places.

Shake That Thing is an excellent record. The choruses are well planned, Hodes taking the first, last, and one in the middle. It gives good balance. Max leads the second chorus while in the next Mezzrow becomes more pronounced. Max has a sharp way of playing the tune that gives it a pronounced down beat. Mezz's chorus is mellow and low, and I wish he had gone on up into the high register for a second chorus. There is a simple one-bar riff chorus. Hodes plays against this riff which is very effective. Riffs have their place in jazz if done simply. Riffs are about the easiest type of music to assimilate and for this very reason become obnoxious when they are "overloaded." After this chorus Kaminsky plays still sharper and more pointedly. This is a very fine chorus of his. One of the best I have ever heard him take in this sharply attacking vein. In the ensemble chorus Mezz plays very beautifully. His tone is warm and clear.

I like Hodes's blues playing and although I like his delicate trill tracery I believe he is beginning to rely on it too much. Once accepted, it does not have great interest in itself. I feel he goes into a chorus waiting for inspiration and when it does not come he substitutes the device of trill tracery. Today, in the recording studio, a player need not be always ready with a spot improvisation. If he is ready, that is fine and good, but if he is not, then he can always borrow some successful device out of his past playing. What comes out on a record is what we judge by, not how it came about. The prolific Mead *remembers* a great deal but on a record we are little concerned with whether he remembered or someone was whispering in his ear.

These comments about Hodes's playing are not specifically directed towards *Shake That Thing* but are general in nature. As a matter of fact *Shake That Thing* is played just right and is on the whole finely conceived.

BUNK JOHNSON'S BAND

Panama	American Music V-255
When You Wore a Tulip	
Walk Thru the Streets of the City	American Music V-256
Darktown Strutters' Ball	

Without even hearing Bunk in the flesh it is quite evident by now how such a band as his functions. They have a pattern which encloses anything they touch. The pattern has qualities of its own that we can abstract from the performance as heard. Whatever they touch melts immediately into this pattern and in addition there is the pure quality of the sound itself. So long as the piece is not too slow it never becomes cheap in their interpretation. With a proper balance of counterpoint and tonal qualities the sentiment and sweetness vanish from music. This goes for all music. Bunk's variations on a tune are inventive but most of the time very close to the tune. The twists he does give are fragmentary and are a vehicle through which to hear Bunk himself although they do not transcend the music as heard. Bunk represents a most critical period between playing straight and the more definite examples of a later jazz.

Lewis does not have such a wide variety as Bunk but his variations are more concise and can transcend the actual performance. The beauty of his figurations, the inventiveness of an ascending passage, and the variety given the descending counterpart are so exquisite that they stand out from the whole band in unmistakable fashion. Lewis with much less of this definiteness of thought would still be a great ensemble player because of his constant musical presence both in figuration and the richest high notes I have ever heard.

The way Jim Robinson kicks himself and the whole band along makes for a more important trombone than I ever heard on the older records. His long notes against the terrific jamming sing out, and at all times he is heard. He has invention but he needs the right moment of excitement to bring it forth.

Baby Dodds plays the most normal drum I can think of. Not by overstressing the rhythm, nor by solo playing, nor by an exaggerated primitive tom-tom does he achieve his effects. Here we have a player who achieves solo variety without playing solo and a true primitive effect without giving the impression that he is imitating African drumming.

Panama has a steady drive all the way through. The end of the record has the greatest variety. Lewis's clarinet has that staccato rhythm that the Chicago school's "all out" used. Bunk plays his low tone wails in the last two choruses. *Walk Thru the Streets of the City* is pretty even. Bunk does some very nice playing after Lewis's solo.

When You Wore a Tulip is the most inventive side of the group. Bunk varies the second chorus just enough. George Lewis's solo is beautiful especially as he takes the melody down on his low register. Robinson has a few rightly placed tailgate slides in this chorus. The next chorus of Bunk's is very inventive. Jim takes a staccato but very rhythmic chorus with Lewis taking the second half. It is a fine side.

Darktown Strutters' Ball is the most successful side. A fine tune and Bunk plays each of his first four choruses quite differently. Lewis has a very beautiful spot in the second half of his chorus. He plays three of his high notes but in a strict rhythm of two to a bar. To hear his clear tone coming out in this fashion is very stirring. Jim has a fine rhythmic solo and then in the next we can hear his wonderful long note work behind Bunk. Bunk plays a descending passage in this chorus that we do not often hear him do. I am continually finding new material in this and the *Tulip* side.

Bunk's band is now in New York City. It is probably the first time that real New Orleans music has been available to a New York audience. It is a fine experience to hear them. We all came to jazz in various ways. The usual introduction is accidental, since we either hear it as the background in some such place as a bar or dance to it. In this sense it becomes a pattern of sound behind other activities. Discrimination can start then, either for good or bad. Even the most undiscriminating can understand and fully enjoy Bunk and this enjoyment may very easily lead into a discrimination both from hearing Bunk in person and listening to his records.

Bunk is old and *sometimes* on off nights, when he gets up to play, the remembrances of the past lead him into what his lips may not be able to execute. If this was the rule no amount of past glory would make me still praise him in the same breath as the younger George Lewis and Jim Robinson. But it is not the rule, and Bunk Johnson brings forth in his own way and in his own time a music equal in its single fashion to the collective feeling of the band.

THE HISTORY OF JAZZ
Vol. 1, The "Solid" South, Capitol Album CE 16

LEADBELLY

Rock Island Line	10021
Eagle Rock Rag	

ZUTTY SINGLETON'S TRIO

Lulu's Mood	10022
Barney's Bounce	

EDDIE MILLER'S CRESCENT CITY QUARTET

Crawfish Blues	10023
Zutty Singleton's Creole Band	
Cajun Love Song	

WINGY MANONE'S DIXIELAND BAND

The Tailgate Ramble	10024
Sister Kate	

NAPPY LAMARE'S LOUISIANA LEVEE LOUNGERS

At the Jazz Band Ball	10025
High Society	

Capitol has chosen to give us the history of jazz in four albums. On the cover of Volume I is a picture of a graveyard with a donkey pulling an old wagon in which is seated a trombone player dressed in his Sunday's best. Following him are cornet, clarinet, and bass drum players.

Oh, if the commercial world could only "tie-up" the idea and the actual thing! They call the album *The "Solid" South.* If one closes one's eyes and thinks of the Solid South, Leadbelly's sides certainly suit the mood, but a terrible shock comes when we hear *Lulu's Mood,* which certainly seems to have no connection with the South.

Of Leadbelly's two sides I like the *Rock Island Line* the best. Probably the dialogue gives extra interest to the singing when it does come. The tune is bright and a few choruses are just about enough. On *Eagle Rock Rag* Leadbelly plays the piano. The whole atmosphere on this side is a little "closed in." He has a fresh way of playing the piano because the idiom he chooses is off the beaten track. But the whole side becomes a little monotonous. His singing, as always, is the real thing for which we can be very thankful in this album.

If only the planners of this album had honestly asked themselves "What comes next?" they could not have answered "52nd Street." But 52nd Street is

The Record Changer, December 1945.

what we get. Barney Bigard's clarinet had its place in Duke's records but the general run of it is far too sweet and smooth. *Barney's Bounce* has "go" to it and when Bigard really gets going the record stops abruptly. Maybe this was the first take.

With the *Cajun Love Song* we get a much finer blues than *Crawfish Blues* or *Lulu's Mood.* We immediately sense something good with Stan Wrightman's piano. Nappy LaMare plays a very good guitar solo, and Miller plays in good blues clarinet style. Irvin Verret's blues singing is excellent and is "Deep South."

Tailgate Ramble has a haunting resemblance to *Willie the Weeper.* It runs along amiably enough but this is about all that can be said. *Sister Kate* strikingly sets its own pace in the very first few bars. Manone's singing has a wonderful neighborhood quality about it. His solos after this are fine. This side of *Sister Kate* is another one of those fine records of this tune which always seems to bring out the best in the men.

Jazz Band Ball lacks any drive in the beginning ensemble. The solos, however, are good with Matlock first, followed by Manone. Manone ends his first half with a fine phrase in real jazz idiom.

The *High Society* side plunges us into the second section immediately. The fanfare leading up to the Picou solo is taken as a trombone solo; giving us a little different slant to the piece. It could have been a little more inspired, though. Matlock plays the Picou item but that is about all. A *High Society* has to be pretty good these days with everyone making it. Eddie Miller also gives us a version of this solo on saxophone. It is the better of the two solos. The "all out" runs along in the even tenor of the whole record.

I can imagine that in *the future* to try to issue new recordings portraying the "history of jazz" might be a difficult problem because unlike classic music the body of historical jazz does not contain notated works of selected composers. It would seem quite foolish to make new sides of the old tunes but *today* with Bunk Johnson's outfit, Sidney Bechet, Louis Armstrong, Jimmy Yancey, Kid Ory, not to mention any number of capable old timers, we could have had two albums of real stuff.

The United States is consciously richer in jazz than it ever was. By that I mean, we now know *who* the great jazzmen are and *where* they are whereas before they existed for us as remote names. It is still up to these large companies to find the great jazzmen and make valid recordings in the real tradition.

Django's Djump HRS 1003
Low Cotton

HRS is back in circulation. Three Duke men and Django play a very sophisticated and later-day jazz. *Django's Djump* is a catchy tune. The ensemble and solos give the tune nice twists and it rides along at a good pace to the end. It is not outstanding but very gracious.

Low Cotton, a re-issue, gives us the very intimate Django. Django exhibits a fertile mind but a mind which has little to do with the jazz of USA. He is interesting to follow and in this number always gives the extra twists to the melody that are not expected. His fine sense of rhythm and strong playing style make his accompaniments very good background for the soloists.

On an instrument that was handled in quite a simple manner before his time, he developed, it seems by himself, his whole outlook on jazz music. Although his style contributes to the development in other jazz guitarists it is so esoteric that it will eventually lose interest for jazz enthusiasts. Bigard falls in very well with suave clarinet giving the right contrast to Django.

"PIGMEAT" ALAMO MARKHAM WITH OLIVER "REV." MESHEUX'S BLUE SIX

How Long—How Long Blues Blue Note 48
Blues Before Sunrise

Back in the 1930s I used to go up to the Apollo in Harlem. They had visiting bands but permanent comedians. One great comedian was Pigmeat. He sang, played at comedy and danced. He was a great dancer. Truckin' and Suzy-Q were his specialties. He says that he started both these dance crazes plus the later boogie-woogie. (Dances are not created quite in such a single fashion but it is very possible that he, in doing them so well, started them on their famous path.) I know that I never saw any one do Truckin' any better than he and the way he would screw his head around to go in another direction I will never forget. He was the perfect example of real *theatre,* actor-dancer-singer. But as many times as I frequented the Apollo I never remember him singing as he does on these Blue Note records and he never featured himself as a blues singer.

Pigmeat, in spite of his active life, is still a young man and has accomplished a great deal as a comedian, dancer, and singer. Besides his legitimate show business (his is the best legitimate theatre) he has been in films and has written for them. He is another one of those who have been discovered late—one of those doing fine stuff but a kind of stuff that has not been preserved. His singing, like Jimmy Yancey's playing, has been a side line and it is this side line which is so easy to preserve and for which there is a market and waiting public.

How Long opens with a very excellent trumpet chorus by Oliver Mesheux. A fairly straight rendition of the eight-bar blues full of warmth and good brass. Pigmeat has great variety from one chorus to the next. From the first

How long
How long

to the last verse

Lord a nickel
Here's a nickel

he keeps your attention by making you visualize the pictures his words intend. Pigmeat has a full voice, in fact in a few places on this side a little too full. There is nothing, however, of the voiced Negro singing spirituals. There is a complete blues attack and a "born to sing" the blues attitude.

Sandy Williams takes two excellent choruses before Pigmeat's last stanza. His trombone with Oliver's trumpet gives great variety to the record both in his accompaniment of Pigmeat and in the instrumental solos.

In *Blues Before Sunrise* Pigmeat has none of the large voice quality noticeable in *How Long.* Here his voice is blues all the way through. I would say that he gives more interest to the words on this side. As you sit and listen you can either take it as fine blues singing backed by Williams and Oliver or you can listen intently to his handling of the words. Each time the different subtleties of his interpretation arouse a different emotion in the listener.

Jimmy Shirley starts this side with a very instrumentalized guitar solo. Behind him is Tommy Benford with a soft cymbal rhythm. Williams backs Pigmeat with some very musical phrases. Oliver takes the chorus before the end. There is plenty to listen to either in the instrumental accompaniment and solos or in the whole authoritative dominance of Pigmeat himself.

I look forward to seeing a lot of his records, a library of the blues and early Negro "pops." It is a great satisfaction to listen to him and to remember that you did listen to him.

"TESCH"—CHICAGO STYLE CLARINETIST
Brunswick B-1017

CHICAGO RHYTHM KINGS
I've Found a New Baby 80063
There'll Be Some Changes Made

JOE "WINGY" MANONE AND HIS CLUB ROYAL ORCHESTRA
Baby, Won't You Please Come Home 80064
Trying To Stop My Crying

ELMER SCHOEBEL AND HIS FRIARS SOCIETY ORCHESTRA
Copenhagen 80065
Prince of Wails

THE CELLAR BOYS
Wailin' Blues 80066
Barrel House Stomp (Second Master)

Records that only could be gotten on UHCA and HRS and some not at all are now re-issues in a Tesch album. Teschemacher stands out on every side however short his solo. Whether it be fast and undulatingly rhythmic or slow and plaintive, his playing has a passionate appeal that is more heated than any of the Chicago group.

In *I've Found a New Baby* he plays a whole chorus in which every section has its own marked interest. The phrases work their way down in descending figurations while the last section sings right out to the accompaniment of Krupa's off-beat drums. From a creative standpoint, creative in the sense of a new instrumental *tune* easily grasped, Mezzrow's sax solo is one of a few great examples. It is the over-all conception that stamps this solo with high merit. Besides this conception Mezzrow has worked out the melodic designs in a manner befitting the use of a reed instrument. In the fourth section of this chorus he repeats the melodic figures of the first section but in a manner denoting the approach of the end. He then breaks it off to swell out into a singing final. The final all-out with Muggsy leading is wonderful. At the very end Teschemacher gives it a special lift by rocking between two notes.

Tesch's solo in *There'll Be Some Changes Made* has a still more pronounced rhythmic way of playing than on the *New Baby* side. He also has a climax on this side although on *New Baby* the solo seems to be better laid out. Both sides

The Record Changer, January 1946.

have their own merits while Mezz's solo adds greatly to *New Baby.* Both all-outs have extreme punch.

Baby, Won't You Please Come Home has fine Tesch on the end of a chorus that he takes and in the "jam" at the end. This is a new side, never pressed before. Tesch's playing in the ensemble at the end would be fitting material for a fine solo as his line is continuous and inventive.

Trying To Stop My Crying has good Teschemacher, besides Manone's singing and playing. There is too much uninspired ensemble and it is Tesch that makes the record worth listening to. On *Copenhagen* and *Prince of Wails* we hear a Teschemacher with a tone not as passionate as usual and with very little to do on these two sides.

Wailing Blues, the most musical of the sides, greatly resembles *King of the Zulus* although really built on a different harmonic structure. Manone's horn is low register and is played with fine tone and beat and with adequate invention. Tesch takes the first release. Manone's high piercing notes form a pick-up before his low-register chorus final, and the contrast is really dramatic. These notes seem "high and piercing" not so much because they are high as because they are so vividly etched against the sober register of the total piece. I find this much more satisfying than it would have been if the whole piece had been played in a higher register. Teschemacher's solo does justice to the name of this piece. It is a great solo and the phrase, before the release, where he plays a figure first in one register and then in a lower one carries with it a tragic touch. Bud Freeman's release with the hurried passage at the end inspires Tesch to finish his next eight bars in a fast cascade of notes.

On *Barrel House Stomp* I especially like Manone's tone. He plays fine horn all the way through. Bud Freeman plays a good solo just before Tesch's. The low register of Tesch comes right out of the Freeman solo, later to strike high and top all the solos as usual.

Frank Teschemacher did not record many records but wherever he is heard, on the few he did make, there is no mistaking that it is he and good "he."

REX STEWART'S BIG FOUR

Solid Rock HRS 1004
Night Wind

Night Wind has three solos typical of the players Rex Stewart, Barney Bigard, and Django Reinhardt. The first two bars of Django's are in his highly sensitive vein while the rest is average running.

Solid Rock is a short version of the twelve-inch record put out by HRS here in America. Rex's introduction and appendage have been shorn off. After Django's opening solo the ensemble theme chorus is played with good down-beat blues intention. It is a striking theme and distinguishes the record immediately. Rex's solo starts at the very bottom of his trumpet with rumbling trills. The use of two registers gives it variety.

Django's second solo has an ending which, although straining away from jazz, clearly shows the musical inventiveness of the man. It always crops out somewhere and always makes one wish that he had really been steeped in jazz over here.

Bigard's solo is vigorous and with Reinhardt's punctuation leads well into the final theme chorus. This is a good side and a fine example of a kind of jazz. Although this side is not too inventive and is tonally not so rich, it strikes forth with good modern blues feeling.

WILL EZELL

Barrel House Woman Signature 910
Heifer Dust

Mixed Up Rag Signature 911
Old Mill Blues

These reissues of Paramount sides have many wonderful choruses by Will Ezell. They reveal great musical depth and should be played many times for the rich reward we will find in becoming well acquainted with them.

Barrel House Woman has a most searching introduction. We are certain to hear a voice after such an introduction but instead we hear a piano blues starting out in bold triplets and revealing the whole tragic intention of the blues as it proceeds. The rest of the record is either triplets or quickly repeated bass notes. It is sprinkled with running boogie bass and the like. It never lives up to its great opening.

Heifer Dust is a very odd blues; a little monotonous at first but a blues revealing in Will Ezell's treatment a most musical passage in the eighth, ninth, and tenth bars of the first two choruses and similar choruses scattered throughout. These choruses with their lovely musical passage are all comprised of thirteen bars except for the fifth which is comprised of twelve. Ezell has some repeated note choruses that are in twelve bars. The sixth chorus has sixteen bars if we include the run-up of the piano within the chorus. Like many others, however, Ezell has a tendency to greatly rush his fast phrases.

Mixed Up Rag contains several old rags, among which is very definitely heard *Milenberg Joys.* After the opening verse *Milenberg* is introduced and it is played again at the end. The introduction of more than one rag gives the record an interest that a too straight continual repetition of one rag would not. In this *Mixed Up Rag* Will Ezell keeps good time. His foot must have been stamping away.

After a good introduction in *Old Mill Blues* we have two very fine and imaginative blues choruses. The next two choruses are highly reminiscent of Yancey. There is certainly a connection between the two men somewhere. The rest of this side is not up to the beginning but is, nevertheless, not monotonous and Ezell has ideas appearing right through to the end.

JELLY ROLL MORTON, Vol. 1
Brunswick Album No. B-1018

JELLY ROLL MORTON (PIANO)
King Porter Stomp 80067
The Pearls

Sweetheart o' Mine 80068
Fat Meats and Greens

Jelly Roll's reputation is continually growing. Nevertheless, however much it grows the genre within which he worked as a pianist is unmistakably a "professor's" music. It is popular to the core and like a great deal of popular music it has its sterling as well as its poor qualities. Taking a large view it would seem that the sterling qualities of the music of this genre have been over-emphasized.

The barrel-house or low-down piano, the slow walking bass, the many versions of the boogie-woogie and such, run into blind alleys because of the lack of imagination in their representation. The unimaginative pianist's use of the monotonous and unmelodic dynamic dissonance in low-down piano and his reiteration of the popularly known boogie-woogie with an ever-increasing dilution of melodic conception makes a refreshing antidote of the great pianist such as Jelly Roll with his rhythmic playing ability and sober use of invention. In jazz, the professional unimaginative catering to the popular conception of what is "low-down" in music is in its own way as nearly unconvincing as is the over-sweetness of "refined" jazz.

In *King Porter Stomp* there is a continuous playing drive slowly but surely building in interest up to the very syncopated last seven bars. Jelly Roll extends the end by two bars thereby giving his syncopation greater play. The old use of the modulation mars the perfect pace of the record and more than anything else dates these tunes. Jelly Roll's well-placed broken chords give great accent and lift to the forward motion of the piece.

The Pearls is played nearly as well as *King Porter* but the tune in spite of its popularity harks back too readily to the old nickelodeon and early movie days with the piano in the small pit. *Sweetheart o' Mine* runs along about the same except for the more pronounced use of fine breaks and the chordal treatment of the trio. *Fat Meat and Greens* does not have the lively playing feeling of Jelly Roll's other pieces nor does it have what may be expected from a piece on the twelve-bar blues.

It is the nickelodeon that runs melodically through all the sides that dates them but Jelly Roll's frank acceptance of the medium is certainly more healthy than the orchestra pianists after him. The frantic Hines, the sweet over-pianistic Wilson, although an advancement, turn out to be not so satisfying. Jelly Roll's playing was a simple but rhythmic statement of the music he liked to play. Without borrowing from the other instruments, a practice so inept in other players, he just carried to a great height the manner of playing of accomplished pianists of the time.

CRIPPLE CLARENCE LOFTON

I Don't Know Session 12–005
Streamlined Train

In de Mornin' Session 10–006
Early Blues

Again Lofton plays his *I Don't Know.* Solo Art some years back issued the same title on a ten-inch disc. This twelve-inch side has the same example of Lofton's drive. In the middle of the piece, on the Solo Art side, the chords seem more interesting in the left hand probably because they are higher up. This music is not simple in its form although Lofton seems to hover very close to between nine and eleven bars to every chorus. I would be very interested in seeing a transcription on paper of this piece. The Session side may seem long but it does have a tremendous power in the last few choruses—an address driving forth with a long musical line in each chorus. It would be difficult to listen any longer if the record continued because of the intensity of this music. This music has sufficient quality to suggest what its value might be were it a passage within a larger work.

The Record Changer, March 1946.

Streamlined Train is much simpler. The last four bars of each blues chorus helps to keep one orientated. Outside of this, this piece has not got the great quality of *I Don't Know.* Although these last four bars are helpful in the beginning, they are, like the chorus endings of a lot of boogie-woogie, a little too much like a sort of tag refrain stamped on the end of every chorus. Lofton has a pattern of his own as most if not all the choruses are eleven and a half bars long. He does not give us much of a lead into what he is going to play next. In other words we are really never at home, although maybe much greater acquaintance with him may reveal an underlying form. There is no mistake of the power inherent in his playing and material but I think he would be more satisfying if more "easier on us."

In de Mornin' is a slow blues. It runs much closer to the traditional blues. Even though Lofton is using the twelve-bar blues structure he varies some of the sections so that he may come out either under or over twelve bars by a half or full bar. The beginning has some extraordinary harmonic changes. The bass is reminiscent of Meade Lux Lewis's *Rising Tide Blues.* This side is a delicate little piece and would, I should think, sound well on the celeste.

One of the most joyous and melodically simple uses of boogie-woogie is in *Early Blues.* It is not a blues but a hymn, and is a perfect example of the way a boogie-woogie player should play such a tune. For those who cannot "go all the way" for Lofton this side will certainly satisfy. His choruses vary between a variation and a repeated passage in the treble. His variations syncopate in a wonderful manner the tune of this hymn. This side was a great surprise to me and I keep wondering what is in store for us next. The simplicity of the tune broken up in such a manner makes this a most ingratiating and "smiling" record.

KID ORY'S CREOLE JAZZ BAND

Maryland	Crescent 3
Oh, Didn't He Ramble	
Down Home Rag	Crescent 4
1919	

Maryland is a sober rendering, in march tempo, of that tune and with the addition of a trumpet *cor de chasse* countervoice and another march tune in the beginning, we get a very interesting and varied piece. Howard stands out with high trills in brass-band style giving a good lift to the piece. Ory's straight solo on the first section has full trombone warmth while his bit with Howard playing the other harmonic voice on the Maryland air is very nice. Carey is a little weak in his rendition of the *cor de chasse* while it is perked up a little when Howard joins him the last time. This side is a little too "marchy." We do not get the fullest satisfaction when a jazz band does play a march straight before it slips into its own style of playing.

Oh, Didn't He Ramble commences with a jazz band's mocking of the Funeral March and then a few words about ashes to ashes, which is all in keeping except for the weird screaming. The transition from this section to the playing of *Ramble* is a good example of how "too modern" the band is sometimes. We quite easily accustom ourselves to various styles, readily accepting them for what they are but there is nothing like a change from one style to another to show us which one we really like. The transition in *Ramble* made one wish for something else. The whole record is weak jamming up to the end. Mutt Carey attempts nothing too difficult so that his fine tone comes out and makes his playing very pleasing. All the players are heard but it is just a case of under playing.

We do not always have to have an arranger to set a piece in order. The men can do it themselves very easily. In *Down Home Rag* each man found his harmonic place in rendering the tune straight. The tune lends itself to what they did, but however it came about, the impression of the music is that of an arranged music. Each phrase is punctuated by a slight glissando on Ory's trombone. This section is repeated much too often. Some of Howard's solo is nice while Ory in his solo takes it mighty easy.

1919—the best of the sides—starts with a little heavy band playing; all right in a dance hall but it does not leave much on a ten-inch disc for something better. After the introduction of the trio the band jazzes it up. Howard plays fine clarinet while the band marches along on the trio tune. Later they more or less all join in for the finale. These four sides are not up to the first four. They are nevertheless very worthy records and contain music reminiscent of early jazz and played by some very great jazz musicians.

BUNK JOHNSON'S BAND

New Iberia Blues American Music V-257
Sister Kate

New Iberia Blues, named after Bunk's home town, is a beautiful blues. It has not got the variety of other slow pieces of Bunk's but this is not much of a reproach considering what we do get. Lewis takes about the only solo, playing low on his clarinet. Baby Dodds backs him with his excellent triplets on the drum rim. Robinson stands out on the last chorus with a passionate blues of his own.

Sister Kate is richly decorated with the musical thoughts of Bunk's men. The variety it *has* got is not so striking as is it absolutely satisfying. The one striking feature is the counter theme Robinson takes near the end. Against the rhythmic jazz background and on Robinson's horn it becomes a great segment of *Sister.* Note Robinson's music after the early breaks. Bunk, Lewis, and Robinson all take fine breaks. Bunk's entrance in the last chorus is very noticeable. This record is a fine representation of Baby Dodds's work. He is touching up *Sister* at every point possible.

The difference between these American Records and the Climax records, on the one hand, and all other records, on the other, is to be found in their absolute purity of folk style at its greatest development. The creation of a new music is complete and the creation of a band treatment is complete. Every instrument has found its correct place in this absolutely free style of playing. Solos, duets, and ensembles have grown out of the simple folk state and have developed an invention and inventive complexity beyond anything we know of in a music coming from the folk. Whatever the tune, the crossing instrumental melodic lines, rich in ideas, carry tremendous import beyond the tune. This kind of music has not reached the straight playing that all other jazz records indulge in. On some of the earliest King Oliver's we do get this feeling but not so consistently on all of them. These AM records are perfect whether the men sustain their peak playing or not. It becomes a matter of degree only whether they are good or bad but a degree within a province that only *they* inhabit. It is difficult, in recording, to ask them to always play the way we might *hope* they will. This makes no real difference when we consider that there is *no* competition from other bands.

The band has a characteristic of its own whether it plays softly, with solos, or a sock ensemble, which is best described as *breadth.* Other bands and especially modern ones sound bunched. Even though they may be all improvising, in later bands there is a singleness in the sum total effect. Bunk's band seems loosely woven. We see the stitch contributing to the total effect. Each instrument in Bunk's band is clearly laid out before us. The omission of this character is a great fault of modern bands and the fault of the records made on Crescent. I hear *one* thing and although I know it is the whole band which is contributing to this *one* thing, there is a stuffiness in this singleness of purpose. For singleness of purpose there must be direction such as is quite evident on the best of contemporary recording. Without this we get the men arranging themselves and, in simple harmonic fashion, following a simple sweet tune. For example I hardly ever hear Ory whereas I always hear Robinson. For such a *blend* of sound the musicians cannot help but repress their own individuality.

At such points we really do not need an Ory. It is in such sections as these that jazz has borrowed from the accepted procedure of the dance orchestra as it has existed in polite society for centuries. My liking of such music is always tempered with some reservation. Unless, of course, it is a masterpiece only coming now and then.

There need be *no* reservation when listening to these Bunk records. We could only wish that all Bunk's recordings might be made when he is most ecstatic. Bunk always plays from the heart rather than from the head but there are times when the heart is warmer.

JOHNNY DODDS
BRUNSWICK ALBUM NO. B-1020, Vol. 1

Weary Blues 80073
New Orleans Stomp

Come On and Stomp, Stomp, Stomp 80074
After You've Gone

JOHNNY DODDS AND HIS BLACK BOTTOM STOMPERS
Joe Turner Blues 80075
When Erastus Plays His Old Kazoo

JOHNNY DODDS WITH THE BEALE STREET WASHBOARD BAND
Forty and Tight 80076
Piggly Wiggly

On every record Dodds, at one time or another, injects a stroke of genius. This is evident in his melodic line which frequently breaks through with real brilliance. Dodds's playing shines above all even though Louis Armstrong is on the first two sides. Dodds's clarinet playing is imaginative and alive throughout but Louis's stomp playing occasionally becomes a little pedestrian. In other records Louis has shown his pre-eminence through the building ability evidenced in some of his solos but Dodds always keeps up a running variety of injected turns and twists.

Weary Blues has plenty of action and a variety of players but as a record it is not too good. Dodds's first solo is excellent and by far the best of this side.

The Record Changer, May 1946.

Louis bursts in on the last twelve-bar section. He then gives us a short eight-bar phrase against an orchestra off-beat. The "unknown" trombonist sounds very good in the legato passage after his cut-note section. A solo by Hines and another by Louis led into the final chorus.

New Orleans Stomp is a fast Louis record. He plays through it all the way. St. Cyr has a chorus but he certainly is not a solo man. His melodic line, although good, does not seem to stand up on a banjo. His alternating notes at the end are interesting. I like Stomp Evans's solo. The rhythmic suspensions stretch to a good length.

Hines's solo on this side is more unique than satisfying. Dodds does not fare quite so well on this side although his two choruses are good Dodds. In the last chorus Louis injects the *Rigoletto Quartet* while the band plays it out nice enough.

Come On and Stomp, Stomp, Stomp is an excellent stomp tune. I wonder that I do not hear it about more. It is one of the jolliest records in this album. In the orchestra's verse and chorus Dodds stands out in great fashion. His solo is superb. His legato notes in the seventh bar with the reaching note in the eighth is great inspiration. After a sad trombone release Dodds plays the end of the chorus. The second chorus by the orchestra is not so good but in the next chorus one especially notes the pick-up just before the last eight bars of the chorus. The heat of this pick-up is kept up all through these last eight bars. It is George Mitchell's lead that *makes* this last eight. Let's have some more of *Come On and Stomp.*

The low spot of this album is *After You've Gone.* Dodds plays the chorus in chalumeau. Chalumeau has come to be a "Tea for two" clarinet register. It is soft and low and has become a lush and saccharin style. Although I do not like this style of playing particularly well I think Dodds "brings it off" better than any other musician. He swells his tone and squeezes into the advancing notes in a vibrant way. His break is excellent.

Baby Dodds has a spot in the midst of the trombone solo to show off his excellent wares.

Joe Turner Blues is a side consisting of excellent breaks by Dodds. Although this tune is not outstanding, this record is very excellent. There is chorus after chorus of collective playing relieved a little by Dodds's breaks at the end of each. Dodds's two-chorus solo is the best blues playing in this album. The ensemble chorus next to the end is a two-bar jam affair with solo breaks by trombone, banjo, and a break by Dodds that is one of the greatest things he has ever done in such a compressed space. The long notes slipping into each other *is* jazz. When we can listen to a fine record and expect such a break at the end we have a two-fold pleasure—the immediate music and that great break to come.

The only thing that saves *When Erastus Plays His Old Kazoo* is the tempo. Dodds plays two-thirds of the verse with good breaks each time. The trombonist, Reeves, plays a little bit of jazz just before the last eight bars of his chorus that shows he *can* play jazz. Dodds leads all the way in the last chorus, again giving us an excellent break.

In *Forty and Tight* and *Piggly Wiggly,* Baby Dodds plays the washboard. This instrument, in spite of a monotony running through it, has an affinity for jazz which is lacking in traps. Baby Dodds's drumming comes closest to a drumming which might be called a "rhythmic beat pattern." While other drumming seems dull and if not dull, arbitrary, Dodds, as unique as he is, does not seem arbitrary. The washboard is never dull nor arbitrary and Dodds plays it with great refinement of rhythm.

Forty is typical of a washboard record although I feel that Johnny Dodds was a little uninspired on this side. He plays four too even solos. He rides along at all times, however, and always keeps a good pace. I like Herb Moran's first trumpet solo. Simple but effective. Melrose is sober good piano.

Piggly Wiggly has about the same good pace as *Forty* without the variety of that piece. Johnny Dodds's breaks are the only high spots with one by Baby Dodds. I like the medium of washboard, trumpet, and clarinet but the inspiration seems to be a little low. The records in this album all have a certain interest of their own, which gives it, as a whole, great variety.

There is a booklet with notes by Eugene Williams which is missing from many albums. Write for it.

SIDNEY BECHET'S BLUE NOTE JAZZ MEN

St. Louis Blues Blue Note 44
Jazz Me Blues

BENNY MORTON'S ALL STARS

The Sheik of Araby Blue Note 46
Conversing in Blue

Conversing in Blue is strictly a modern record. In it the musicians have achieved the fullest in improvisation without any of the stereotype of arranged jazz. The richness of intonation and the *running* machinery of a band have great sustaining qualities. When we do away with the best in *sound* quality and reduce the rhythmic section to an implied beat we are left with only the melodic line. *Conversing* may not have a highly significant run of solos but they do represent in their outline the workings of melodic emotion. *Conversing* reveals some bad taste on the part of Webster and Bigard. Bigard is not happy in his choice of the long piercing note followed by a velocity descent into chalumeau. This side is unique in certain moods that it achieves and in the degree of melodic interest sustained.

A flexible trombone rhapsodic passage leads into an interesting introduction by Crosby. Morton starts the blues with a long pick-up. His solo keeps up great activity while Bigard's two choruses are more relaxed. Webster enters his section with energy, and plays two very fine choruses except for his forced vibrato.

The last chorus is quite unique in many respects. Although at first it seems as though Webster were finishing this side, later a "far away" feeling is induced by a nostalgic spirit in the ensemble playing. Benskin's blues piano is the modern tremolo blues associated with Avery Parrish. It is not as satisfying as his playing on the next side.

Sheik of Araby consists of two-chorus solos for the most part. Although Morton's solo is constructed well as a whole, Benskin shows something more than fast 52nd Street piano. He has some good phrases that stand out. There must be something really harmonically different to make this brand of playing *really* satisfying. Webster uses some atonal high notes that detract from his playing.

When Handy published his *St. Louis Blues* the second or sixteen-bar section was written in the Habanera rhythm. Abbe Niles in his introduction to the *Blues* by Handy says, "To Handy also is to be credited the introduction in the accompanying bass of some of his blues, of the Habanera or tango rhythm. . . ." Granting that this rhythm originated in Africa and that there may have been some trace of it in early or pre-jazz music, its deliberate use in the published form, the accentuating of it in bands since then, has always given this section a non-jazz quality. The Blue Note men play the theme fairly straight in order to set it. On a one-chorus blues this is like a stated theme. The *b a a b* arrangement used here lengthens the whole theme too much, especially when the Habanera rhythm is so pronounced in the repeat of *b.*

There is a great change with Bechet's reedy clarinet solo. He plays his two choruses with invention, great tone, and an over-all feeling for register. The band behind him keeps up a good subdued busyness. While on Bechet, it must be noted that his work before his solo is excellent. His improvising on the harmonic part gives great interest to this secondary voice.

Dickenson's solo is very good although a little easy after Bechet's intensity. Hodes's piano comes through with some interesting material while at the end of his solo we can hear Pops Foster plucking away at the bass. It is a good touch. The last two choruses led by De Paris are good but a little under the best of this record.

Jazz Me Blues, that old jazz tune, is certainly a healthy specimen. The band plays it with great feeling for its style. In the second section Bechet takes some of the most wonderfully intriguing breaks I have ever heard him take.

De Paris's solo is in his best manner of playing. He twists the tune just enough to give it a wonderful lilt. When he plays this way I feel that his whole musical spirit is coming forth. It is a very rhythmic style of playing and is one of the best off-straight styles there is.

Dickenson's solo is good and varied but he is between two very great players so that he is a little dwarfed. Bechet's solo is most plaintive. It is legato in feeling as a contrast to that of the whole piece. Again he has a wonderful break and the whole solo is Bechet in his best form. Hodes's solo is especially good in the second half. He has a nice phrase that he repeats in different registers. The ensemble is fine at the end again with Bechet playing his wonderful break.

All through this side Bechet can be heard. He is behind every solo and his playing gives the whole thing a wonderful contrapuntal "punch."

MEZZROW-BECHET QUINTET

Gone Away Blues King Jazz No. 140
Deluxe Stomp

Bowin' the Blues King Jazz No. 141
Old School

Ole Miss King Jazz No. 142
Out of the Gallion

MEZZROW-BECHET SEPTET

Blood on the Moon King Jazz No. 143
House Party

PLEASANT JOE (Vocal)

MEZZROW-BECHET SEPTET

Saw Mill Man Blues King Jazz No. 144
Levee Blues

JIMMY BLYTHE, Jr. (Piano Solo)

Boogin' with "Mezz" King Jazz No. 145
I Finally Gotcha

Mezzrow has started ambitiously by issuing twelve sides. They range from ensembles and vocals to piano solos. *Gone Away Blues* starts with two choruses

Jazz Forum, January 1947.

by Mezz. Bechet can be heard in the second chorus followed by a striking solo of his own. From then on they both build for tonal quality and get a great effect of sustained sound.

De Luxe Stomp is played by both Bechet and Mezz all the way through. They weave around each other throughout in a good medium rhythmic style. Fitz Weston, piano, Pops Foster, bass, give good rhythmic background.

Bowin' the Blues has a nice feeling in Pops Foster's bowing. It sounds greatly like an organ and probably would have sounded more like a bowed *bass* had he given us a few accented scrapes on the bow. Bechet and Mezzrow play closely together again in an effort to get straight tonal quality.

Old School runs along about the same as *De Luxe Stomp* with Bechet and Mezzrow weaving all the way through. *Out of the Gallion* starts with a nice mellow tune played by both. Bechet plays excellent solos, the second of which has an exciting climax. This side is very good.

Ole Miss is the sixteen-bar strain that is usually played at the end of *Bugle Call Rag*. It is a trio in a composition by W. C. Handy called *Ole Miss* named after the fastest train between Memphis and New Orleans. This side is the familiar *Bugle Call Rag* but has been renamed after the last part. This side is pleasantly familiar and nicer to listen to than some of the others. The third break by Bechet certainly shows how he can come through in such spots. Kaiser Marshall gives good punctuation to Bechet's two phrases in this break.

Blood on the Moon, a number by Oran "Lips" Page, is sung by Poppa Snow White, who also plays the trumpet chorus at the end. Poppa's singing is good but a little the same all the way through. Bechet's solo is excellent in the middle of this side.

House Party starts with a fine simple piano introduction by Jimmy Blythe, Jr. The best solo on this side is by Poppa Snow White. Blythe, Jr., repeats his intro at the end where it sounds especially good, coming as it does after the strong tone built up on the record.

Levee Blues and *Saw Mill Man Blues* are singing sides by Pleasant Joe. He has a good folk voice but he has not too much variety. The background on these two sides could have pointed up the music more.

The piano solos are by Jimmy Blythe, Jr., not to be confused with the original Jimmy Blythe. The pieces are credited to Sammy Price. Blythe, Jr., plays *I Finally Gotcha* in a slow boogie mood. Blythe, Jr.'s playing is reminiscent of Avery Parrish in the beginning, some of Jimmy Yancey's bass in the middle and sounds a great deal like Isreal Crosby's bass figure playing later on. I do not mind hearing old friends but I feel Blythe, Jr., falls too little short of being an original boogie-woogie player. The last chorus sounds the most original.

Boogin' With "Mezz" is fast but still less inspired. Blythe, Jr., is a strong player but playing boogie-woogie is too easy, unless he shows inspiration.

The orchestral sides are excellent except that in the slow numbers Bechet and Mezz build too much for tonal effect. Such continuous building should have been relieved by a little rhythmic variation. These records sharpen the interest and make one wonder what his next sides will contain.

Note: Poppa Snow White and Jimmy Blythe, Jr., are Oran "Hot Lips" Page and Sam Price.

GEORGIA PEACH & SKYLIGHT SINGERS

The Road Is Mighty Rugged — Manor 1008
Does Jesus Care

FAMOUS GEORGIA PEACH WITH THE HARMONAIRES

Shady Green Pastures — Apollo 103
Here Am I, Do Lord Send Me

Who Is That Knocking? — Apollo 107
Where the Sun Will Never Go Down

Some time ago I went to the Golden Gate in Harlem to hear Negro religious music. The wonderful voices spotted here and there throughout all the quartets were a revelation as to the present condition of the great tradition of Negro singing. Concessions to radio technique, the pat arrangements of voices marred to a certain extent a complete pure feeling of song but their timbre of voice is a timbre that is lost in the theatre or secular song. The current blues singing of today has become so tainted that we speak of blues singing as of the past. The ecstasy of religious singing brings out the best of musical utterance and where it has not been led into devious paths, there is evidence that folk singing is still plentiful.

There were two old women from Georgia, the Two Keys. One of them played a minor role in this duet by playing the guitar and singing in an undertone. The other singer has a tremendous voice of the deepest religious passion. Her delivery was direct and straightforward with a sense of a continual onwardness to it. Georgia Peach, a famous singer, sang very little but that very little showed that she did have a voice with the modulating subtlety of the great blues singers. There are three records of hers current now.

Does Jesus Care shows her great range and emotion. The record starts off with a few chords on a guitar after which Georgia Peach enters with her warm low tones to be joined immediately by the Skylight Singers. She winds her way through their steady background with a voice that seems to be capable of anything. The little I have heard of her has revealed to me her rich imagination and the knowledge of its existence keeps one in a continual state of expectancy. The quartet behind her has a little too conspicuous a manner of the "singers'" accompaniment. *The Road Is Mighty Rugged* does not speak out in musical art as *Does Jesus Care.*

In *Shady Green Pastures* her voice is clearly outlined all the way through while the Harmonaires are not too intrusive, although their style is far too polished and is absolutely lacking in rapport with her voice. In *Here Am I, Do Lord Send Me,* which she recorded some time ago for Decca, she reveals more of her great voice and artistry. The background, this time, assumes a pat rhythmic accompaniment, modulating around in a technique so popular on the radio.

Who Is That Knocking? commences in a slow legato fashion, later to gather momentum. This faster section shows the Peach's great feeling for swinging along. She does not altogether let out but this is still a worthy side. Needless to say her background is out of keeping with her voice and I hope some day to hear her alone with either piano or organ.

A melodic line sung by the Peach is song drama of the greatest depth. Its continuity, its long line with the telling harmonic changes, is a music which certainly, when delivered in such a manner, has qualities of great music.

PEE-WEE RUSSELL'S RHYTHMAKERS

Baby Won't You Please Come Home HRS 1000
Dinah

There'll Be Some Changes Made HRS 1001
Zutty's Hootie Blues

PEE-WEE, ZUTTY AND JAMES P.

Everybody Loves My Baby HRS 1002
I Found a New Baby

HRS has re-issued some of its first waxings. *Baby Won't You Please Come Home* has fine Pee Wee, a little of Dicky Wells, and a typical chorus of James P. Johnson. It is a fine side. *Zutty's Hootie Blues,* a new name in this re-issue for *Horn of Plenty Blues,* is the best side. Dicky Wells plays a beautiful trombone solo followed by Zutty Singleton singing some verses of the blues. He has a subdued moaning delivery that has great appeal. Kaminsky's solo after the singing is excellent against the chordal background.

On *I Found a New Baby* and *Everybody Loves My Baby* Pee Wee plays some beautiful clarinet. His tone *makes* these solos and these sides in general.

James P.'s solo is best on *Everybody.* The long drum solos of Singleton are quite meaningless.

Altogether these three records have good jazz on them.

"HOT LIPS" PAGE AND HIS ORCHESTRA

You'd Be Frantic Too Commodore 571
Rockin' at Ryans

You'd Be Frantic Too is an interesting record. It is well planned but the piano and sax are a little too modern. Ace Harris's introduction is not blues piano but it has an extremely pleasant feeling. Page's singing has a nice hoarse quality but it is more a passionate statement of fact than it is blues singing. The sax weaving behind Page's singing gives a fluid quality to the background and its coming forward for a solo in the same manner gives a nice feel to this sequence. Page's trumpet chorus is good but he seems a little too interested in forcing the "blues" feeling. I like the general set-up of the sequences as they give something to the record as a whole.

Rockin' at Ryans is typical of a fast and furious jazz. I see no reason for Page extending himself so on a Commodore label. I think here he could well afford to take his time.

BERTHA "CHIPPIE" HILL, vocal with LOVIE AUSTIN'S BLUES SERENADERS

Trouble in Mind Circle J-1003

BERTHA "CHIPPIE" HILL, vocal with BABY DODDS STOMPERS

How Long Blues

BERTHA "CHIPPIE" HILL, vocal with LOVIE AUSTIN'S BLUES SERENADERS

Careless Love Circle J-1004

BERTHA "CHIPPIE" HILL, vocal with BABY DODDS STOMPERS

Charleston Blues

After a period of about twenty years Bertha Chippie Hill has been recorded again. Here she has made no concession to cheap popularity and has kept her integrity as a singer.

This *Trouble in Mind* is a little faster than her first recording of it. I would say that the older version has a little more subtlety to it. This eight-bar blues sounds better at a much slower tempo. The repeated tune does not come around so often. Each chorus has more of a life of its own. In this version we hear for the first time, I think, the singing of the verse. It is twelve bars, an odd length for a verse.

Chippie Hill's voice has a wonderful metallic ring to it. It is a strong folk singing that comes right at you. It is not altogether a pleasant sound, in the accepted sense, but its strength, vitality, and stark quality are unmistakable. She has a tendency to shout a little too much and I am sure her voice would sound more musical if she subdued it a little.

How Long Blues gives Lee Collins a chorus. His playing is in good style but melodically it does not seem to stand out. Shayne's piano is stronger than is Lovie Austin's. This fact makes *How Long* a better record from the instrumental standpoint. Lindsay's bass is quite prominent. Baby Dodds's playing is subdued as is proper for the accompaniment to a singer. In Chippie Hill's humming chorus it is wonderful to hear her humming with closed lips and to actually visualize her parting her lips which she does for the latter part of the chorus.

Careless Love shows her voice off the best. On some of her registers her voice, although a little hard, has great feeling. There are few singers that can extract from one unaltered note so much pathos. She changes the tune a little as she does in *Trouble in Mind.*

Charleston Blues is a rollicking old blues. What a story it tells. One of the verses runs:

> I'm going back to the fish house, baby, and get me some shrimps
> I'm going back, baby, and get me some good shrimps
> I've got to feed, baby, two or three hungry old pimps

Shayne is on this side again playing very well.

Chippie Hill is head and shoulders above most blues singers. The tempo of these records is all about the same. If she had one or two very slow sides I am sure they would have given the set more variety. This kind of singing is not much longer to be with us. There is *nothing* to take its place when it is gone. The voice cannot fall back on melodic line or invention, it is purely itself and *must* be there. Chippie Hill has it.

DUKE ELLINGTON AND HIS FAMOUS ORCHESTRA
HMVC 3504–5

Black, Brown and Beige
i. *Work Song*
iv. *Three Dances*
a. *West Indian Dance*
b. *Emancipation Celebration*
c. *Sugar Hill Penthouse (Beige)*
Black, Brown and Beige
ii. *Come Sunday*
iii. *The Blues*

Victor has issued the much talked about *Black, Brown and Beige* on four twelve-inch sides. To go into a close examination of each subtitle in view of the unfavorable impression given by the style of this music would have no real meaning. A few remarks, therefore, will suffice. In the first three movements Ellington is obviously striving towards a development of a style to hold this material together. In the last movement, however, he obviously reverts to the usual Ellington.

The first movement of *Black, Brown and Beige* is subtitled *Work Song.* At times there is a contemporary approach to such an idea but it is spoiled by the constant intrusion of a style associated with "arranged jazz." The use of Tricky Sam's art is most disappointing. His muted manner, so fine in his own improvisations, sounds truly "corny" when applied to straight quarter notes. All the polyrhythms and inventive lines that he improvises are missing. For a composer to turn his back on this contribution of Nanton's shows that the right time for *the* jazz composer has not arrived.

When I think of the name *Come Sunday* I am reminded of *Early Blues* by Lofton. This *is* church music for jazz composition. Instead the second movement is a long sobbing wail of Hodges resembling nothing we might associate with religious music.

The modernity of the movement called *The Blues* makes one wonder what use was served by the whole era of Bessie Smith as a contribution to the blues in jazz. The whole feeling of the lyrics in *The Blues* is that derived from *Porgy and Bess* and such writing cannot even be called derivations of the "blues" type.

The last movement consists of three dances. *West Indian Dance* could be any lesser arrangement of the Duke's. In *Emancipation Celebration* Taft Jordan unfortunately takes on the fouled-note mantle of Rex Stewart. Tricky Sam takes a break, the only bit of jazz in this work.

With *Sugar Hill Penthouse* we come to the end of this work. The Duke says in the notes that it is "representative of the atmosphere of a Sugar Hill penthouse in Harlem . . ." It not only is "representative" but it *is* a facsimile of a certain style of jazz today. Herein lies the fault of such compositions. At one moment they *represent* and at another they *are.* There seems to be a mania in both folk music and folk dance to portray a folk epic. Either an epic concerning the growth of the art or of the people. If the whole work is couched in one style, then such a theme does not disrupt the homogeneity necessary in a large work of art. The Duke's composition does keep to a style removed from his general output but on the last side and especially the *Penthouse* dance, he reverts to an Ellington mean. Either subordinate the work song and blues to what we know as Ellingtonia or keep the whole work in the style established in the beginning.

We have had improvisation, then arrangement, and lastly composed pieces in a sort of jazz idiom. A work of art should be evaluated within its own category. The validity of the category within which the Duke is functioning invites questioning.

I believe that notation in jazz is one means of contributing to the further development of jazz. Notation will lead improvisation through various stages of development and will finally enable jazz to reach a new category, that is, composed jazz. Although the works in early composed jazz will show some change, they will still have a strong bond with improvised jazz. Except in very advanced musical compositions which have "come out" of a *tried* period before them, the early composed piece should stand on the same ground as the improvisation. This the Duke's composition cannot do. His compositions cannot do it because he has not developed the jazz idiom but has *borrowed* from both the "arrangement" and the last echo of Debussy. It can hardly be said that he has developed jazz nor can it be said that he displays good taste. This Duke composition should be listened to as a piece which exemplifies the path that jazz should not take.

Of all the ambitious compositions created, none except the *Rhapsody in Blue* seems to have made a name for itself or would seem to be worth hearing again. I do not think that *Black, Brown and Beige* will survive any longer than the other "ambitious compositions." There are, however, several tuneful sections which might become popular and survive.

Duke Ellington in his own way is a unique personality in American jazz. Just as Gershwin was a unique personality growing out of Tin Pan Alley, who wrote a composition that has been popular for these last twenty-two years, so Ellington is a man who has done something no other man could have done. Unfortunately neither of them used the rich and fertile field of jazz improvisation nor as composers can we compare them with great composers in other styles. However, they both have occupied a niche which in a strictly musical sense would be hard to estimate at this time.

EARL HINES, piano solo

Off Time Blues	HRS 1009
A Monday Date	
Chimes in Blues	HRS 1010
Blues in Thirds	
Panther Rag	HRS 1011
Stowaway	
Chicago High Life	HRS 1012
Just Too Soon	

Some years back, HRS put out the old QRS sides of Earl Hines's piano solos. They have now re-issued them.

For some collectors, these sides have been the tops in piano jazz. Earl Hines was a unique piano player and was head and shoulders above everyone else in the style that he was solely responsible for. Possessed with great talent he made of jazz piano a highly personalized art. His execution gave a sparkle that the content of the piece did not possess. The sparkle was extremely light, however, and not in keeping with the jazz of the other instruments. He is known for his trumpet style of playing, but if the notes did come from the trumpet, all the trumpet solidity is gone. I think William Russell was a little kind to him when he said he derived from Chopin. Although the two may have a lightness in common, the content of Hines is more of the MacDowell-Nevin school; *To a Wild Rose* jazzed up.

A Monday Date is mostly wonderful pianistics and the lesser of all these sides. In *Off Time Blues,* Hines has a nice walking bass, but it does not stand out as it should. By means of accentuation he gives an element of syncopation. The last two choruses are extremely nice. It is the singing blues treated in a Hines fashion. It is the one expressive spot on this side. His delicate tremolos fit in well with his version of the blues here. The *Wild Rose* effect finishes the last chorus.

Blues in Thirds is slower, but goes along in the Hines fashion without saying too much. *Chimes in Blues* uses the chimes effect, possible on the piano, for a beginning. Along the middle there is a lovely section of delicate notes against pacing chords. Hines did this sort of thing to perfection.

Stowaway has the best elements of some of his other records besides giving us a cadenza from *Manhattan Serenade.* It seems to me these Hines sides were first.

Panther Rag is a hurried piece, like so much of Hines. There is no great solidity in Hines's fast playing. He is supposed to have a great sense of suspended

Jazz Forum, Autumn 1947.

rhythm in which he does not lose the beat over passages that do not have the established beat. This may be so, but he gives no great feeling of power when *in* the beat. He introduces *Milenberg Joys* in the middle which he plays with great gusto.

Chicago High Life and *Just Too Soon* both have bits of good Hines. *Too Soon* has odd passages up and down the piano while *Chicago* builds to a solid ending.

Altogether these HRS re-issues represent a good cross-section of Hines's various treatments. If one is in the mood for Hines they are satisfying, but time will tell for how long one will be in the mood.

MEADE LUX LEWIS, piano solo
Disc Album 502

BOOGIE WOOGIE AT THE PHILHARMONIC, pt. 1 6020
Medium Boogie
BOOGIE WOOGIE AT THE PHILHARMONIC, pt. 2
Fast Boogie

BOOGIE WOOGIE AT THE PHILHARMONIC, pt. 3 6021
Slow Boogie
BOOGIE WOOGIE AT THE PHILHARMONIC, pt. 4
Honky Tonk Train Blues

These four sides comprise a disappointing album not because Meade Lux has deteriorated, but because he has added an ingredient of flashiness which spoils his playing. Unlike other outstanding performers who, through constant improvisation, may have changed for the worse over the years, Meade Lux is still playing the same good music except that his rendition has changed.

Medium Boogie, pt. 1, the first title on the label, is really *Yancey Special* played at a great clip. To come upon it even in such a playing guise is always a great musical experience. *Fast Boogie,* pt. 2, is a short bit, actually one minute 36 seconds, that has all been heard before. *Slow Boogie,* pt. 3, has a beginning which I have never heard him play. The right hand sounds something like *Yancey.* It is a very beautiful variation. The rest is music one can hear all through his Blue Note records. The piece is well rounded and holds together as a composition.

Honky Tonk Train Blues is faster than even his tremendous Blue Note version. It has also one variation that was used on that recording. It now looks as though Meade Lux were rushing to get through. The left hand has lost all the notes which are impossible to play at this tempo.

We hear the applause after the numbers and Meade Lux's splashing away at some cheap tune through this applause. It is quite shocking to hear this gravy before he breaks into *Honky Tonk Train.* Even though we have earlier Meade Lux records that show him at his best this has the particular interest that comes from an actual performance for an audience which indeed this was.

DINK'S GOOD TIME MUSIC

Grace and Beauty	American Music 515
Stomp de Lowdown	
So Dif'rent Blues	American Music 516
Take Your Time	

ORIGINAL CREOLE STOMPERS

Up Jumped the Devil	American Music 513
Eh, La-Bas!	

Dink Johnson's music is the most infectious piano since the advent of boogie-woogie. The complete satisfaction it invokes through its very simplicity is hard to explain. When I say simplicity, it must be understood that I only speak of the appearance of the music. It is void of piano exhibitionism and even invention, but Dink Johnson's rhythmic sense as a performer is unique. The marked syncopation he injects plus his great mastery of a performer's jazz touch makes the music jump with life.

Stomp de Lowdown has an opening theme which is just what early jazz should have been but never was on any recorded piano playing. The guitar and bass give great support. Especially noticeable is their support when Dink later repeats the first theme. There is a typical ragtime theme which leads into his quick note variation on the main theme. The piece acquires variety through just the correct emphasis and changes.

Grace and Beauty is a typical fast ragtime. In it Dink does not fare so well. It is nice to listen to, but he is not the pianist to play runs and the like. It is always gratifying, however, to get ragtime played by a performer who is a great jazzist, and this we have.

Take Your Time, another ragtime, but at a slower tempo. It fits more in with Dink's temperament and fingers. It is a fine tune and not to be easily dismissed. Dink can be heard singing under the music, so to speak. On this side he brings his voice out to sing the title of the song. It is a very natural way for a jazz player to sing and play. He is not singing a song for you with piano accompaniment but is playing piano while the very complete character of his performance spills out in an undertone of song.

So Dif'rent Blues is a real twelve-bar blues with the most Dinkish pick-up and beginning you ever heard. In the second chorus the bass and guitar come in softly to give the greatest feeling of change and rhythmic satisfaction to the second chorus without the use of any other device of musical change. This whole side is great blues playing without, as I have said before, any obvious invention and the like. It is baffling in the same way a Strauss waltz is baffling in that it is hard to put one's finger on *the* thing that makes it what it is. The guitar has a part of a chorus which is more effective as guitar color in the right place than as an outstanding solo.

There could have been no better title for this music than *Dink's Good Time Music.* It is nothing else but that in the best sense. The variety of stomp, ragtime and blues give these sides great interest but Dink is above all a *Stomp King.* It will be interesting to see wherein these sides do fit in the jazz picture as we look back upon them in the future.

From just two sides it is hard to estimate Wooden Joe Nicholas's playing but what is heard on *Up Jumped the Devil* is the kind of trumpet that one could always listen to. From just the first four bars in the introduction we can see beat, timbre, and all that goes to make up great jazz brass. He has a very direct beat in the chorus and what seems a simplification of the tune which might as well be called *Sister Kate.*

George Lewis's weaving is in here too with Jim Robinson's trombone. Robinson takes a break in the second chorus that is exceptional and quite different from what one would expect from him. Jim plays his *Sister Kate* counter melody in some of the choruses he takes. Baby Dodds, besides backing the band up as he always does, also gives us an outstanding break near the end.

Everything considered this is a surprising side and in addition gives us the pleasure of hearing Wooden Joe.

Eh, La-Bas! a song they all sing in New Orleans, is sung here by the clarinetist Albert Burbank. It is sung in French with great feeling for the singing of just such a lively song. His clarinet is excellent and soars away at the music instead of weaving as does Lewis. Wooden Joe can be heard to great advantage after the singing. He shows he has lots to say by giving rhythmic passages, long notes, and under-the-lead playing.

SIDNEY DE PARIS' BLUE NOTE JAZZ MEN

Who's Sorry Now Blue Note 41
Ballin' the Jack

Who's Sorry Now has some very nice material in it. The beginning ensemble opens up with Sidney De Paris's wonderful lead. His melodic phrase right on the tune is a most happy beginning. The tune may be just a popular tune, but De Paris seems to have lifted it beyond what we are accustomed to when singing this tune. Arthur Shirley's guitar solo is most striking. Vic Dickenson, James P., Edmond Hall, and De Paris all have a complete chorus which they play in excellent fashion. Take notice of the fine syncopation Hall gets at the end of his chorus. The final is a tightly woven, strong "all out." It is fine what the men do with this popular tune. From beginning to end it keeps "well up."

Ballin' the Jack has the most terrific beat. From De Paris down to Sidney Catlett the emphasis is on "beat." This is a record made at the same session as another great record, the *Call of the Blues.* After two choruses of ensemble, Hall gives us a little relief, so far as beat is concerned, with two choruses in his best manner. He has long lines and plenty of melodic invention.

Sidney De Paris plays a very rhythmic chorus. He emphasizes the beat with abrupt phrases very reminiscent of the tune. After playing a chorus, which one would think was all that could be said in this very pronounced vein of his, he goes on to play another chorus in the same vein but with more beat, less tune, a tightening all round that makes this chorus so charged that you feel something must "give way" somewhere. It is the greatest chorus of this kind I have ever heard De Paris or any other trumpet player play. Extra punch and stringency is gotten by the "just under statement" of melody which De Paris has in every note of this chorus. These two choruses of his make a wonderful solo. The first strongly exhibit a virile type of playing and the second chorus is the same in character but is notched up, so to speak, a peg or two.

Catlett gives great support to De Paris all the way through. Abrupt phrases need a sympathetic drummer and Catlett is right on the beat with De Paris. In fact, Catlett shines all the way through this record.

Naturally anything after this is an anti-climax but nevertheless Dickenson and James P. play their solos well. James P.'s second chorus gets into something of the groove that De Paris established which sounds good on the piano.

This is a great side and one of the best *Ballin' the Jack*s to be made in this vigorous style.

ART HODES AND HIS CHICAGOANS

Maple Leaf Rag	BN 505
Yellow Dog Blues	
She's Crying for Me	BN 506
Slow 'em Down Blues	
Doctor Jazz	BN 507
Shoe Shiner's Drag	
There'll Be Some Changes Made	BN 508
Clark and Randolph	

It is always gratifying to review records by players who are capable of the best jazz, who know what it is and who at the same time make an honest attempt to give us the best. Such a group are Art Hodes and His Chicagoans, not to mention Alfred Lion's supervision.

It is not always a genuine pleasure to be placed in the position of *trying* to find something nice to say about a record, but with records such as these, the reviewer is challenged to so grade his remarks as to evaluate them without in any way detracting from the high plane maintained throughout. There is a richness and balance of tone in the ensembles which we usually do not find except when lush arrangement is resorted to. Blue Note manages to get balanced tone in improvised ensembles.

The records have material to please both those who like ensemble and those who like solos. *Maple Leaf Rag* is practically all ensemble whereas *Slow 'em Down Blues* is a series of fine solos. The records show Rod Cless to better advantage than do any others. He has tone, invention, and fire. Kaminsky plays lead in all ensembles and the tone of his horn is rich with flavor. A new trombonist, Ray Conniff, is excellent. His breaks, ensembles, and solos keep his end of the music on a high level. Both he and Kaminsky have an intonation that is reminiscent of very early records. Kaminsky sounding like Armstrong when he accompanied Bessie, and Conniff sounding like the general tone of the Hot Five. Hodes plays his typical style, which in a very slow blues has developed a very original manner—the Hodes-blues.

In *Maple Leaf Rag* the tempo is not too fast and, because of this moderate pace, is reminiscent of the feeling we get from the way a jazz band plays the old marches. Cless takes the first break while Conniff has an exceptional break on the repeat of the first section. Cless is heard all the way through the second section up in the register a clarinet should play when in ensemble.

There are very simple devices which in an improvising group are purely subconscious so far as any meditation on the matter goes. To illustrate, take the first section which the ensemble plays each time. The first two times there is no instrument predominant, but when later it is repeated, Kaminsky plays the tune of *Maple Leaf,* bringing it out more distinctly. Had he done it the first two times the effect would have been lost, but hearing it not too distinctly it becomes a great pleasure to hear him play it with that wonderful intonation he has when playing fairly straight. Conniff has a solo in the third section—very fine. It is probably the best *Maple Leaf* of any put out for some time.

Yellow Dog Blues is a record that grows on you. The beginning is very ominous. There are forebodings of all sorts. More than any other instrumental *Yellow Dog,* this one sounds much like Bessie Smith singing. And nothing better could be said of a band than that they sound like Bessie. The usual run of records on this tune do not seem to bring out that wonderful mysterious verse and relaxed chorus inherent in the tune itself. Hodes makes a nice mixture of the chorus and of what is typical in his playing. Cless is next with a very inventive solo riding right along in double time—a double time, however, with rests between the phrases. Kaminsky leads with a wonderful tone—that early tone I spoke of before. Cless weaves with invention in this ensemble up high where he can be heard. His descending phrase right after Kaminsky's ensemble work is the greatest thing in clarinet jazz. It is familiar to me but off-hand I cannot remember where I heard it. Such music is the greatest thing in jazz.

She's Crying for Me, made so famous by the New Orleans Rhythm Kings, stands up in these 1940s as well as it did in those early 1920s. In the second bar after the familiar introduction we hear Rappolo's high clarinet note. There is always something missing in this piece when I do not hear it. Cless must have felt the same way too. Conniff has a wonderful break at the end of this chorus. Cless's two choruses are very stirring. Behind them we hear Danny Alvin playing simple but fine drums—*One,* two, three, four, accenting the one for the first twelve bars and a mild off-beat on the second twelve. The drums have many telling rhythmic devices but only too often we hear either just a continuous vamp or fireworks protruding through the music, saying nothing. The last ensemble is as thrilling as ever. Cless and Conniff are distinctly heard. Cless weaves on his high register with fine material. The usual corny interpolation in the middle of the tune is gladly missed.

Slow 'em Down Blues is the best blues of the lot. It is a great blues record. Hodes presents us to the ensemble with a short introduction. They are all playing, but in such a manner as to permit each one at a time to come to the surface and be heard. They are listening to each other also; we can be sure of this when we hear them imitate each other as does Cless when he imitates a phrase of Conniff's. Hodes's solo is a real Hodes-blues. It is a good blues style pushed just far enough. Any further and he would be over the border of jazz. Notice Haggart's gentle running bass at the beginning of this solo. Cless's chorus in chalumeau is rich in heavy vibrato. Along about the middle we can hear Hodes-blues in the background, moving along behind him. Besides being a good solo blues style it is a very befitting accompaniment. Conniff plays a very sensitive and agitatedly expressive solo. This is the last chorus with the others coming in for a build-up at the end. There is a lot of material on this record with no dead spots to mar it.

Dr. Jazz starts with an ensemble which is too straight for my taste, in fact more so than in Jelly Roll's original record. The reverse should be the case as we develop jazz. The first half of Cless's solo is too straight also but in the second he rises up in register to give us a good solo. Conniff's solo is full of invention and rides along in good trombone style. It is fine horn and saves the record from being a little monotonous.

Shoe Shiner's Drag starts with the familiar introduction introduced by Jelly Roll. It is a good record although there are no really outstanding solo passages. The blues itself is good enough, however, to stand up in simple ensemble rendition. The general pattern throughout is solo with ensemble joining in. The overall tonal quality of the music is rich in instrumental timbre.

In *There'll Be Some Changes Made* the men play the first chorus straight as did the early Chicagoan innovators. I know others take it for granted and like to hear the boys just playing so long as they have tone and show some spirit but for something that I am to play *over* I want a little more. Hodes is really going when he bursts in with a descending passage in quarter notes. This whole solo is spirited and sets a fine pace for the others to follow. Cless is reminiscent of Teschemacher. He plays with intensity and in an atmosphere of restlessness, not content with just going along but with a desire to say plenty in the smallest space of time. Conniff finishes his solo with a sustained note that leads us into the final ensemble. This note, not much in itself, but in such a place is always so effective. It seems that the whole balance between solo and ensemble is kept intact by just so simple a device as this one long lusty note. The second section of the chorus is rich in a double rhythmic background. The very ending is one of the best ensembles I have heard. Cless is wailing on the very top and the rhythm below carries to the last beat.

The last of these eight sides is a great Art Hodes piece, *Clark and Randolph.* The theme is a very good blues theme and inspires a low-down strong beat. Kaminsky's solo is his best. His intonation and musical intervals are, as I have said, reminiscent of Louis when with Bessie. I do not know of any trumpet solos that have this in the early records. We only find it when the trumpet accompanied blues singers. Kaminsky seems to incorporate both the singer and the old trumpet accompaniment. A great way to play the blues.

When I play the records over, the slow tempo of *Maple Leaf,* the wonderful solos in *Slow 'em Down,* and the great chorus at the end of *There'll Be Some Changes* seem to stand out. Eight such good sides all sounding very authentic besides being inventive and at the same time vibrant in sound is quite a treat these days.

CRIPPLE CLARENCE LOFTON

The Fives Session 10–002
South End Boogie

Lofton played *Pinetop's Boogie Woogie* and, although Lofton played it his own way, it certainly stemmed quite directly from Pinetop. In *The Fives,* however, I see no relation to Yancey's *Fives.* Maybe it is just coincidence that they are both called *The Fives* and if this is so I should like to know from what the name derives. It is a great piano piece whatever the question of names. Lofton really has only two themes. He plays the first theme about six times, varying it the last two. After the fourth we hear the breaks that Romeo Nelson uses on *Head Rag Hop.* Listen to Lofton's right hand in this break. The musical idea and rhythmic insistence of this right hand against the passage that the left hand is carrying on makes this break really great. He repeats the break but changes the right hand, giving it variety this time. Lofton is certainly fertile. The theme that he starts with is a good boogie theme while retaining some of the qualities of the blues as played by a wind instrument.

South End Boogie is slow. The introduction barely establishes the tempo but the old foot stomping away can be heard on the record. The piece proper starts with Lofton's left hand deep in the bass. The rhythm of this piece is heavy and rhythmically sonorous. His inventions are very musical and in the medium of the greatest piano jazz. Some of his inventions are beautifully delicate, such as the second, which, however, ends up in a far from delicate mood. This whole boogie blues has a majesty in its slew-foot pace that few boogie pieces have. The whole tonal feeling of Lofton's music is one of the most refreshing experiences. When we hear choruses repeated we also realize the paucity of improvisation in a great deal of jazz, and I use improvisation as meaning extemporaneous playing. The themes, however they came about, are as set as are the bound works of music on Schirmer's shelves.

De PARIS BROTHERS ORCHESTRA

Change o'Key Boogie Commodore 567
The Sheik of Araby

EDDIE CONDON AND HIS BAND

Singin' the Blues Commodore 568
Pray for the Lights To Go Out

Singin' the Blues is all ensemble except for a fine solo of Pee Wee's. The ideas that he shows, the clarity with which they stand out, all go to make this side something "special."

The reverse side, *Pray for the Lights To Go Out,* has much better ensemble. After two solos, one by Bushkin and one by Pee Wee, the boys break through with something to hear. Pee Wee is outstanding in this ensemble.

Change o'Key Boogie has a fine charge of instrumental boogie all the way through. As I expected from the name, the change of key was annoying, but upon repeated hearings the changes take on musical value. One change of key may give a nice lift to a piece but too many changes puts it into the "novelty" class. Clyde Hart's boogie-woogie playing is even and adequate in so far as only a few piano choruses are used. Edmond Hall's and Wilbur De Paris's are the most striking entrances in the change of key. The three choruses of Sydney De Paris are a little evenly conceived although coming, as they do, in the early part of the piece they set the pace for Hall's more sparkling solo. Wilbur De Paris has great imagination in the first few bars of his solo. It is the high musical spot of the record.

"PIGMEAT" ALAMO MARKHAM WITH OLIVER "REV." MESHEUX'S BLUE SIX

See See Rider Blue Note 509
You've Been a Good Old Wagon

EDMOND HALL'S SWINGTET

It's Been So Long Blue Note 511
I Can't Believe That You're in Love With Me

ORIGINAL ART HODES TRIO

Blues 'n' Booze Blue Note 512
Eccentric

It's Been So Long and *I Can't Believe That You're in Love With Me* are both on the popular side. Edmond Hall's clarinet has that wonderful crackling tone and melodic twist that he always manifests. Harry Carney's deep and rich sax tone is always pleasant to hear. His solo is especially nice on *It's Been So Long.* Benny Morton comes through nicely on *I Can't Believe. It's Been So Long* is the better side giving everyone a good chance to come forth.

In *Eccentric,* Kaminsky has an interesting little figure running through the first section. It is very musical but he "wears it a little thin" playing it so often. Hodes's first solo is very good in the beginning but lacks interest in the end. His familiar rapid alternating hand technique with which he usually begins his solos is here sprinkled throughout the entire chorus which in this case comprises his second solo. It's very good.

In *Blues 'n' Booze* there is a little stressing for tonal effects on the part of Kaminsky. It is a slow blues with three choruses sung by Fred Moore. His singing makes a good interlude but three verses are a little too long. Kaminsky's first solo is a simple and melodic statement of the blues.

See See Rider is a joyous singing side of Pigmeat Markham's. The movement of the whole band behind Pigmeat added to which is Pigmeat's happiness of singing—happiness to a point of barely containing himself—makes a record that sparkles all the way through. Vivian Smith's bold piano and Isreal Crosby's walking bass are notable. There is a veritable tumult of exciting sounds behind Pigmeat. Vivian Smith's solo is fresh and finds itself at perfect ease in a blues "stepped up" in tempo. Her glissando at the end has good continuity with the triplets issuing from Jimmy Shirley's guitar. These triplets seem to come like a bolt from the sky. Although these triplets have a "closing" feeling, Shirley does not end the piece there but eases on to finish with a fine chorus. Oliver takes the next solo playing with his fine tone and imagination. These three solos contribute to a fine and interesting instrumental section. Pigmeat ends this fine piece with two more choruses.

You've Been a Good Old Wagon, one of those double meaning songs so popular in the Negro shows, now gets a rebirth in Pigmeat's hands. This song is full of emotional content in which the words lose the vulgarity of the usual obvious double meaning song and become instead rather tragic symbols.

I think it a mistake to compare Pigmeat too much with Bessie Smith. That whole era whether simply stated by the lesser lights or richly stated by Bessie Smith is quite different in feeling. Pigmeat is great in his own way whether that way is equal to the other or not. We can "go" all the way for Pigmeat without comparing. His singing, as objective as it is, is far more subjective than Bessie's. When a singer, like Bessie, can at the same time be so objective and so close to our emotion, then we are perilously near the end of what *can* be said in song. Pigmeat is a great singer living the part he sings with every inch of his body and voice. These two sides are fine examples of the rolling quality of a song when in the hands of a singer who is so alive to the tempo of a piece.

 Georgia Peach (Clara Hudman), *ca.* 1953.

GEORGIA PEACH: GOSPEL IN THE GREAT TRADITION

DANNY BARKER, guitar and banjo
JAMES FRANCIS, piano (Side "A")
JOHN EPHRAIM, piano (Side "B")

Religion has been the greatest influence in creating the music we associate with the Negro people in the United States. Two distinct types have come forth from this musical expression, the spirituals and the blues. It is the blues together with instrumental practices which led to the creation of jazz. This entire secular expression has had an exhaustive experience leading into the demise of blues singing and an instrumental practice far removed from its virile beginning. Only in the religious manifestation do we find a musical art that still has vitality and an unspoiled character. It is a music little known outside of its habitat, the secular audience only becoming slightly aware of this music in the last few years. The term *Gospel Songs* as a rule covers all the religious music sung by the Negroes today. In gospel music we will find some old spirituals as well as modern religious songs closely resembling the older spirituals.

Only in the Negro churches do we hear unspoiled voices with a plaintive character not found in the professional world today. It is the attitude towards singing held by these singers which is responsible for this character. Deep within this entire school of singing is an emotional utterance that has pervaded both its secular and religious expressions, while the up-tempo numbers are sung with all the hand-clapping necessary to insure a rhythmic beat of great power. When a great voice is brought up in this tradition the remarkable result is the singing we find in Georgia Peach.

Added to this we have the embellishments which are an integral part of gospel singing. Embellishments are part of all music, *becoming* the music when they are written in. We find critical statements throughout the 17th century, and earlier, berating the liberties taken by singers. Such criticism naively tells us that this practice is bad when it is a matter of the individual performer's bad taste in embellishment. I find such bad taste even in gospel singing today, but where embellishment is suited to the occasion and the style we have great art in a fitting façade.

What is remarkable about Georgia Peach's singing is that, without losing any of its fervor, it fits in with our musical life on records, as distinguished from an on-the-spot recording or recorded imitation of revival hysteria. This kind of worship is a phenomenon of the greatest vitality, but it is a little out of place when we are not within the context of the church itself.

Acquaintance with Georgia Peach's voice reveals that she, in the manner of

Liner notes to *Georgia Peach: Gospel in the Great Tradition,* Classic Editions # CE5001.

all folk singers, uses more than one registral tone. Folk singers usually concentrate on a tone emanating from the throat or front palate. These tones they greatly qualify by diction emphasis, giving their singing many different colorings. It is this wide variety of choice which makes it possible for the singing of one culture, nation, or period to sound so different from another. Happy rhythmic and word emphasis is achieved by a just intonation, bringing into prominence high points of the melodic or scanned word line. Georgia Peach adds to this folk procedure the deep tones of the chest usually associated with academic study and practice. Whereas in the academic singer these chest tones are relied upon to do what would be better accomplished up nearer the mouth, Georgia Peach's extraordinary flexibility in this region and her innate readiness to run from chest tones to palate tones makes, in her hands, for the most natural delivery of the singer's art.

Georgia Peach, or for that matter any other gospel singer, should not be compared to the traditional blues singers such as Bessie Smith. The blues singers developed an idiom much further removed from European delivery than the singing of Georgia Peach. When a singer sings within an unspoiled school and has great emotional fervor, she expresses this feeling without necessarily going very far from the norm. If one listens to *Tired,* one hears a magnificent song with suggestions of the blues idiom, achieved through a relatively simple delivery.

What is very remarkable about Georgia Peach's singing is the phrasing and meaning she puts into the words. They take on word meaning (as opposed to purely melodic meaning) found in the speech delivery patterns of the Negro. Her alteration of the staid nature of the words coincides with a rhythmic alteration making more meaningful the entire context, words and music. Without losing the finer effects of the Negro folk-singing style, Georgia Peach brings her unique contribution, a vibrant and tremendous voice, a precious rarity in singing.

Georgia Peach was born Clara Hudman in Georgia. At an early age she sang in the junior choir of her church, later forming a trio with two of her brothers. She sang in revival meetings, the churches, and in religious concerts, covering a good portion of the United States. Her first recordings were with the famous Rev. J. M. Gates. Later, in 1931, she came to New York to record for Okeh. She has recorded for the Decca, Manor, Apollo, and Signature labels. She came to New York in 1938 and has lived here ever since. In private life she is Clara Gholson Brock. Once a year she gives an anniversary concert either in Georgia or New York. Her thirty-ninth anniversary concert occurs this year (1954).

Index

Smith, Charles Edwards, 65, 66, 123